BIOTECHNOLOGY BY OPEN LEARNING

Technological Applications of Biocatalysts

PUBLISHED ON BEHALF OF :

Open universiteit   and   **University of Greenwich (formerly Thames Polytechnic)**

Valkenburgerweg 167
6401 DL Heerlen
Nederland

Avery Hill Road
Eltham, London SE9 2HB
United Kingdom

Butterworth-Heinemann

Butterworth-Heinemann Ltd
Linacre House, Jordan Hill, Oxford OX2 8DP

A member of Reed Elsevier group

OXFORD LONDON BOSTON
MUNICH NEW DELHI SINGAPORE SYDNEY
TOKYO TORONTO WELLINGTON

First published 1993

© Butterworth-Heinemann Ltd 1993

All rights reserved. No part of this publication may be reproduced in any material form (including photocopying or storing in any medium by electronic means and whether or not transiently or incidentally to some other use of this publication) without the written permission of the copyright holder except in accordance with the provisions of the Copyright, Designs and Patents Act 1988 or under the terms of a licence issued by the Copyright Licensing Agency Ltd, 90 Tottenham Court Road, London, England W1P 9HE. Applications for the copyright holder's written permission to reproduce any part of this publication should be addressed to the publishers

British Library Cataloguing in Publication Data
A catalogue record for this book is available from the British Library

Library of Congress Cataloguing in Publication Data
A catalogue record for this book is available from the Library of Congress

ISBN 0 7506 0506 5

Composition by University of Greenwich
(formerly Thames Polytechnic)
Printed and Bound in Great Britain

The Biotol Project

The BIOTOL team

OPEN UNIVERSITEIT, THE NETHERLANDS
Prof M. C. E. van Dam-Mieras
Prof W. H. de Jeu
Prof J. de Vries

UNIVERSITY OF GREENWICH (FORMERLY THAMES POLYTECHNIC), UK
Prof B. R. Currell
Dr J. W. James
Dr C. K. Leach
Mr R. A. Patmore

This series of books has been developed through a collaboration between the Open universiteit of the Netherlands and University of Greenwich (formerly Thames Polytechnic) to provide a whole library of advanced level flexible learning materials including books, computer and video programmes. The series will be of particular value to those working in the chemical, pharmaceutical, health care, food and drinks, agriculture, and environmental, manufacturing and service industries. These industries will be increasingly faced with training problems as the use of biologically based techniques replaces or enhances chemical ones or indeed allows the development of products previously impossible.

The BIOTOL books may be studied privately, but specifically they provide a cost-effective major resource for in-house company training and are the basis for a wider range of courses (open, distance or traditional) from universities which, with practical and tutorial support, lead to recognised qualifications. There is a developing network of institutions throughout Europe to offer tutorial and practical support and courses based on BIOTOL both for those newly entering the field of biotechnology and for graduates looking for more advanced training. BIOTOL is for any one wishing to know about and use the principles and techniques of modern biotechnology whether they are technicians needing further education, new graduates wishing to extend their knowledge, mature staff faced with changing work or a new career, managers unfamiliar with the new technology or those returning to work after a career break.

Our learning texts, written in an informal and friendly style, embody the best characteristics of both open and distance learning to provide a flexible resource for individuals, training organisations, polytechnics and universities, and professional bodies. The content of each book has been carefully worked out between teachers and industry to lead students through a programme of work so that they may achieve clearly stated learning objectives. There are activities and exercises throughout the books, and self assessment questions that allow students to check their own progress and receive any necessary remedial help.

The books, within the series, are modular allowing students to select their own entry point depending on their knowledge and previous experience. These texts therefore remove the necessity for students to attend institution based lectures at specific times and places, bringing a new freedom to study their chosen subject at the time they need and a pace and place to suit them. This same freedom is highly beneficial to industry since staff can receive training without spending significant periods away from the workplace attending lectures and courses, and without altering work patterns.

SOFTWARE IN THE BIOTOL SERIES

BIOcalm interactive computer programmes provide experience in decision making in many of the techniques used in Biotechnology. They simulate the practical problems and decisions that need to be addressed in planning, setting up and carrying out research or development experiments and production processes. Each programme has an extensive library including basic concepts, experimental techniques, data and units. Also included with each programme are the relevant BIOTOL books which cover the necessary theoretical background.

The programmes and supporting BIOTOL books are listed below.

Isolation and Growth of Micro-organisms
Book: *In vitro* Cultivation of Micro-organisms
       Energy Sources for Cells

Elucidation and Manipulation of Metabolic Pathways
Books: *In vitro* Cultivation of Micro-organisms
       Energy Sources for Cells

Gene Isolation and Characterisation
Books: Techniques for Engineering Genes
       Strategies for Engineering Organisms

Applications of Genetic Manipulation
Books: Techniques for Engineering Genes
       Strategies for Engineering Organisms

Extraction, Purification and Characterisation of an Enzyme
Books: Analysis of Amino Acids, Proteins and Nucleic Acids
       Techniques used in Bioproduct Analysis

Enzyme Engineering
Books: Principles of Enzymology for Technological Applications
       Molecular Fabric of Cells

Bioprocess Technology
Books: Bioreactor Design and Product Yield
       Product Recovery in Bioprocess Technology
       Bioprocess Technology: Modelling and Transport Phenomena
       Operational Modes of Bioreactors

Further information: Greenwich University Press,
University of Greenwich, Avery Hill Road, London, SE9 2HB.

Contributors

AUTHORS

Dr R.D.J. Barker, De Montfort University, Leicester, UK

Professor M. Griffin, Nottingham Trent University, Nottingham, UK

Dr D. Griffiths, Cranfield Biotechnology Ltd, Newport Pagnell, UK

Dr E.J. Hammonds, Nottingham Trent University, Nottingham, UK

Dr S.H. Kirk, Nottingham Trent University, Nottingham, UK

Dr C.K. Leach, De Montfort University, Leicester, UK

Dr O. Misset, Gist-brocades, Delft, Netherlands

Professor T. Palmer, Nottingham Trent University, Nottingham, UK

EDITOR

Dr R.D.J. Barker, De Montfort University, Leicester, UK

SCIENTIFIC AND COURSE ADVISORS

Prof M.C.E. van Dam-Mieras, Open universiteit, Heerlen, The Netherlands

Dr C.K. Leach, De Montfort University, Leicester, UK

ACKNOWLEDGEMENTS

Grateful thanks are extended, not only to the authors, editors and course advisors, but to all those who have contributed to the development and production of this book. They include Mrs A. Allwright, Ms H. Leather, Mrs A. Liney and Miss J. Skelton.

The development of this BIOTOL text has been funded by **COMETT, The European Community Action Programme for Education and Training for Technology**. Additional support was received from the Open universiteit of The Netherlands and by University of Greenwich (formerly Thames Polytechnic).

Contents

How to use an open learning text	viii
Preface	ix
1 Introduction to the technological applications of biocatalysts R.D.J. Barker and C.K. Leach	1
2 Industrial production and purification of enzymes T. Palmer	19
3 The use of soluble enzymes in industrial processes M. Griffin, E.J. Hammonds and C.K. Leach	51
4 Enzyme immobilisation M. Griffin, E.J. Hammonds and C.K. Leach	75
5 Immobilised enzyme reactors C.K. Leach	119
6 Use of enzymes in large-scale industrial applications M. Griffin, E.J. Hammonds and C.K. Leach	149
7 Use of enzymes in analysis M. Griffin and E.J. Hammonds	181
8 Biosensors D. Griffiths	209
9 Use of enzymes in molecular biology and biotechnology: restriction and associated enzymes S.H. Kirk	247
10 Use of enzymes in molecular biology and biotechnology: enzymes other than restriction endonucleases S.H. Kirk	275

11 The application of protein and genetic engineering to industrial enzymes
    O. Misset and C.K. Leach   299

12 Industrial enzymology using non-aqueous systems
    C.K. Leach   327

    Responses to SAQs   339
    Suggestions for further reading   377
    Appendix 1 -Abbreviations used for the common amino acids   378

How to use an open learning text

An open learning text presents to you a very carefully thought out programme of study to achieve stated learning objectives, just as a lecturer does. Rather than just listening to a lecture once, and trying to make notes at the same time, you can with a BIOTOL text study it at your own pace, go back over bits you are unsure about and study wherever you choose. Of great importance are the self assessment questions (SAQs) which challenge your understanding and progress and the responses which provide some help if you have had difficulty. These SAQs are carefully thought out to check that you are indeed achieving the set objectives and therefore are a very important part of your study. Every so often in the text you will find the symbol Π, our open door to learning, which indicates an activity for you to do. You will probably find that this participation is a great help to learning so it is important not to skip it.

Whilst you can, as an open learner, study where and when you want, do try to find a place where you can work without disturbance. Most students aim to study a certain number of hours each day or each weekend. If you decide to study for several hours at once, take short breaks of five to ten minutes regularly as it helps to maintain a higher level of overall concentration.

Before you begin a detailed reading of the text, familiarise yourself with the general layout of the material. Have a look at the contents of the various chapters and flip through the pages to get a general impression of the way the subject is dealt with. Forget the old taboo of not writing in books. There is room for your comments, notes and answers; use it and make the book your own personal study record for future revision and reference.

At intervals you will find a summary and list of objectives. The summary will emphasise the important points covered by the material that you have read and the objectives will give you a check list of the things you should then be able to achieve. There are notes in the left hand margin, to help orientate you and emphasise new and important messages.

BIOTOL will be used by universities, polytechnics and colleges as well as industrial training organisations and professional bodies. The texts will form a basis for flexible courses of all types leading to certificates, diplomas and degrees often through credit accumulation and transfer arrangements. In future there will be additional resources available including videos and computer based training programmes.

Preface

The specificity and efficiency of enzymes in catalysing a wide array of chemical reactions are of crucial importance to metabolism. The importance of enzymes is however, much wider than simply their ability to fulfil their roles in metabolism. Increasingly enzymes are becoming tools of industry in which their catalytic properties are harnessed to bring about changes to materials to produce higher valued products. They are also being employed increasingly as analytical and therapeutic reagents.

A knowledge of enzymes and their applications is, therefore, essential to many in the scientific and industrial communities. As you might anticipate, discussion of enzymes permeate throughout the BIOTOL series of texts. Two BIOTOL texts have, however, been specifically designed to describe the technological applications of enzymes. The first of these, 'Principles of Enzymology for Technological Applications' deals with the underpinning principles of enzymology and aims to provide a thorough understanding of the structure, properties, isolation and analysis of these important molecules. This, the second BIOTOL text devoted to enzymes, builds upon this foundation and examines the ways enzymes may be utilised.

Chapter 1 provides a brief introduction to the technological application of enzymes. It elaborates on the strategy used in the BIOTOL series to explain the principles and practices of industrial enzymology and includes some helpful reminders of the basic principles of enzymology. It, therefore, provides readers with a context in which the rest of the text may be studied and enables readers to quickly judge whether or not they have the appropriate background knowledge so that they will benefit fully from subsequent chapters.

Chapter 2 examines the production and purification of enzymes on an industrial scale. It described the sources used and the different strategies employed in the preparation of secreted and intra-cellular enzymes. In Chapter 3, the advantages and disadvantages of using soluble enzymes are discussed and some application of soluble enzymes are described.

The growing importance of enzyme immobilisation and immobilised enzyme reactors is reflected by the inclusion of two chapters on these topics. The potential advantages and disadvantages of using immobilised enzymes are discussed in Chapter 4 which also includes details of the immobilisation process, the nature of the supports used and the consequences of immobilisation on the properties and behaviour of enzymes. This discussion is extended in Chapter 5 into an examination of the design and performance characteristics of immobilised enzyme reactors. The material covered in Chapters 3-5 is integrated in Chapter 6 which includes more detailed accounts of specific large-scale processes.

In Chapter 7, we introduce the concept of using enzymes as analytical reagents and explain how kinetic and end-point (equilibrium) methods may be used to assay a wide range of materials. The chapter includes discussion of assay systems based on single and on coupled enzyme systems. In Chapter 8 we examine the design and application of biosensors especially those which employ the use of enzymes. Since the key to the success of producing biosensors is the application of appropriate transduction methodologies, this chapter includes some discussion of non-biological aspects of biosensor technology especially emphasising how biological signals are transduced into electronic signals. It is not, however, our intention here to provide an indepth theoretical treatment of the physical processes of signal transduction. This chapter also briefly examines the manufacturing techniques used to produce biosensors. In this way, the reader is made aware of the range of techniques and processes available for the design and production of biosensors.

In Chapters 9 and 10 we examine the enzymes used to manipulate genetic material. In Chapter 9, the properties and activities of restriction enzyme are explained whilst the other key enzymes are covered in Chapter 10. Although used in small quantities, their potential practical and commercial significance is tremendous. One needs hardly to emphasis the commercial significance of genetic engineering, but these enzymes also have applications in many quality control processes as well as in a range of forensic science techniques.

This molecular biological approach is extended in Chapter 11 where we describe the application of genetic engineering to the production of modified enzymes of greater industrial value. The final chapter briefly examines the use of enzymes in non-aqueous systems.

This text is a valuable learning resource and its qualities reflect the skills and knowledge of the author:editor team as enzymologists and as teachers. The knowledge they provide through this text will enable readers to contribute to the development and operation of processes which employ the use of enzymes.

Scientific and Course Advisors: Professor M.C.E. van Dam-Mieras
Dr C.K. Leach

Introduction to the technological applications of biocatalysts

1.1 Introduction	2
1.2 The BIOTOL strategy for discussion of enzyme technology	2
1.3 Assumed knowledge	3
1.4 Arrangement and contents of chapters	16
Summary and objectives	18

Introduction to the technological applications of biocatalysts

1.1 Introduction

In preparing a text on the technological application of enzymes, authors are faced with the difficulty of judging the knowledge base of readers. Equally, readers, when they first encounter a text dealing with technical issues, may have doubts about whether or not they have sufficient knowledge and experience to cope with the material covered by the text. The main purpose of this brief introductory chapter is to explain to readers what assumptions have been made regarding the knowledge needed to understand the issues described in subsequent chapters. To do this, the BIOTOL strategy for discussing enzyme technology will be described and the organisation and topic coverage of subsequent chapters will be explained.

1.2 The BIOTOL strategy for discussion of enzyme technology

The ability of enzymes to fulfil their roles in living organisms largely resides in their specificities and their powers of catalysis. These same properties make them attractive as reagents in a wide variety of industrial, medical and research activities. The potential of enzymes to play a part in a wide range of processes is reflected by the fact that enzyme technology has become one of the most rapidly expanding areas of biotechnology. A major part of contemporary biotechnology now employs the use of enzymes *in vitro*. Consistent with this importance, the discussion of enzymes permeates throughout the BIOTOL series of texts. The technological application of enzymes is arranged into three levels of study. The first level deals with the fundamental properties of enzymes and the knowledge needed to successfully apply enzymes. These issues are covered by the BIOTOL text, 'Principles of Enzymology for Technological Applications'. The second level, the topic of this text, builds upon this underpinning knowledge to show how enzymes may be used to achieve a wide variety of practical objectives. The third level is provided for by in-depth case studies of specific examples of enzymes in a variety of business sectors. These are contained within relevant volumes of the BIOTOL 'Innovations' series, for example, chymosin and specific amino acid metabolising enzymes are discussed in, 'Biotechnological Innovations in Food Processing,' whilst the use of specific enzymes in the diversification of antibiotics is described in, 'Biotechnological Innovations in Chemical Synthesis'. This general organisation of the discussion of enzyme technology is shown diagrammatically in Figure 1.1.

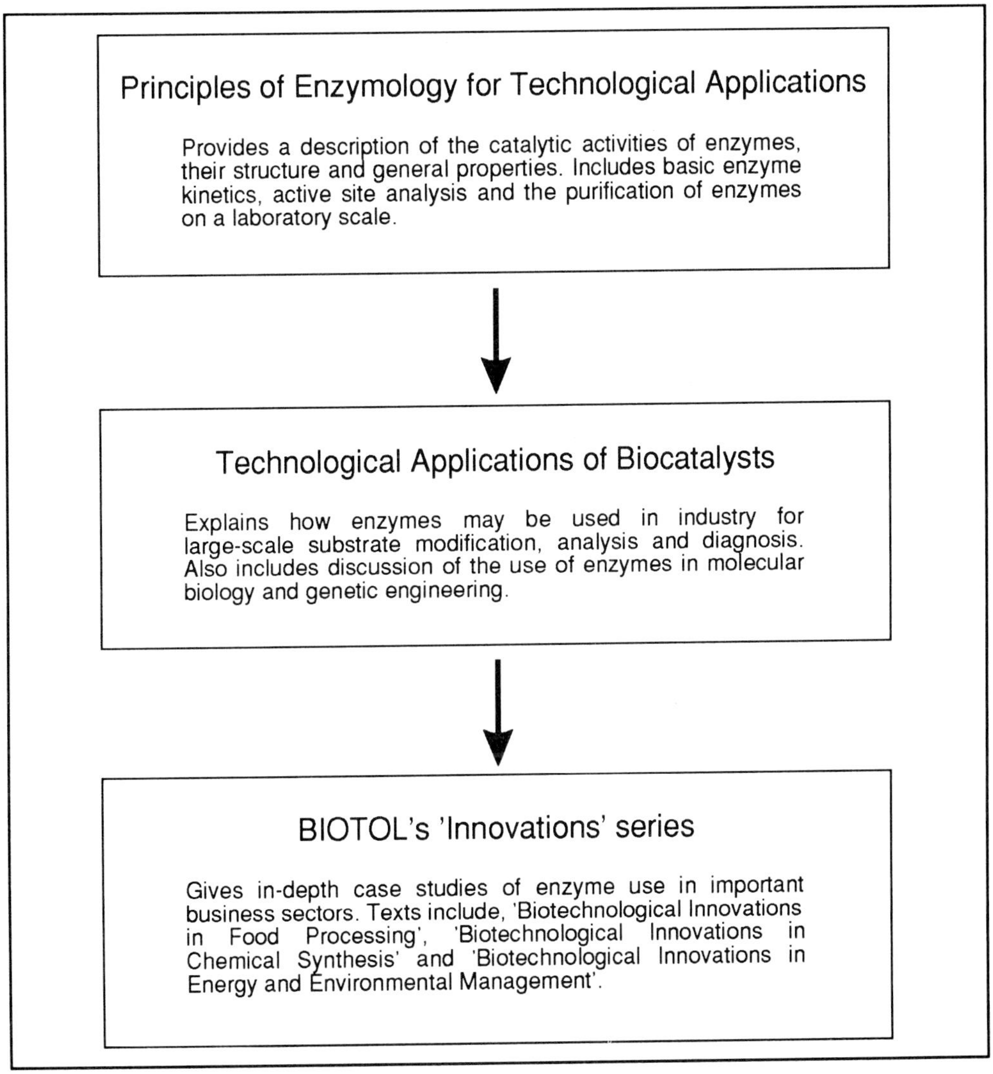

Introduction to the technological applications of biocatalysts

---

Principles of Enzymology for Technological Applications

Provides a description of the catalytic activities of enzymes, their structure and general properties. Includes basic enzyme kinetics, active site analysis and the purification of enzymes on a laboratory scale.

Technological Applications of Biocatalysts

Explains how enzymes may be used in industry for large-scale substrate modification, analysis and diagnosis. Also includes discussion of the use of enzymes in molecular biology and genetic engineering.

BIOTOL's 'Innovations' series

Gives in-depth case studies of enzyme use in important business sectors. Texts include, 'Biotechnological Innovations in Food Processing', 'Biotechnological Innovations in Chemical Synthesis' and 'Biotechnological Innovations in Energy and Environmental Management'.

Figure 1.1 Arrangement of BIOTOL texts dealing with enzyme technology.

1.3 Assumed knowledge

Throughout the text, the assumption has been made that readers are familiar with the basic properties of enzymes and the methods used to analyse these properties. Despite these assumptions, authors have provided many helpful reminders about essential issues and readers with only limited experience of enzymology should encounter little difficulty in understanding the material covered. The inclusion of a variety of in text activities and self assessment questions (SAQs) provides vehicles for reinforcing this understanding.

In this section, we briefly outline what knowledge gained from previous studies, readers will need to be able to fully comprehend the issues discussed in subsequent chapters.

The ideal background to this text is provided by the BIOTOL text, 'Principles of Enzymology for Technological Applications,' but there are many other routes by which the essential underpinning knowledge may have been gained. Here we have divided the discussion of the essential pre-knowledge into a number of sub-sections including:

- enzymes as proteins;
- enzymes, cofactors and coenzymes;
- mechanisms of enzyme catalysis;
- enzyme kinetics.

1.3.1 Enzymes as proteins

influence of physical parameters and enzyme purification

Enzymes are proteins and many of their properties reflect their proteinaceous nature. Thus the influence of environmental parameters (eg pH, temperature, salt concentration) can largely be explained on the basis of the properties of proteins. Similarly, the methods used to purify enzymes (eg ion exchange chromatography, gel filtration, selective precipitation, electrophoresis, iso-electric focusing) are the methods that are generally used to purify proteins.

In this text, the assumption has been made that the readers are familiar with the general properties of proteins in terms of their structures, the effects of environmental parameters on these structures and the methods used to purify proteins from cell extracts. It is also assumed that the reader is familiar with the methods used to determine the primary and higher order structures of proteins. A number of SAQs have been included that enable you to judge whether or not your knowledge of these is sufficient.

SAQ 1.1

Each of statements 1-3 contains two alternative terms. Which is correct?

1) The sequence of amino acid residues in a protein is referred to as the primary/secondary structure of the protein.

2) Hydrophobic sidechains of amino acid residues are predominantly found on the outside/inside of the proteins.

3) If the pI value (isoelectric point) of a protein is 7, at a pH of 8 the protein will carry a net positive/negative charge.

4) Proteins absorb UV light at 280 nm because of the presence of which amino acids?

SAQ 1.2

Decide whether each of the following statements is true or false.

1) SDS (sodium dodecyl sulphate) gel electrophoresis is a good way of purifying active enzymes.

2) In gel filtration column chromatography, larger molecules will be eluted faster from the column than smaller ones.

3) In ion exchange chromatography, proteins bound to the column can be eluted by changing the pH of the eluting medium.

4) In ion exchange chromatography, proteins bound to the column can be eluted by applying a salt gradient.

Introduction to the technological applications of biocatalysts

SAQ 1.3

A liver homogenate was shown to contain 50 mg ml^{-1} protein. Samples of the homogenate were assayed for the presence of three different enzymes A, B and C and the results obtained were as follows:

enzyme A:45 units ml^{-1}; enzyme B:120 units ml^{-1}; enzyme C:85 units ml^{-1}.

The homogenate was then heated to 80°C for 10 minutes, and precipitated protein removed by centrifugation. The protein content of the clear supernatant was shown to be 15 mg ml^{-1} protein. The three enzymes were again assayed and the results obtained were as follows:

enzyme A:2 units ml^{-1}; enzyme B:60 units ml^{-1}; enzyme C:83 units ml^{-1}.

Would this heat treatment step be of any use in purifying any of the three enzymes? When assessing your reasons, use the terms 'specific activity' (ie units of enzyme per mg total protein) and 'yield'.

SAQ 1.4

Solid ammonium sulphate was added to a tissue homogenate to give 35% saturation and the resultant precipitate was centrifuged down. The supernatant was collected and further ammonium sulphate was then added to the supernatant to give 45% saturation. The resultant precipitate was again collected by centrifugation. Continuing in this way a series of precipitates were collected at 35, 45, 55, 67 and 75% ammonium sulphate saturation. Each pellet was re-dissolved in buffer solution and assayed for both total protein, and the presence of the enzyme aspartate aminotransferase (AAT). The data obtained are tabulated below. Using the data, devise a purification step for AAT using ammonium sulphate fractionation.

% saturation of ammonium sulphate	total protein in precipitate (mg)	total enzyme units in precipitate
35	200	0
45	450	15
55	500	270
67	400	20
75	150	5

SAQ 1.5

Sephadex G-50 fractionates proteins in the molecular mass range of 1500 to 30 000 Daltons. If a mixture of the following proteins was passed through the column, how many peaks of protein would you expect to elute from the column?

Protein	Mol. mass
phosphorylase	92 500
transferrin	78 000
hexokinase	45 000
ovalbumin	43 000
lactate dehydrogenase	36 000
trypsin	23 800
soya bean trypsin inhibitor	20 100
lysozyme	14 400
insulin	5 800

Would a better separation be achieved if a column of Sephadex G-75 (fractionation range 3000-70 000) was used?

1.3.2 Enzymes, cofactors and coenzymes

Many enzymes require special factors in order to carry out their catalytic functions. These may include particular metal ions and a range of organic molecules. These non-proteinaceous components are generally referred to as cofactors. Such cofactors may be tightly bound to the enzyme either covalently or by strong ionic interaction. Others are more loosely associated with the enzyme and may be more easily separated from the enzyme-cofactor complex. Organic members of this latter group are sometimes referred to as coenzymes and are usually converted into a different chemical form during catalysis.

apo- and holo-enzymes

The removal of cofactors from enzymes renders the enzymes inactive. The inactive form of an enzyme is referred to as an apo-enzyme whilst the active form, containing the relevant cofactor, is referred to as the holo-enzymes. We can represent these in the following way:

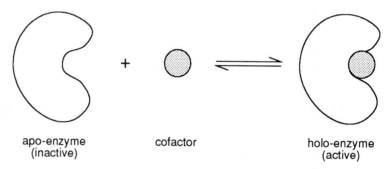

apo-enzyme (inactive)   cofactor   holo-enzyme (active)

It is anticipated that readers will be familiar with a wide range of cofactors/coenzymes and can, in general terms, explain their roles in enzyme reactions. (Try SAQ 1.6 to test your knowledge).

SAQ 1.6

Match the following cofactor (coenzyme) with the appropriate reaction types. (Note: more than one cofactor may be involved with each reaction type).

Cofactor/coenzyme

nicotinamide adenine dinucleotide (NAD^+)

pyridoxal phosphate

biotin

tetrahydrofolate

flavin mononucleotide (FMN)

flavin adenine dinucleotide (FAD)

thiamine pyrophosphate

coenzyme A

Reaction types

cleavage or formation of C-C bonds adjacent to carbonyl carbon atoms

oxidation - reductions involving hydride anion (H^-) transfer

transfer of acyl groups

transamination, decarboxylation and racemisation

carboxylation, transcarboxylation

transfer of one-carbon units

1.3.3 Mechanisms of enzyme catalysis

The general mechanism of enzyme catalysis is to reduce the activation energy barrier that needs to be overcome to enable a thermodynamically favourable reaction to proceed. There are many mechanisms by which an enzyme may achieve this. These include:

- proximity effects. By bringing substrates closer together reactions proceed at a faster rate. Effectively the localised concentrations of substrates are much higher at the enzyme active site than they are in free solution;

- orientation effects. By binding substrates in particular orientations, reactive groups are orientated in such a way that reactions can proceed faster;

- strain. Enzymes may distort (strain) substrates which effectively raises the energy level of the reactant thereby leading to faster reaction rates;

- donation/acceptance of protons. This is usually referred to as acid-base catalysis. Functional groups on enzymes may donate or accept protons (or electrons) to and from substrates. This may contribute to the reaction and may be an important component of rate enhancement;

- covalent bond formation. The enzyme may form covalent bonds with reaction intermediates. This may stabilise what would otherwise be unstable reaction intermediates thereby promoting the reaction rate.

active site

All of these mechanisms imply an association between substrate(s) and a particular part of the enzyme, the active site.

It is this association with the active site that imposes specificity on enzymes. Only the 'correct' substrate(s) will bind with the enzyme. In analysing the mechanisms of enzyme catalysis it is essential to be able to identify the active site and the functional groups within the enzyme that are involved in the catalytic processes. It would be a bonus if you knew how such analysis is carried out and could cite some specific examples.

SAQ 1.7

Comment on the truth, or otherwise, of the following statements. In each case write no more than 2 sentences to support your conclusion.

1) Enzymes alter the equilibrium position of the reaction they speed up.

2) An enzyme-substrate complex must be formed before enzymatic catalysis can occur.

3) When an enzyme binds substrate, binding sites for 2 groups on the substrate confer stereospecificity on the enzyme (ie the ability to distinguish stereoisomers).

4) Enzymes are specific to only one compound and hence only one reaction.

5) Substrate specificity is a consequence of the particular shape and charge distribution of the active site.

1.3.4 Enzyme kinetics

The study of the kinetics of enzyme-catalysed reactions is a major topic which many students find quite difficult. Nevertheless these kinetics are of vital importance to the design and operation of enzyme-catalysed processes used in industry. For this reason, a thorough treatment of enzyme kinetics has been presented in the partner BIOTOL text, 'Principles of Enzymology for Technological Applications'. Here we briefly summarise some of the major aspects of enzyme kinetics which we have assumed you have encountered previously and which are an important foundation to the study presented in this text.

Most students of enzymology will be aware that physical parameters such as pH and temperature may influence the kinetics of enzymatic processes. They are also aware that substrate concentrations also have a marked influence on the rate of reaction. It is assumed that readers are familiar with Michaelis-Menten kinetics in which the initial velocity of an enzyme-catalysed reaction is related to substrate concentration $[S_0]$, the maximum velocity of the reaction (V_{max}) and the affinity of the enzyme for its substrate (as related by the Michaelis constant, K_M) in the following way:

Michaelis-Menten kinetics

$$v_0 = \frac{V_{max}[S_0]}{[S_0] + K_M}$$

K_M is the value of $[S_0]$ which gives a $v_0 = \frac{1}{2}V_{max}$ and the value V_{max} is the reaction rate in the presence of saturating amounts of substrate. It is also assumed that readers are familiar with the determination of K_M and V_{max} using double reciprocal plots (Lineweaver-Burk plots) of $\frac{1}{v_0}$ against $\frac{1}{[S_0]}$.

The Michaelis-Menten relationship is built upon the assumption that the reaction follows the pathway:

$$E + S \underset{k_{-1}}{\overset{k_1}{\rightleftharpoons}} ES \underset{k_{-2}}{\overset{k_2}{\rightleftharpoons}} P + E$$

where E = enzyme, S = substrate, ES = enzyme:substrate complex and P = product.

It also assumes that $k_{-2}[P] = 0$. In other words the back reaction, $P + E \rightarrow ES$ is not taking place. The Michaelis-Menten relationship, therefore, only holds for the initial reaction rate (v_0) for reversible reactions. The velocity of the forward reaction is in fact given by $k_2[ES]$. The maximum velocity (V_{max}) is given when all of the enzyme present is in the form of an enzyme:substrate complex (that is $[ES] = [E_0]$, where $[E_0]$ is the concentration of enzyme used).

Thus $V_{max} = k_2[E_0]$.

turnover number Often k_2 is substituted by k_{cat} (thus $V_{max} = k_{cat}[E_0]$) where k_{cat} is known as the turnover number of the enzyme. This represents the maximum number of substrate molecules that the enzyme can 'turnover' to product per unit time.

specificity constant Another useful term is the ratio k_{cat}/K_M which determines the relative rate of reactions at low substrate concentrations. It is then known as the specificity constant (or catalytic efficiency).

Although strictly speaking the Michaelis-Menten relationship only applies to the initial phase of reactions, in practise it is more widely used. (We shall apply this relationship quite extensively in subsequent chapters). It, in principle, only applies to single substrate reactions, but many reactions involve more than one substrate. If, however, all except one of the substrates are present in saturating amounts, then we can then usually apply a Michaelis-Menten approach.

Π Is the hydrolysis of sucrose by the enzyme sucrase (invertase) likely to exhibit Michaelis-Menten kinetics?

The enzyme catalyses the reaction:

$$\text{sucrose} + H_2O \rightleftharpoons \text{glucose} + \text{fructose}$$

It would appear to be a second order reaction. However, in an aqueous medium, water is present in saturating amounts and the reaction follows what is referred to as 'pseudo first order' kinetics. Under these circumstances, it is likely that the Michaelis-Menten relationship holds.

In many two-substrate reactions of the general type:

$$A + B \rightleftharpoons C + D$$

the reaction is genuinely second order. The precise kinetic relationships which hold depend upon the mechanisms of the interaction between substrates and the enzyme. We can broadly divide reactions into three types. These are:

- random order reactions;
- compulsory order reactions;
- 'ping pong' reactions.

In random order reactions, the two substrates, A and B may be added to the enzyme in any order, the products C and D are also released in any order.

Thus we may represent the mechanism by:

$$E + A \rightleftharpoons EA \searrow^B \quad\quad ^C\searrow ED \rightleftharpoons E + D$$
$$EAB \rightleftharpoons ECD$$
$$E + B \rightleftharpoons EB \nearrow^A \quad\quad ^D\nearrow EC \rightleftharpoons E + C$$

In compulsory order reactions, there is a defined order in which the substrates associate with the active site of the enzyme and also the products are released in a fixed order. A typical example could be:

$$E + A \rightleftharpoons EA \overset{B}{\rightleftharpoons} EAB \rightleftharpoons ECD \overset{C}{\rightleftharpoons} ED \rightleftharpoons E + D$$

In 'ping pong' reactions, the enzyme combines with the first substrate to form an enzyme:substrate complex. The substrate is modified to form a product and, at the same time, the enzyme is modified. The modified enzyme then forms a complex with the second substrate to form a new product and the original form of the enzyme is regenerated. We can represent this by:

$$E + A \rightleftharpoons EA \rightleftharpoons E^*C \overset{C}{\rightleftharpoons} E^* \overset{B}{\rightleftharpoons} E^*B \rightleftharpoons ED \rightleftharpoons E + D$$

where E* represents the modified form of the enzyme.

∏ Consider the following reaction sequence. Is this a random order, compulsory order or a ping pong reaction?

```
        A -Ⓟ                              A
         ↓                                ↓
enzyme ⇌ enzyme A -Ⓟ ⇌ enzyme -ⓅA ⇌ enzyme -Ⓟ
                                          ⇅ B
                                       enzyme ⓅB
                       B -Ⓟ               ⇅
                        ↓
              enzyme ⇌ enzyme B -Ⓟ
```

You should have concluded that this is a ping pong reaction. Notice that a modified form of the enzyme (enzyme -Ⓟ) is formed as an intermediate.

The influence of substrate concentrations on reaction rates is different for each of these reaction types. Indeed, the mechanisms of these type of reactions can be determined by measuring reaction rates at different concentrations of the two substrates. We do not propose to reiterate these kinetics here as they are adequately covered in the partner BIOTOL text, 'Principles of Enzymology for Technological Applications'. Our intention here is to alert you to the fact that in designing industrial processes using enzymes we must consider the influence of substrate concentration on process kinetics. We have provided a summary of the key relationships in Table 1.1. We do not anticipate that you will know all of these relationships off by heart. Read through this table and see if it makes sense. Do not despair if you have difficulties with some of these, we have largely confined our discussion in subsequent chapters to using Michaelis-Menten kinetics. Authors have also provided many helpful reminders in the text. You may also find Table 1.1 a useful reference source. The SAQs at the end of this section will help you to decide whether or not your knowledge is sufficient.

If after attempting these SAQs, you feel that you need to study enzyme kinetics a little more, we recommend the BIOTOL text, 'Principles of Enzymology for Technological Applications'. However, many good biochemical texts also provide adequate coverage.

types of inhibition

Also included in Table 1.1 are kinetic relationships for enzymes that are inhibited. Enzymes may be inhibited in a variety of ways including competitive inhibition, non-competitive inhibition and irreversible inhibition. The various mechanisms of inhibition have important consequences on the kinetics of enzyme-catalysed reactions and can be distinguished by their effects on K_M and V_{max}. It is anticipated that you will have previously encountered this aspect of enzymology. Substrates too may inhibit their enzymes when present in high concentrations.

Laws of mass action:

$$v = k[A] \quad \text{1st order reaction}$$
$$v = k[A][B] \quad \text{2nd order reaction}$$
$$v = k[A]^2 \quad \text{2nd order reaction}$$

Single substrate and pseudo first order enzyme catalysed reactions

for the reaction:

$$E + S \underset{k_{-1}}{\overset{k_1}{\rightleftharpoons}} ES \overset{k_2}{\longrightarrow} E + P$$

Michaelis-Menten kinetics:

$$v_0 = \frac{V_{max}[S_0]}{K_M + [S_0]}$$

where: v_0 = initial reaction rate = $k_2[ES]$
V_{max} = maximum velocity = $k_2[E_0]$
K_M = Michaelis constant

Lineweaver-Burk equation:

$$\frac{1}{v_0} = \frac{K_M}{V_{max}} \frac{1}{[S_0]} + \frac{1}{V_{max}}$$

plots of $\frac{1}{v_0}$ against $\frac{1}{[S_0]}$ give straight lines with intercepts at $\frac{1}{V_{max}}$ and $-\frac{1}{K_M}$ and slope of $\frac{K_M}{V_{max}}$

Eadie-Hofstee relationship:

$$v_0 = -K_M \frac{v_0}{[S_0]} + V_{max}$$

plots of v_0 against $\frac{v_0}{[S_0]}$ give graphs of slope = $-K_M$ and an intercept on the $\frac{v_0}{[S_0]}$ axis of $\frac{V_{max}}{K_M}$; the intercept on the v_0 axis = V_{max}

turnover number:

$$V_{max} = k_2[E_0] = k_{cat}[E_0]$$

k_{cat} = turnover number of the enzyme

specificity constant:

$$\frac{k_{cat}}{K_M}$$

note that evolution leads to maximising the specificity constant while keeping K_M in the same order as the naturally-occurring substrate concentration

Inhibition

competitive inhibition (assuming Michaelis-Menten kinetics):

$$v_0 = \frac{V_{max}[S_0]}{K_M' + [S_0]}$$

competitive inhibitors do not affect V_{max}, but increase K_M

where $K_M' = K_M \left(1 + \frac{[I_0]}{K_i}\right)$

$[[I_0]]$ = inhibitor concentration

Continued .../...

Table 1.1 Some key relationships used in enzyme kinetics.

uncompetitive inhibition:

$$v_0 = \frac{V'_{max}[S_0]}{K'_M + [S_0]}$$

both V_{max} and K_M are changed

where $V'_{max} = \dfrac{V_{max}}{\left(1 + \dfrac{[I_0]}{K_i}\right)}$; $K'_M = \dfrac{K_M}{\left(1 + \dfrac{[I_0]}{K_i}\right)}$

non-competitive inhibition:

$$v_0 = \frac{V'_{max}[S_0]}{K_M + [S_0]}$$

V_{max} is changed but K_M remains unchanged

where $V'_{max} = \dfrac{V_{max}}{\left(1 + \dfrac{[I_0]}{K_i}\right)}$

substrate inhibition:

$$v_0 = \frac{V_{max}[S_0]}{K_M + [S_0]\left(1 + \dfrac{[S_0]}{K_s}\right)}$$

relationship which accounts for inhibition of some enzymes at high substrate concentrations

K_s = binding constant for substrate

Two-substrate reactions

general rate equation for the reaction: $AX + B \rightleftharpoons A + XB$

$$v_0 = \frac{V_{max}[AX_0][B_0]}{K_M^B[AX_0] + K_M^{AX}[B_0] + [AX_0][B_0] + K_s^{AX}K_M^B}$$

where: V_{max} = maximum velocity when AX and B are both present in saturating concentrations
K_M^{AX} = concentration of AX which gives $v_0 = \frac{1}{2} V_{max}$ when B is present in saturating concentrations
K_M^B = concentrations of B which gives $v_0 = \frac{1}{2} V_{max}$ when AX is present in saturating concentrations
K_s^{AX} = dissociation constant for $E + AX \rightleftharpoons EAX$

for 'ping pong' reactions the relationship simplifies to:

$$v_0 = \frac{V_{max}[AX_0][B_0]}{K_M^B[AX_0] + K_M^{AX}[B_0] + [AX_0][B_0]}$$

Table 1.1 (continued) Some key relationships used in enzyme kinetics.

Finally in this section, we must mention allosteric enzymes which show sigmoidal relationships between velocity and substrate concentration. Two models have been proposed to explain allosteric behaviour. These are:

Introduction to the technological applications of biocatalysts

- the model of Monod, Wyman and Changeux (MWC or concerted symmetry model);
- the model of Koshland, Nemethy and Filmer (KNF model).

MWC model In the MWC model, the subunits of a multiunit enzyme are assumed to be able to exist in two forms, a low-affinity form (T-form) and a high affinity form (R-form). Although the low-affinity T form predominates in the absence of substrate, binding of a substrate molecule to one subunit causes this subunit to switch to the high affinity R form. This, in turn, causes the adjacent subunits to adopt the high affinity R form. These subunits now bind substrate with higher affinity and results in a sigmoidal relationship between velocity and substrate concentration. We can represent this in the following way:

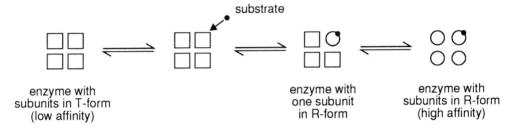

enzyme with subunits in T-form (low affinity)   enzyme with one subunit in R-form   enzyme with subunits in R-form (high affinity)

In the KNF model, it is proposed that both R and T forms can co-exist within a multisubunit enzyme. This is a more complicated model but it is more generally applicable. For example, it can explain cases where initial substrate binding impedes additional binding. It would be helpful, but not essential, for you to understand these models in order to tackle the remainder of this text.

SAQ 1.8

1) Calculate K_M and V_{max} for an enzyme, for which the following data are provided:

	[S], mmol l^{-1}	initial velocity, μmol min^{-1}
(A)	0.6	1.14
(B)	1.0	1.70
(C)	1.5	2.36
(D)	3.0	3.64
(E)	6.0	5.00
(F)	15.0	6.46

2) Assume that you have to accurately measure the K_M of an enzyme, for which preliminary estimates indicate that the K_M is about 4×10^{-3} mol l^{-1}. Recommend what concentrations of substrate should be used for the careful analysis.

SAQ 1.9

Which one of the following statements is correct? Jot down reasons for rejecting the other statements as incorrect. A non-competitive inhibitor:

1) lowers V_{max} and raises K_M;
2) lowers both K_M and V_{max};
3) raises K_M and leaves V_{max} unaltered;
4) lowers V_{max}, whilst K_M is unchanged;
5) results in complete abolition of enzyme activity.

SAQ 1.10

A compound, is known to inhibit an enzyme. However, the type of inhibition is not known. Enzyme assays were conducted in the presence of a fixed concentration of the inhibitor at various substrate concentrations. In all other respects, the assays were identical. The data given below show the [S_0] used and the initial velocity in the absence (column A) and presence (column B) of inhibitor.

[S_0] (mmol l^{-1})	initial velocity (μmol min^{-1})	
	A (- inhibitor)	B (+ inhibitor)
0.02	31.3	16.1
0.03	41.7	22.2
0.04	47.6	27.8
0.06	57.1	35.7
0.12	71.4	52.6

1) Using whatever procedure you consider to be most appropriate, determine what type of inhibition is occurring.
2) Are there any additional experiments which you would recommend to confirm your conclusions? Briefly outline these.

SAQ 1.11

Consider the following statements. Write down, with brief reasons, whether they are true or false.

1) Allosteric enzymes are likely to be more rigid than enzymes which display Michaelis-Menten kinetics.
2) Allosteric enzymes give linear double reciprocal (Lineweaver-Burk) plots, from which the Michaelis constant can be deduced.
3) Allosteric enzymes are ideally suited to a regulatory role.
4) A common feature of allosteric enzymes is possession of more than one subunit.

Enzyme kinetics and the physical environment

In addition to the effects of substrate concentration and the presence of inhibitors on enzyme kinetics, the physical environment also has an effect. Changes in pH cause changes in the charged groups on the enzyme. This may have an effect on both the binding of substrate and on the reaction mechanism itself. Thus we must anticipate that pH will have an effect on the velocity of the reaction and may also have an effect on K_M. Furthermore if substrates and/or products are also ionised, changes in pH may also affect the equilibrium position of the reaction.

changes in v, K_M and K_{eq}

effects of ionic strength

The ionic strength of a solution is also an important parameter affecting enzyme activity. This is particularly true when catalysis depends upon the interaction of charged molecules (for example charged substrates binding to an enzyme or movement of charged groups in the catalytic site). An approximate relationship may be written:

$$\log (k) = \log k_0 + Z_A Z_B \sqrt{I}$$

where k is the actual rate constant, k_0 is the rate constant at zero ionic strength and Z_A, Z_B are the electrostatic charges on the reacting entities. I is the ionic strength of the solution.

In many circumstances, especially with enzymes in which the nature of interactions between substrates and the active site are unknown, the effects of pH and ionic strength on reaction kinetics cannot be predicted from theoretical considerations and have to be determined experimentally. In practice, data from such determinations throw some light on the nature of the interactions involved in the catalytic process.

Students of enzymology are also aware that temperature has a marked effect on enzyme reaction kinetics. Generally as the temperature rises, reaction rates increase according to the Arrhenius relationship:

Arrhenius relationship

$$k = Ae^{-\Delta G^*/RT}$$

where k is the kinetic rate constant; A is the Arrhenius constant; ΔG^* is the standard Gibbs Energy Function change (free energy) of activation; R is the gas constant and T is the absolute temperature.

Q_{10} typically = 2

Typically for many reactions the rise in rate of reaction for every 10°C (Q_{10}) is by a factor of about 2. This factor is usually denoted by Q_{10} (ie Q_{10} is the relative increase in the rate of reaction over a 10°C rise in temperature).

However, as the temperature rises, enzymes may be irreversibly denatured and lose activity. The ΔG^* for the denaturation is often about 200-300 kJ mol^{-1} which gives Q_{10} values of about 5-35. This means that above a critical temperature, enzymes may rapidly lose activity.

In some instances, the inactivation of enzymes follow a simple first-order reaction. Thus:

$$[E_t] = [E_0]e^{-k_i t}$$

where $[E_t]$ = active enzyme concentration at time t; $[E_0]$ = initial concentration of active enzyme; k_i = inactivation rate coefficient and t = time.

In practice, the kinetics of inactivation of enzymes may be quite complex since the inactivation may involve a series of intermediates. Because of the importance of this topic to the successful operation of enzyme-based processes, we will deal with it again in appropriate sections of the text. In an industrial context, the half-life ($t_{\frac{1}{2}}$) of an enzyme is often quoted. This is the time it takes for the activity to reduce to half its original value under specified conditions. The thermal stability of an enzyme is an important consideration in the long term storage and use of enzymes. In practice, the thermal stability of enzymes is usually determined experimentally rather than from theoretical considerations.

1.4 Arrangement and contents of chapters

Although the activities of specific enzymes may, in some instances, be used in intact organisms, in most cases there is a requirement to purify the enzyme of interest before it can be used for commercial or medical purposes.

∏ See if you can list two or three reasons why enzymes need to be purified before they can be used in many practical circumstances.

diverse products

physical barriers to the delivery of substrate

low enzyme levels

The most obvious reason is that organisms contain many enzymes arranged into a variety of metabolic pathways. Thus if we used whole organisms instead of a single enzyme, we are likely to generate a variety of products and the yield of the desirable product may be low or even negligible. By using a single, purified enzyme we can specify the reaction. Another reason for using purified enzymes, rather than intact organisms is than the plasma membranes of the organism may act as a barrier to delivery of substrate to the enzyme thereby reducing reaction rates. Furthermore, the enzyme catalysing the desired reaction represents only a small portion of the total enzyme complement of cells thus, even if we pack cells close together, we still have only a relatively low concentration of the desirable enzyme.

Apart from these difficulties, it is also difficult to study the properties of the enzyme *in vivo* (for example the effects of environmental parameters on stability and activity, the influence of substrate concentration and the effects of inhibitors on activity). This information is vitally important if we are to get the best out of our enzyme. You may well have thought of additional reasons why we need to use purified enzymes.

sources of enzymes and purification

For these reasons, we begin Chapter 2 by considering the sources of enzymes we may use and the methods used to purify them on a commercial scale. These methods have many similarities with the processes used to purify enzymes in the laboratory. They make use of the same molecular properties (for example, differences in solubility, size, ionic charge, etc) but the equipment used is somewhat different.

use of soluble enzymes

In Chapter 3 we examine the use of soluble enzymes in large-scale industrial processes. We consider the advantages and disadvantages of using enzymes in a soluble form. The main disadvantage is that the enzyme is usually lost at the end of the process. This may be overcome by attaching the enzymes to suitable supports or carriers in a process usually referred to as enzyme immobilisation.

Introduction to the technological applications of biocatalysts

production and properties of immobilised enzymes

We examine the processes and materials used to immobilise enzymes in Chapter 4. This chapter also considers aspects of the effects of immobilisation on the properties and behaviour of enzymes. This is extended in Chapter 5 where we describe the design and operational characteristics of immobilised enzyme reactors.

In Chapter 6, we draw together the information given in Chapters 3-5, by examining some specific applications of enzymes used in large-scale processes.

use of enzymes in analysis

Enzymes, however, do not only find use as agents to mediate chemical changes on a large scale. They find extensive use as tools for analysis and diagnosis. Chapters 7 and 8 examine this aspect of the technical application of enzymes. In Chapter 7, we describe the basic principles of enzyme-based analysis whilst in Chapter 8 we examine their uses in the rapidly expanding business area of biosensors.

use of enzymes in genetic engineering

In addition to being tools used for analytical purposes, specific enzymes are vitally important in the manipulation of genetic information. In Chapters 9 and 10 we examine the use of enzymes in the areas of genetic engineering and molecular biology. In Chapter 9 we predominantly focus on a vitally important group of enzymes, the so called restriction enzymes, which enable us to cut DNA molecules into specific fragments. In Chapter 10, we examine a range of other nuclease enzymes, ligases and DNA modifying enzymes of vital importance in contemporary biotechnology and biomolecular research.

protein engineering

Enzymes are not only of vital importance to genetic engineering, they are themselves targets for genetic engineering. Developing novel (modified) enzymes with more desirable properties (for example, improved stability, higher catalytic efficiency, changed substrate specificity) often employs the techniques of the genetic engineer. This aspect of enzyme technology is examined in Chapter 11 in which we include some specific examples to illustrate the kinds of outcomes which may be achieved.

In our final chapter, we explore the use of enzymes in non-aqueous systems. Traditionally we think of enzymes as operating in an aqueous milieu. However, there are many reasons for considering employing enzymes in non-aqueous media, for example because of the low water-solubility of substrate, changes in equilibria (especially of hydrolytic reactions) and potential improvements in product recovery. In this final chapter, we explain these reasons in greater depth and describe the strategies used in this relatively new and exciting application of enzymes.

Summary and objectives

The main purpose of this brief introductory chapter is to ensure that readers are able to judge whether or not they are properly equipped to tackle the material covered in the remainder of the text. We also described, in outline, the BIOTOL strategy for discussion of enzyme technology and the structure and context of the text.

Now that you have completed this chapter you should be able to:

- judge whether your previous knowledge of enzymes and enzymology is sufficient to enable you to fully benefit from this text;
- understand the organisation of material covered by this text.

Industrial production and purification of enzymes

2.1 Introduction	20
2.2 Sources of enzymes for industry: the importance of micro-organisms	20
2.3 Comparison of the criteria applied to the extraction of enzymes for scientific and industrial purposes	25
2.4 Developing a strategy for extracting an enzyme on an industrial scale	27
2.5 Large-scale extraction - separating particular matter	29
2.6 Disruption of cells on a large scale	35
2.7 Large-scale purification	42
2.8 Production of artificial enzymes	48
Summary and objectives	50

Industrial production and purification of enzymes

2.1 Introduction

Biotechnological applications often involve the use of purified enzymes, and in this chapter we shall consider how enzymes may be obtained in large amounts in a purified form. However, we shall not be restricting ourselves to a catalogue of purification procedures used in industry. We shall take a brief look at general strategies by which a required enzyme may be extracted from cells and then purified. In the main, we shall focus on those procedures which are particularly relevant to enzyme purification on an industrial scale.

The most suitable enzymes for commercial purification are those which can be obtained in large amounts from a convenient source by simple procedures. Some enzymes which would be of considerable value in a purified form are not easily accessible so in consequence are currently of little industrial interest. From a commercial point of view, the value of the product has to be judged against the cost of obtaining it, whereas a purely scientific investigation might have different priorities. In either case, however, a knowledge of the general principles of extraction and purification is important. The first requirement, therefore, is to ensure that you are familiar with these topics, for only when the general principles have been assimilated can we go on to consider the particular requirements of industry.

As we described in Chapter 1, this text has been written on the assumption that you have studied previously aspects of enzymology and are familiar with their basic properties. It has also been assumed that you are familiar with the general principles of protein (especially enzyme) purification. In this chapter, we will focus attention on the processes used to produce and purify enzymes on an industrial scale. We have, however, provided a brief revisionary review of the general principles of enzyme extraction and purification which apply regardless of scale. Further details of the general principles of enzyme extraction and purification can be obtained from the BIOTOL texts, 'Principles of Enzymology for Technological Applications', and, 'Analysis of Amino Acids, Proteins and Nucleic Acids'.

2.2 Sources of enzymes for industry: the importance of micro-organisms

The sources of enzymes used in industry are mainly commonly cultivated plants and animals and a variety of micro-organisms. In the case of plants and animals, the preparation of useful enzymes often only involves part of the organism. For example, chymosin was traditionally isolated from the stomach of calves. Similar other animal-derived proteolytic enzymes (eg trypsin, chymotrypsin) only involved extraction from specific organs. Likewise plant-derived enzymes (eg ficin, papain) are extracted from specific organs. Often enzyme extraction is only a small part of the motivation behind raising the original crop and is of only minor importance to those cultivating the appropriate plants and animals. Furthermore, the large-scale extraction of enzymes from plant and animal sources poses technical, economical and, sometimes, ethical problems.

Thus, although there are a number of plant and animal enzymes with important industrial uses, some going back many centuries (for example chymosin in cheesemaking), there is an increasing tendency to use microbial sources. It is for this reason, therefore, that we will predominantly focus on the preparation of enzymes from microbial sources. However, as we will explain later, not all enzymes may be prepared from such sources and so some consideration will be given to animal and plant cells as potential providers of industrial enzymes.

increasing tendency to use microbial enzymes

2.2.1 Replacing animal and plant sources by micro-organisms

There are many attractions to replacing animal and plant sources by micro-organisms. The characteristics of microbial cultures (fast growth rates in defined conditions) and the ability to produce pure (cloned) strains make micro-organisms particularly attractive as biochemical resources.

microbial enzymes may resemble those from plants and animals

Because of the large number and wide range of micro-organisms available, a species can usually be found which contains an enzyme bearing a reasonable resemblance to one from a plant or animal source. Moreover, sometimes one is found which is more advantageous than traditional plant or animal enzymes for a particular application. For example, a more heat-stable enzyme may be found in a micro-organism whose natural habitat is hot springs.

use of genetic engineering

If no suitable microbial enzyme can be found, it may be possible to engineer one by means of recombinant DNA technology. This technique, which is considered in a later chapter, enables genetic material to be inserted into micro-organisms so that they will produce new enzymes, including ones normally synthesised by eukaryotic organisms. This approach is largely at the development stage as far as industrial applications are concerned, but it offers much promise for the future.

increase enzyme yield by genetic engineering

One successful application of genetic engineering has been to increase the yield of penicillin-G-amidase from *E. coli* by inserting multiple copies of the relevant gene into the host, using the plasmid pBR322 as a vector. The increased yield of the enzyme at this stage is important from a commercial point of view because it helps to reduce downstream processing costs (ie the costs of further purification).

Particular problems are encountered if an attempt is made to produce a eukaryotic protein by inserting the relevant gene into a micro-organism.

Π From your previous studies, what would you consider to be the main one, and how may it be overcome? (Hint: think about the structures of prokaryotic and eukaryotic genes).

problems with introns

The main problem is that the genes of eukaryotic organisms contain non-expressed intron regions, which are normally excised at the transcription stage, but microbial hosts are unable to do this. The solution is to insert the appropriate cDNA, which contains no introns, instead of the whole gene. cDNA is copy (or complementary) DNA that has been made using mRNA (from which the introns have been removed) as a template. A single-strand cDNA can be synthesised complementary to the relevant mRNA by the use of a viral reverse transcriptase.

post-translational modifications eg glycosylation

However, problems may still remain if a cDNA such as this is inserted into *E. coli* via a suitable vector. The final structure of the protein produced may still not be quite correct, because of the inability of the bacterium to bring about the required post-translational modifications such as the addition of sugar units. A partial solution is to clone DNA into shuttle vectors, (for example plasmids which can replicate both in *E. coli* and in a yeast). Since yeast cells are eukaryotic, they are more likely to be able to achieve a reasonable (if not total) degree of success with the post-translational modifications. For example, calf chymosin, an enzyme used traditionally in cheesemaking, has been produced by inserting the appropriate DNA into the yeast, *Kluyveromyces* spp., an organism acceptable to the food industry.

protein engineering

A further refinement is that the structure of a gene may be carefully altered prior to insertion into a micro-organism, so that an enzyme of modified characteristics (for example, with greater stability or easier to purify or greater catalytic efficiency) will be produced. This process is called protein (or enzyme) engineering. We will return to these issues in a later chapter.

However, despite the undoubted potential of techniques involving recombinant DNA technology, currently it often makes better commercial sense to search for a suitable enzyme source in more conventional ways. Much effort, for example, has gone into developing systems for cloning glucose isomerase, but with little impact so far on industrial applications.

Π If a micro-organism is found which is capable, naturally or otherwise, of synthesising an appropriate enzyme, can it be assumed that it would make a suitable source for the industrial extraction and purification of that enzyme? If not, what other factors need to be taken into account?

The assumption cannot be made automatically. For example, if the organism is to be used in an industrial process, it should be safe to handle and should not produce, in addition to the required enzyme, significant amounts of substances that might be hazardous to health. Enzyme yield is another factor that has to be taken into account and so a strain is often sought which will have a minimum of negative controls restricting the synthesis of the enzyme.

Π From your previous studies, write down what factors are likely to regulate enzyme synthesis, assuming that an adequate supply of nutrients is available to the micro-organism.

regulation of enzyme synthesis

In general, enzyme synthesis in prokaryotic cells is regulated by the processes of induction and repression. Although differences occur from system to system, synthesis of an enzyme is often induced in the presence of its substrate and repressed in the presence of a product (or a further metabolite of the product).

mutagenesis

strains selection

If a population of cells of a particular micro-organism is exposed to, say, X-rays, or a chemical mutagen, a variety of genetic mutations will occur. It may then be possible, by trying to culture the cells under conditions where the substrate of an enzyme of interest is present in only limited amounts, to isolate and culture a mutant strain which does not require induction, as this could be the only one to thrive under the conditions used. An alternative way to produce such a mutant would be by means of recombinant DNA technology. In this context, it should be noted that penicillin-G-amidase is produced

Industrial production and purification of enzymes

constitutively, ie without requiring induction, by the system utilising the inserted plasmid genes mentioned previously. This is not the case in the original organism.

use of fermentation to produce enzymes

When it is considered that a suitable strain of a suitable micro-organism has been obtained, by whatever means, it may then be grown in culture to produce large amounts of the relevant enzyme. This process is often loosely termed fermentation and the vessels used for the cultivation of large amounts of microbial biomass are referred to as fermenters. Strictly speaking fermentation is an anaerobic process. Since microbes are often cultivated under aerobic conditions, we prefer to call the vessels in which they are cultivated bioreactors.

SAQ 2.1

Which of the following facts encourage the replacement of animals and plants by micro-organisms as sources for enzymes? Record + to indicate those features which encourage their use and - to indicate those which discourage their use.

feature	encourage/ discourage
1) great diversity of microbes available;	
2) fast growth rates;	
3) can be grown in pure culture;	
4) some microbes can grow in extremes of physical conditions;	
5) many are unicellular;	
6) micro-organisms are often associated with disease;	
7) many micro-organisms carry plasmids;	
8) post-translational modification of proteins;	
9) RNA processing (eg removal of introns);	
10) metabolic regulation.	

2.2.2 Plant and animal cell cultures as sources of enzymes

In the same way, plant or animal cells could be cultivated to enable large-scale extraction of their enzymes to be carried out.

animal cell culture

With animal cell culture, the first stage is often to treat isolated tissue with collagenase or other proteolytic enzymes to break down the matrix and enable the individual cells to be dispersed. The conditions of proteolysis should, of course, be sufficiently mild so that each isolated cell retains most of its original characteristics, but it is very difficult to prevent some damage occurring to surface proteins. Nevertheless, under favourable conditions, isolated cells in culture may grow and divide, similar to unicellular micro-organisms. The main technical problem in long-term culture is the prevention of contamination, particularly by micro-organisms, but there is also a tendency for cells to lose their precise function (ie to de-differentiate) as time passes.

plant cell culture

Similar difficulties are found with plant tissue culture. Hence, despite the advances that have taken place, the extraction of enzymes from cultured plant or animal cells has yet to find widespread application in industry, and microbial fermentation continues to be the main method for the large-scale production of enzymes.

2.2.3 Production of cells

Some fungal enzymes, eg *Aspergillus* proteases, may be obtained from a semi-solid culture, where the low water content and high degree of aeration at the surface apparently favours their production. However, the vast majority of microbial enzymes are produced in aseptic, submerged culture, where conditions such as temperature, pH and degree of aeration can be finely controlled. The synthesis of most of the enzymes of commercial significance is related to cell growth, but can lag behind it. This is particularly true in the case of hydrolytic enzymes, and enzymes whose production is controlled by plasmids. Generally, in a single-batch procedure, the number of cells present may increase for about a day, but enzyme production is likely to continue at significant levels for several days after this.

surface and submerged cultivation

∏ A vessel contains a single bacterium which has a replication time of one hour. Thus, 2 bacteria will be present after 1 hour and 4 bacteria after 2 hours. Assuming that replication continues at this rate, what is the likely number of bacteria after 24 hours?

After 24 hours there will be $2^{24} = 1.7 \times 10^7$ bacteria present. In practice, of course, it is impossible for growth to continue indefinitely at the same rate in a finite volume of medium. Nevertheless, you can see how tremendously fast the biomass could, in principle, increase in a microbial culture.

The culture medium must contain the essential nutrients needed by the micro-organism, together with any inducer necessary for the synthesis of the required enzyme. However, if available, a strain which does not require induction is used. Regardless of induction, the synthesis of many catabolic enzymes is repressed if the cells are grown rapidly on readily-utilisable carbohydrate or protein.

catabolite repression

∏ How do you think, in general, such catabolite repression might be overcome?

One way is to avoid the use of a readily-utilisable carbon source, introducing instead a source which can only be utilised if the enzyme in question is synthesised. Another way is to control the availability of the carbon source causing the repression, either by feeding small amounts into the system at regular intervals, or by adding it in a form (for example as an ester) which is slowly hydrolysed to yield the nutrient. Similarly, in the case of biosynthetic enzymes (and some others), feedback repression may occur in the presence of metabolic end-products, so the build-up of these should be avoided. Alternatively, the problems of repression may be overcome by the use of mutants.

We do not intend to discuss the technical issues of cultivating cells on a large scale in this text. These aspects are covered elsewhere in the BIOTOL series (see for example, in 'Operational Modes of Bioreactors', or, '*In Vitro* Cultivation of Micro-organisms').

We do, however, need to mention oxygen. Oxygen is often a limiting nutrient in cell cultivation procedures, and it is useful to assess the capacity of the system for absorbing oxygen.

importance of bioreactor design

For controlled bioprocessing, it is important that the culture medium is well-mixed, so that pO_2, pH, temperature and nutrient concentrations are the same throughout. Hence the mixing procedure, which usually involves a propeller or flat-bladed turbine, must be adequately designed. This presents one of the main problems in converting a small-scale procedure to a large-scale one.

Industrial production and purification of enzymes

many commercially important microbial enzymes are extracellular

Many of the enzymes obtained commercially by means of microbial cultures are extracellular ones, ie they pass into the culture medium, so they do not have to be extracted from the cell. Hydrolases generally fall into this category. Increased yields of extracellular enzymes may be obtained by introducing surfactants, such a Tween 80, into the culture medium, although the precise reason for this is not known.

Extracellular enzymes may be separated from the cells by simple procedures once the cell cultivation is complete. Continuous culturing processes could be employed, but these are more difficult to design and control than are single-batch ones. In contrast to extracellular enzymes, intracellular enzymes must, of course, be released by cellular disruption before separation can take place.

An intermediate situation can exist with Gram negative bacteria such as *E. coli* in which some enzymes are found in the periplasmic space between the plasma membrane and the outer membrane. These can be released into the medium by relatively mild treatments, such as osmotic shock. To release enzymes which are truly intracellular, harsher treatments are required.

Now we are at the position of selecting and producing our source of enzyme, we now need to consider how we are going to extract and purify the enzyme. Your previous experience of enzyme extraction and purification of enzymes probably relates to the purification of enzymes from a scientific viewpoint. The industrial perspective is somewhat different. We will examine this in the next section.

2.3 Comparison of the criteria applied to the extraction of enzymes for scientific and industrial purposes

The requirements for enzyme extraction and purification as part of a commercial operation may well be different from those where it forms part of a scientific study.

Π Apart from the obvious difference in scale, what do you think the main differences in requirements might be?

The aim of enzyme purification as part of a scientific study is usually to obtain a specimen in as pure a form as possible, disregarding to a certain extent yield, effort and cost. In contrast, the important requirements of an industrial process will be high yield and low cost. The enzyme will often only be purified as far as is necessary to remove contaminants which would interfere with the function for which the enzyme is intended. Purification steps will be reduced to a minimum, since each will contribute to the overall cost, effort, and reduction in yield. If just one of the steps involved has a low yield, the overall recovery of the enzyme will be low. We have illustrated this in Figure 2.1. In this figure we have plotted the overall yield against the number of process steps for different step yields. Thus the upper line represents the yield when the yield for each step is 95%. The lower line represents the yield when each step yields 80%. The data reported in this figure can be calculated using the following equation:

overall yield = yield of step 1 (as a decimal) x yield of step 2 x yield of step 3 etc.

Then coverting the overall yield to % by multiplying by 100.

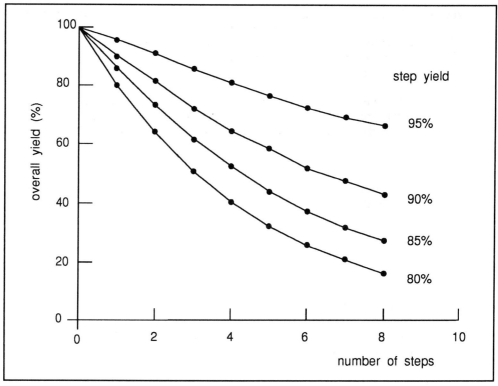

Figure 2.1 Overall yield as function of the number of process steps.

In practice, however, it is uncommon for each step in a process to give the same yield. Attempt SAQ 2.2 to give yourself some practice.

SAQ 2.2

Two processes for enzyme purification, each of five steps, were tested. In the first process, each step gave a 90% yield of the active enzyme. In the second process, four of the steps gave a 100% yield of active enzyme, and the fifth a 60% yield. Which process gave the best overall yield of active enzyme?

In some cases, where the enzyme being produced is extracellular, the culture medium (fermentation broth) itself may be suitable for marketing as an enzyme preparation, after appropriate concentration and addition of stabilisers. If, however, downstream processing is considered necessary to enable the product to meet required specifications, then this is made more straightforward if the concentration of the relevant enzyme after cultivation is high, forming something of the order of 10% of the total protein content.

Π It is also important that the required enzyme does not lose activity at this (or indeed any) stage, if it can be avoided. Summarise the factors which might cause loss of enzyme activity.

From your knowledge of enzymes you should have realised that enzyme activity may be lost as a result of heat, oxidation, or an inappropriate pH. The effect of proteolytic enzymes might also be significant, as may the presence of heavy metal ions (for example, Hg^{2+}).

Industrial production and purification of enzymes

control of conditions during purifications

These and other factors must be taken into account during extraction and purification. In particular, it is important to maintain a low temperature throughout (unless the enzyme in question is particularly heat-stable), to adjust and maintain the pH as necessary, and to remove (or inhibit) unwanted proteolytic enzymes (proteases) as soon as possible.

Thus we can conclude that for an industrial process, it is usual to design a process with a minimum number of operational steps consistent with producing a product which meets the specifications needed for its subsequent use. Each operational step should be optimised to give high yields of products with low costs of operation.

We will now move on to developing a strategy for extracting and purifying enzymes on a large scale. We begin by considering the importance of the cellular location of the required enzyme.

2.4 Developing a strategy for extracting an enzyme on an industrial scale

In this section we will begin by considering the importance of the location of the desired enzyme within the cell. We will then examine the strategies and techniques used in extracting the desired enzyme. In the subsequent section we will examine how the extracted enzyme might be purified.

2.4.1 Importance of the cellular location of enzymes

Microbial enzymes may be extracellular, intracellular, or, in some instances in Gram negative bacteria, located in the periplasmic space between the plasma and outer membranes (note that periplasmic enzymes may also be found in yeast cells). Intracellular enzymes may be soluble or membrane-bound, but the latter type, especially when the enzymes are integral proteins, have not so far been of much commercial interest.

∏ Why do you think this is?

It is because of the extra difficulties associated with the extraction, purification and stabilisation of membrane-bound enzymes. (Note that these are discussed in the Appendix). Hence we shall not be considering membrane-bound enzymes in the present section.

inclusion bodies

Enzymes produced in bacteria by the use of recombinant DNA techniques may be secreted out of the cell altogether, like natural extracellular enzymes; they may accumulate in the periplasmic space, or they may remain within the cell as components of insoluble membrane-free aggregates of protein and nucleic acid, called inclusion bodies. The reason for the formation of inclusion bodies is not clear, but one factor may be the inability of bacterial systems to add appropriate sugar units to eukaryotic proteins. The location of a required enzyme within an inclusion body has advantages as well as disadvantages for its purification. Additional steps are required to separate the enzyme from other components of the inclusion body, which might not be straightforward, but on the other hand, the enzyme is likely to be more concentrated than it would be if it was simply dissolved in the culture medium (fermentation broth), and also it could be protected from the effects of proteolytic enzymes.

Basic approaches to obtaining initial cell-free preparations of extracellular, soluble intracellular and periplasmic enzymes from the products of microbial cultures are shown in Figure 2.2. Also indicated in Figure 2.2 is what, in general, is likely to be found in each type of preparation, in addition to the enzyme of particular interest.

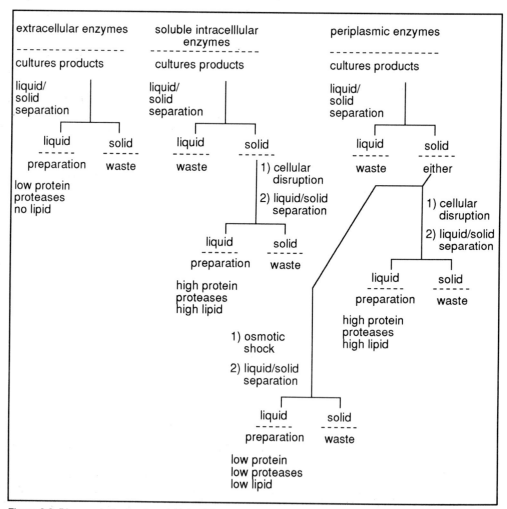

Figure 2.2 Diagram indicating how initial cell-free preparations of various types of enzymes may be obtained from microbial culture. The diagram also indicates what these preparations are likely to contain, in addition to the enzyme of interest.

Π In the case of periplasmic enzymes, alternative approaches to extraction are given in Figure 2.2. What do you see as the main differences between them?

extraction of periplasmic proteins

Extraction by osmotic shock (eg by suspending bacteria in 20% buffered sucrose, centrifuging, and re-suspending in cold water) can give an enzyme preparation with low levels of contaminants, since the plasma membrane should remain intact and hold back most intracellular components. More robust treatment, disrupting the plasma membrane, will give an enzyme preparation with high levels of contaminants. Unfortunately, extraction by osmotic shock tends to give lower yields of periplasmic enzymes than are obtained by complete cellular disruption, so a choice has to be made

Industrial production and purification of enzymes

between a procedure giving a relatively pure preparation with a low yield of the relevant enzyme, and one giving a higher yield, but with much more contamination.

extraction from inclusion bodies

Enzymes present in insoluble inclusion bodies would be released from cells in the same way as soluble intracellular enzymes, but as stated above, additional steps would be required to separate inclusion bodies from both soluble material and cell debris, and then to liberate the required enzyme in a soluble form.

Approaches along the same lines as those outlined for microbial systems in Figure 2.2 are also applicable to plant and animal systems. Intracellular plant and animal enzymes must, of course, be released by cellular disruption, whereas extracellular enzymes are secreted naturally by the living cells. Secretions from cells forming part of a tissue (in contrast to secretions from individual cells in culture) often give rise to a straightforward liquid/solid separation. Hence, as, for example, with enzymes in plant latex, no other liquid/solid separation may be required.

SAQ 2.3

In a pilot experiment a suspension of bacterial cells was divided into two samples. Sample A was disrupted using ultrasonic disintegration. After removal of intact cells and cell wall debris by centrifugation, the total activity of an enzyme in the liquid supernatant was determined to be 1000 units and the supernatant contained 2.0g of protein.

Sample B was subjected to osmotic shock and then cells and liquid were separated by centrifugation.

The total activity of the same enzyme in the liquid supernatant was determined to be 550 units and the supernatant contained 0.1 g protein.

Suggest the likely cellular location of the enzyme and select which of the two strategies looks most promising to use to produce the enzyme for commercial use.

2.5 Large-scale extraction - separating particular matter

liquid/solid separations by centrifugation, filtration or biphasic partitioning

Liquid/solid separations in microbial systems may have to be carried out in the initial separation of cells from the culture medium; in the separation of debris of disrupted cells from soluble contents which have been released, and in the separation of precipitates from soluble material. Such separations on an industrial scale often involve:

- centrifugation;
- filtration;
- aqueous biphasic partitioning.

In general, centrifugation is particularly useful when the solid fraction is the one containing the enzyme of importance, whilst the other two procedures provide convenient ways of removing unwanted solid when the important enzyme is in solution.

2.5.1 Centrifuges

The types of centrifuges commonly seen in laboratories, involving tubes in fixed-angle or swing-out rotors, are not suitable for scaling up for use in industrial processes.

∏ Can you suggest reasons for this?

One reason is that the mechanical stress on the centrifuge rotor is proportional to the square of the radius, so problems (including ones involving safety) arise if bigger centrifuges and rotors are made to the same basic design. Also, starting-up and slowing-down procedures are time-consuming, so single-batch operations are inefficient, particularly if the capacity is low. Hence, systems which allow material to be fed in continuously without stopping the centrifuge are preferred for industrial applications. Some of these are illustrated in Figure 2.3.

use of continuous centrifugation

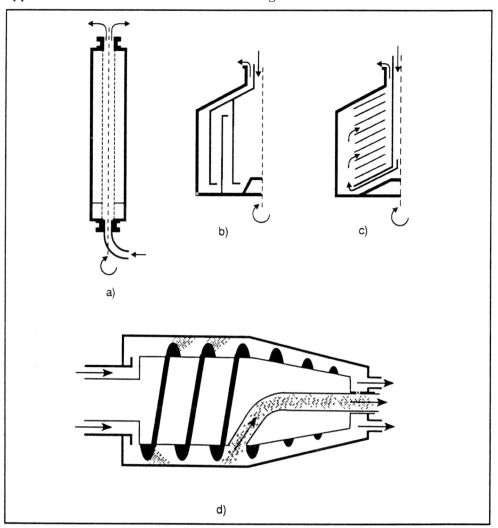

Figure 2.3 Continuous centrifuges: a) tubular centrifuge, b) multichamber, c) disk stack centrifuge, d) decanter centrifuge (see text for description).

Industrial production and purification of enzymes

tubular bowl centrifuge

One common design is the tubular bowl centrifuge, (Sharples centrifuges are well-known examples, see Figure 2.3 a). The rotor of a tubular bowl centrifuge is essentially a long tube, usually placed vertically, with constrictions at the top and the bottom. The feed material comes in continuously at the bottom and the clarified liquid passes out at the top, whilst the sludge collects at the side of the tube (as a result of centrifugal effects), concentrating in the lower half (because of gravity). Such a system is only semi-continuous, because centrifugation has to be stopped at intervals to enable the solid material to be removed. However, because of the narrowness of the tubular rotor, stopping and accelerating to full speed again can be accomplished quickly. Tubular bowl centrifuges can be effective in sedimenting debris and precipitates, but only on a relatively small scale. With larger-scale operation, mechanical factors limit the centrifugal field achievable. A similar principle is applied in the multichamber type (Figure 2.3b)

disc bowl centrifuge

An alternative design is the disc bowl (or disc-stack) centrifuge (Figure 2.3 c). Here the rotor contains a stack of discs which effectively increases its length, and so increases the effectiveness of the separation process. Such a system may be operated semi-continuously, like a tubular bowl centrifuge, or alternatively the sludge may be forced out continuously through a valve.

scroll (decanter) centrifuge

Another design which can be operated continuously is the scroll or decanter centrifuge (Figure 2.3 d). This is something like the tubular bowl centrifuge, but it is orientated horizontally and incorporates an Archimedes screw to scrape the sludge from the walls of the rotor and force it in the direction opposite to that of the flow of clarified liquid, enabling both fractions to be removed continuously. Disc bowl and scroll centrifuges (especially the latter) have restricted accelerations, so they are best suited to the collection of large particles.

The selection of the type of centrifuge to be used is based on the size of the particles to be separated and the volume of particles in the feed. Figure 2.4 gives some guidelines.

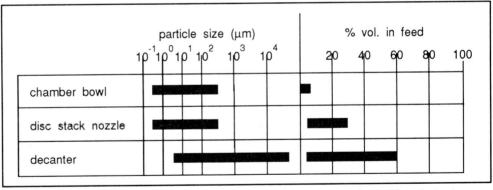

Figure 2.4 Selection criteria for centrifuges and decanters. % vol in feed = % v/v of solids in the feed.

SAQ 2.4

From the guidelines given in Figure 2.4, which type of device is best suited for collecting larger particles and which type can only handle dilute suspensions?

use of coagulation or flocculation to increase particle size

It will be apparent from the above that the collection of small particles on an industrial scale is much more difficult than the collection of large particles. However, one way out of this difficulty is by coagulating or flocculating particles to give rise to bigger particles. Coagulation may be brought about by pH adjustment, to remove electrostatic charges which would otherwise prevent particles from aggregating together. Flocculation results from the addition of substances that help form bridges between charged groups on different particles.

2.5.2 Filtration

size and shape are important

Filtration is another important method for bringing about liquid/solid separation on an industrial level. Essentially this technique separates on the basis of particle size, but shape may also be a relevant factor. The viscosity of the liquid component, the maximum allowable pressure and the compressibility of particles are other factors which need to be taken into consideration. Particles which are easily compressed can cause filter blockage as the cake (ie the solid material) builds up, so a filter aid such as celite may be added to the suspension before filtration to minimise this problem. However, this could in turn make recovery of the solid phase more difficult, so filter aids are generally added only when the important enzyme is in the liquid phase. Even then, activity may be lost as a result of enzyme being held back in the filter cake. One way to avoid the build-up of filter cake, and hence to dispense with the need for filter aids altogether, is to have the flow of feed material at right angles to the filter, and hence to the direction of filtration. The general principle of cake filtration is illustrated in Figure 2.5.

use of filter aids

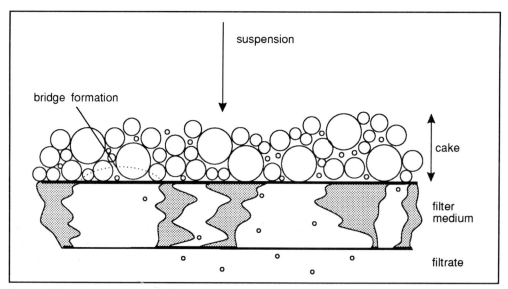

Figure 2.5 Basic principles of cake filtration. Large particles are retained; the very small particles are not, aperture 3-7 times particle size.

use of centrifugal filters

In addition to straightforward systems where the filter remains stationary, large-scale filtration may involve centrifugal or rotary vacuum filters. In the former, a suspension is fed into a perforated bowl centrifuge, and debris collects as liquid is forced through a filter cloth by centrifugal effects. The debris can be skimmed off as required by some suitable device but, once again, safety considerations limit the angular velocity which can be used, making the procedure effective only for the collection of large particles. Note, however, particles smaller than the aperture of the filter media may be collected because of the formation of 'bridges' (see Figure 2.5).

use of rotary vacuum filters

Rotary vacuum filters act essentially in the opposite direction to centrifugal ones and are illustrated in Figure 2.6. The feed comes into contact with a rotating drum from the outside, and liquid is sucked through a filter cloth covering the drum because there is

Industrial production and purification of enzymes 33

a vacuum on the inside. The filter cake may be scraped off and collected as the drum rotates.

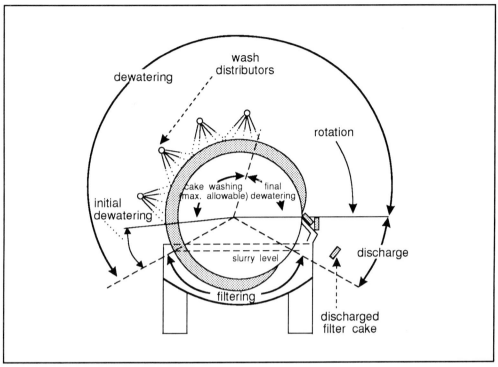

Figure 2.6 Schematic drawing of a rotary drum vacuum filter. (Adapted from Bosley R, 1965, 'Vacuum Filters' in 'Process Engineering Technique Evaluation', Morgan Grampian (Pub) Ltd, London, pp40-55).

filter presses

An alternative form of filtration is the filter press. The basic unit of a filter press is a series of perforated plates alternatively with hollow frames mounted on suitable supports. Each face of these plates is covered with a filter medium (cloth, membrane) to create a series of lined chambers. Slurry is forced into these chambers under pressure. The plate and frame are held together by hydraulics or screw rams. The solids are retained in the chambers while the filtrate discharges into hollows on the plate surface and then to drain points.

Figure 2.7 shows a plate and frame assembly. Washing of the cake can be achieved by using the slurry inlet to inlet washing liquid (water/buffer). In comparison with rotary vacuum filters, filter presses often give higher yields and a drier cake. These features, however, depend upon the nature of the lining material and filter presses tend to be expensive.

Note that the mathematical relationships for filtration efficiency are given in the BIOTOL text, 'Product Recovery in Bioprocess Technology'. In practice, however, efficiency has to be determined experimentally because of differences in the compressibility of particles and in the shape and size of the particles to be separated.

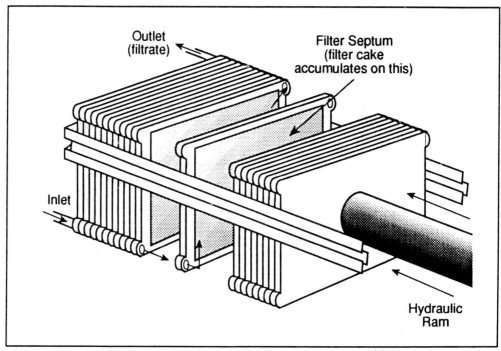

Figure 2.7 Plate and frame filter assembly.

∏ Earlier in this section you learned that there are three important methods for large-scale liquid/solid separations. We have just considered two of them. What is the third?

It is the use of aqueous biphasic partition, ie the separation of components between two phases, both essentially aqueous in character.

2.5.3 Aqueous biphasic partition

two-phase systems

Many two-phase aqueous systems are now known, but the one that has been studied in the greatest detail is the dextran/polyethylene glycol one. Various concentrations of each polymer can be used, with dextran generally forming the lower phase, which is the more hydrophilic of the two, and polyethylene glycol the upper phase.

∏ What do you think will be the main advantage of a two-phase aqueous-aqueous system rather than a two-phase organic-aqueous system being used in a procedure for the extraction and purification of an enzyme?

Organic solvents can easily cause denaturation of enzymes, so a procedure which avoids the use of an organic solvent must be preferable, if everything else is equal.

factors influencing distribution

The distribution of material between the two aqueous phases will depend on polymer concentration, pH, temperature and ionic strength. Cells and cell debris will, of course, generally remain in the lower phase, since they will not dissolve in either.

∏ If a particular enzyme is being extracted in a biphasic aqueous system, conditions may be selected so that the enzyme passes almost entirely into one of the two phases. Which phase do you think should contain the important enzyme?

Industrial production and purification of enzymes

It is preferable for the enzyme being extracted to concentrate in the upper phase, because it may then be more easily separated from the cell debris in the lower phase. Ideally, many other enzymes, including unwanted proteases, will concentrate in the lower phase.

Partition in such a two-phase system can be achieved in a matter of minutes, providing an opportunity to remove proteases before they can do much damage to the enzyme being extracted. Also, it is easy to design a continuous process involving a biphasic system, for example using a disc bowl centrifuge to separate the phases from each other. In the case of extracellular enzymes, as summarised in Figure 2.2, removal of cells by liquid/solid separation leaves an enzyme preparation which, apart from the elimination of any unwanted protease activity, is ready to undergo whatever concentration and/or purification procedures may be thought necessary. With intracellular enzymes, on the other hand, the initial liquid/solid separation (usually by centrifugation) leaves the important enzyme in the solid fraction. Hence, after the cells have been harvested, they must then be disrupted to release intracellular enzymes. After this, those enzymes which are soluble in the extraction medium under the conditions used may be isolated from the cell debris by means of another liquid/solid separation, and then processed as required.

SAQ 2.5

1) A microbial culture contains spherical cells with an average diameter of 10 μm, yet it can be successfully collected by filtration through a filter material containing pores that are 30 μm in diameter. Explain why this is so.

2) Given that the cell suspension described in 1) consists of 20% v/v cells, what type(s) of centrifuge could be used to successfully separate the cells from the suspending media?

2.6 Disruption of cells on a large scale

Π From your earlier studies you have probably encountered several techniques that may be used for the disruption of micro-organisms. Note down as many of them as you can.

You probably included drying (in air or vacuum), lyophilisation, solvent extraction, grinding, applying pressure, sonication, the use of specific enzymes and osmotic shock, or combinations of these. (If these are unfamiliar to you, we suggest you refer to the appendix).

suitability of cell disruption procedures for large-scale use

Some of these are suitable for scaling-up to use in industrial processes and some are not. Egg-white lysozyme, for example, is effective on a small-scale in weakening bacterial cell walls as a prelude to disruption, but at the present time it is too expensive for large-scale use. Procedures involving drying micro-organisms in air or under vacuum can be, and are, used in large-scale processes, but are very slow compared to techniques based on mechanical disruption. Techniques involving forcing frozen cell pastes (at around -20°C) through narrow orifices at high pressure (around 200 MPa) are effective in disrupting cells, mainly by solid shear, but such freeze-presses can generally only be used for small-scale, single-batch applications. Although some designs (eg the X-press) are suitable for semi-continuous operation on an industrial scale, they have not been applied in enzyme-extraction processes to any great extent. Similarly, sonication (or, to be more precise, ultrasonication, since the frequencies used are in the ultrasonic range, above 18 kHz) is widely-used in small-scale procedures for disrupting cells, and has been scaled-up for some industrial applications, but not generally for ones concerned with the extraction of enzymes.

At the present time, the main techniques used in industry for the mechanical disruption of cells in order to release enzymes are grinding in bead mills and high-pressure homogenisation. Both of these techniques involve disruption by liquid shear, although other factors may be involved.

2.6.1 Bead mills

use of bead mills

Bead mills are essentially containers in which cell suspensions can be agitated in the presence of glass or steel beads, these usually being less than 1 mm in diameter. In general, large cells will be disrupted more easily than small bacteria, and a large number of small beads (for example, of 0.2 mm diameter) will be more effective than a small number of larger beads. Bead mills such as the Mickle shaker have been widely-used for small-scale work, and systems that can be operated continuously, such as the Dyno-Mill, are suitable for industrial use. Such systems can handle up to 200 kg wet yeast per hour, but rather smaller amounts of bacteria.

Π Why do you think there is this difference in performance between handling yeast and bacteria?

As we saw above, large cells are disrupted more easily than small ones in bead mills, so bacterial cells are disrupted at a lower rate than yeast.

An illustration of a typical bead mill is given in Figure 2.8.

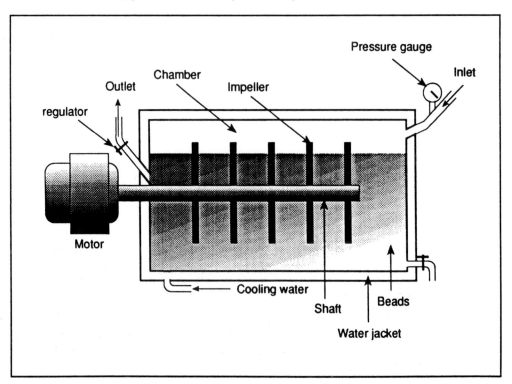

Figure 2.8 Horizontal bead mill. Note the inlet at the base for cooling water to fill the jacket surrounding the bead mill.

You can see from this illustration that bead mills consist of horizontal or vertical grinding cylinders. These cylinders are fitted with a central shaft bearing a number of impellers. The shaft is driven by an electromotor. Usually the grinding cylinder is made of stainless steel or, for smaller devices, of glass.

In operation, cylinders are partially filled with small beads. These beads are moulded out of wear-resistant materials such as zirconium oxide, zirconium silicate, titanium carbide, glass or alumina. The impeller is designed to give an efficient energy transfer to the beads often with a tip speed of the order of 10-15 ms^{-1}.

SAQ 2.6

Before reading on, make a list of as many factors as you can think of that will influence the efficiency of releasing enzymes from cells by a bead mill. (Consider the organisms used, the location of the derived enzyme, the design of the bead mill and how the mill is used). Then check your list against ours.

bead mill efficiency

Because of the complexity of the factors which influence bead mill efficiency, relationships between these variables and enzyme release are generally unknown. In practice, the optimal conditions for the release of enzymes have to be determined experimentally.

It can be shown that in many cases the release of enzymes from many cell types is by first order kinetics. The concentration of released enzyme [E] is proportional to the concentration of disrupted cells.

If we assume that $[E^{max}]$ represents the maximum concentration of enzymes that can be released from the biomass, then the rate of release of the enzyme $\frac{d[E_t]}{dt}$ is proportional to the concentration of unreleased produce ($[E^{max}]-[E_t]$) where $[E_t]$ is the concentration of released enzyme at time t.

In other words:

$$\frac{d[E_t]}{dt} = k_b ([E^{max}] - [E_t])$$

where k_b is the first order release constant for bead milling under the conditions used.

Integrating this equation gives:

$$[E_t] = [E^{max}] (1 - e^{-k_b t})$$

Note that these equations only apply for bead mills that are being used in batch mode. In the case of continuous mode the mathematical description is much more complex. Nevertheless you should release that a key feature will be the mean residence time of the biomass in the mill.

In selecting a bead mill or in selecting operating conditions, it is clear that we should be attempting to maximise k_b. k_b depends upon the factors we described earlier as influencing the release of enzymes, (impeller type, bead size, bead load, impeller speed, temperature).

SAQ 2.7

From the information presented in the figure below, determine as accurately as possible, the values of k_b for the two types of impellers. Note that the two types of impellers are able to achieve different maximum concentrations [E^{max}] of enzyme from the biomass.

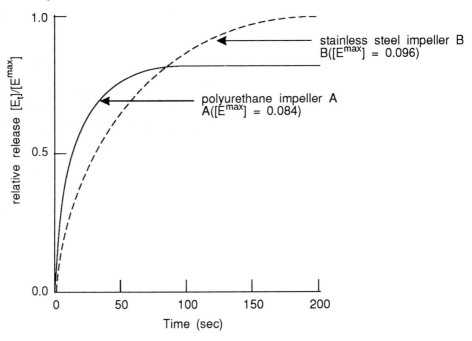

A problem with bead mills is that they tend to warm up. Almost all of the power input via the impeller is dissipated into heat. This is lost through the mill walls. The bigger the mill the greater is the problem because:

- the ratio of heat transfer area (walls)/volume decreases;

- the greater the volume, the greater the power input to achieve the same impeller speed.

Because of these problems an alternative approach is sometimes needed. Homogenisers constitute such a commonly used alternative.

2.6.2 Homogenisers

Homogenisers are basically high-pressure displacement pumps which pump the cell suspension through an adjustable orifice discharge valve.

The pressures used vary between 200-1000 bar depending on the nature of the cell suspensions. A typical homogeniser discharge valve is illustrated in Figure 2.9.

Industrial production and purification of enzymes

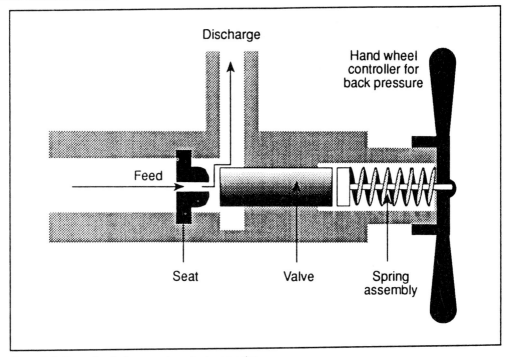

Figure 2.9 Example of a homogeniser discharge valve.

During discharge, the suspension passes between the valve and the seat. The back pressure is controlled by a hand wheel which provides the pressure on the seat via a spring mechanism. In order to overcome the irregular discharge from the homogeniser, homogenisers are usually equipped with multiple (3-5) cylinders and dampers.

As you may well imagine, the valves and seats of homogenisers are both subject to abrasion and are, therefore, made from resistant materials such as tungsten carbide. A variety of discharge designs are available (Figure 2.10) which give high efficiencies of discharge.

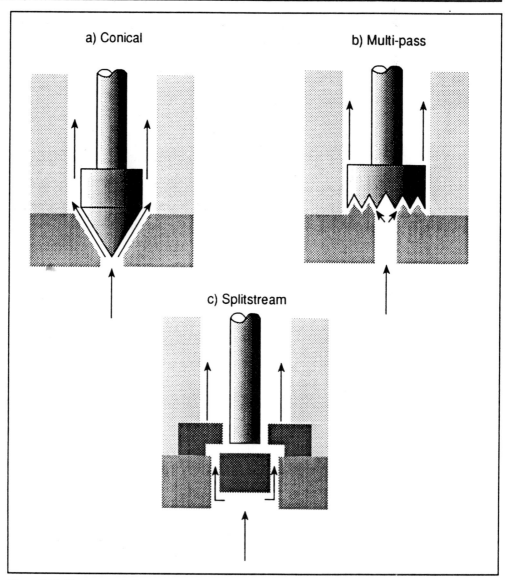

Figure 2.10 Discharge valves and seats.

Π The efficiency of cell disruption depends upon many of the same factors as cell disruption in bead mills (nature of the cells, location of the enzyme in the cells). There are, however, a variety of factors which influence the efficiency of the disruption which are specific to homogenisers. See if you can list these.

The sort of factors we hoped you would list are:

- the pressure used;

- type of valve and seat;

- the number of passages through the homogenisers.

Industrial production and purification of enzymes

Experimentally, it can be shown that the release of an enzyme from many cell suspensions follows the relationship:

$$[E] = [E^{max}](1 - e^{-k_h \Delta p^\beta N})$$

where [E] is the concentration of enzyme released; $[E^{max}]$ is the maximum concentration of enzyme that can be released; k_h is a first order rate constant characteristic of the homogeniser and the operating conditions; Δp is the pressure difference (in bars) used in the homogeniser; N is the number of passages through the homogeniser. The exponent β is a characteristic of the cell suspension used and has to be determined experimentally. For bakers' yeast β has the value of 2.9.

Figure 2.11 shows the relationship of the relative release of soluble enzyme ($[E]/[E^{max}]$) against the number of passages through a homogeniser.

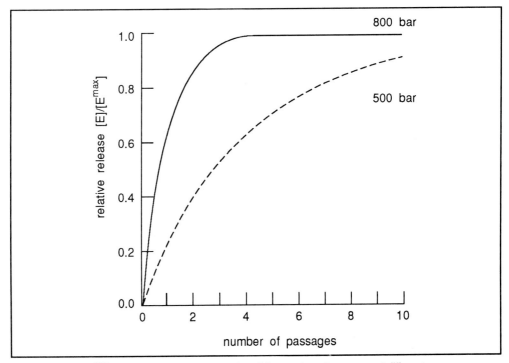

Figure 2.11 $[E]/[E^{max}]$ for baker's yeast as a function of the number of passages at different pressure differences in an homogeniser (the relationship is as described by the equation given in the text). Temperature 5°C.

Π From Figure 2.11, is a single passage sufficient to liberate the enzyme from yeast cells?

You should have concluded that a single passage is not sufficient. Even at high pressures, two or more passages are necessary.

Π Explain how an increase in pressure differences influences the performance of the homogeniser.

effects of pressure difference

We can see from the data presented in Figure 2.11 that the increase in pressure difference increases both the rate of release of the enzyme and the final amount of enzyme released.

effects of temperature

In principle, temperature also influences the efficiency of homogenisation by influencing the viscosity of the cell suspension. Usually, however, the temperature is kept low to prevent denaturation of the product.

The scale-up of homogenisers is relatively simple. All we need to do is to increase the size of the plunger pumps and discharge valves. Temperature rise is not a particular problem of scale-up. Since temperature rise is only a function of the applied pressure difference and the properties of the cell suspension, it is not a function of scale.

Once the cells have been disrupted, we can separate the released enzyme from the cell debris and unbroken cells by one of the solid/liquid separation procedures described earlier. Thus whether we are attempting to prepare an extracellular enzyme, a periplasmic enzyme or a cytoplasmic enzyme, we are ultimately left with a solution containing the enzyme and a variable amount of contaminating molecules. In the next section, we will examine ways of purifying these preparations.

2.7 Large-scale purification

Once an initial enzyme preparation has been obtained, by whatever means, various characteristics need to be determined to give an indication of how it should be processed. These include the activity (and, ideally, the concentration) of the important enzyme; the total protein concentration; the nucleic acid concentration; the conductivity; the pH; and an assessment of the non-protein contaminants present. We will explain the importance of these and the methods of circumventing problems which may arise in the following sections.

2.7.1 Increasing the concentration of the enzyme

enzyme concentration procedures

If it is thought that the enzyme concentration is too low (say, less than 1 gl^{-1}), it may be increased by ultrafiltration. Ultrafiltration membranes can be obtained with known molecular mass cut-offs covering the range 1000-100 000 D. Enzyme concentration (as well as purification) may also be increased by means of chromatography, for example using ion-exchange or affinity chromatography. A short, wide column would generally be used at this stage, under conditions where the important enzyme would bind but many contaminants would pass straight through. Alternatively, it might be more convenient with a large-scale operation to carry out essentially the same process on a batch basis rather than passing solutions through columns. In either case, the bound enzyme would later be released by changing the conditions.

ammonium sulphate precipitation not commonly used on a large scale

Ammonium sulphate precipitation, which is widely used in laboratories in procedures for the concentration and purification of enzymes, is far less important in large-scale processes. This is because ammonium sulphate is corrosive; it forms dense solutions which are not easily separated from precipitates using industrial centrifuges; and it gives off ammonia gas at alkaline pH.

2.7.2 Reducing viscosity and total protein content

The total protein content of the initial enzyme preparation, together with the nucleic acid content, can affect its viscosity. If the viscosity is high it may be necessary to reduce it before chromatography is carried out to prevent problems arising from increased column back-pressure. This may be done by simple dilution, or by using nucleases to break down nucleic acids. Precipitation of nucleic acids, for example by streptomycin sulphate or protamine sulphate, tends to be avoided in large-scale processes because of the costs of the materials.

purification by chromatography

If the total protein content is very much greater than that of the important enzyme, the first chromatography step will generally aim to hold back the enzyme and let many of the protein contaminants pass straight through the column. If, however, the ratio of total protein to enzyme is much closer to one, and there is a well-defined contaminant, it might be better to try to absorb this on the column in the first instance, letting the enzyme pass through.

2.7.3 Importance of conductivity measurements

A conductivity value of greater than $5 \text{ m}\Omega \text{ cm}^{-1}$ for the initial enzyme preparation at neutral pH would indicate potential problems if the intended first chromatography step involved ion exchange.

∏ What would a high conductivity value indicate about the ion concentrations and how would they interfere with a chromatographic step?

A high conductivity indicates the presence of a high concentration of ions. These small charged molecules might quickly fill up most of the available sites on the resin. The answer could be to dilute the preparation or, better, to remove the small molecules by gel filtration or some other filtration procedure (this is known as 'desalting').

2.7.4 Importance of determining pH

The importance of determining the pH of the initial enzyme preparation is that the pH value can then be adjusted as required.

∏ What are the likely requirements for the pH?

It must be at a value suitable for the preservation of enzyme activity. Also, it must be at a carefully selected value if the first chromatography step involves ion exchange or hydrophobic interaction chromatography.

2.7.5 Importance of the presence of contaminants in the enzyme extract

Some possible contaminants have already been mentioned, including proteins, nucleic acids, and small charged molecules.

∏ What other contaminants do you think might be present in the initial enzyme preparation?

As we saw in Figure 2.2, some preparations will contain lipids. Other contaminants which might be present include carbohydrates, pigments and any cell debris which has not been removed.

Some of these contaminants might fill available sites on a chromatographic column, as mentioned above, or simply cause clogging. Depending on circumstances, a preliminary 'desalting' step, and removal of lipids and small molecules by filtration, might be appropriate.

∏ From what you have learned in this section, can you suggest an alternative approach?

It might be possible to hold back the enzyme of importance, say by affinity or ion exchange chromatography, using a column which would allow many of the contaminants to pass straight through without binding or clogging. Clearly, there can be no general rule.

2.7.6 Developing an overall scheme for the downstream processing of an enzyme

use of pilot scale trials

Now that we have considered some of the features of our initial extract and shown how these may influence subsequent purification steps, we can turn our attention to developing an overall scheme for the purification of an enzyme after its initial extraction. We need to emphasise that this is normally done on a pilot scale first before attempting to establish a really large-scale process with all its encumbering capital costs.

As we have seen, an initial enzyme preparation, particularly of an extracellular enzyme, may sometimes be of commercial value without requiring a great deal of downstream processing. However, in the vast majority of cases, a series of purification steps are necessary, and generally these involve chromatography. A wide range of types of chromatography may be used in large-scale processes. In general, a pilot scheme will be developed, and then scaled-up for industrial use.

We have already noted that there are no clear-cut rules stating that certain chromatographic procedures are always used before others. However, in general, a sequence is chosen such that the products of one step are suitable to be the feed for the next. So, for example, a sample can be applied at high salt concentration to a column for hydrophobic interaction chromatography and then eluted in low salt, which makes the eluate suitable for subsequent application to an ion exchange column. Alternatively, the high salt eluate from an ion exchange column can be used as feed for a hydrophobic interaction system. Gel filtration causes sample dilution, so (except when used for 'desalting') it is applied near the end of a sequence, when the process volume is low. The eluate from gel filtration may then be made more concentrated, for example, by ion exchange or affinity chromatography, if required.

We can perhaps give you a general rule:

> try to alternate purification steps which lead to a dilution of a product with ones that lead to its concentration.

Although this is not a rigid rule, it will make sense.

∏ Think about the following. If you designed a process which involved three steps all of which led to dilution of the desired enzyme, what would be the outcome?

Industrial production and purification of enzymes

We would, of course, end up with a very large volume of a dilute enzyme. Handling large volumes costs money and does not allow the sample to be used in certain procedures. For example, we would need an enormous gel filtration column to satisfactorily separate two enzymes present in a large volume of liquid.

Chromatofocusing, which utilises an ion exchanger with many ionisable groups covering a wide range of pKa values (eg polyethyleneimine agarose), and ampholyte buffers to set up a pH gradient down the column, as with isoelectric focusing (see Appendix if you are unfamiliar with this technique), is sometimes used as the final step in the sequence, because of its high resolving power.

Some possible sequences of chromatography steps are shown in Figure 2.12. The sequence chosen will depend to a large extent on the characteristics of the initial preparation.

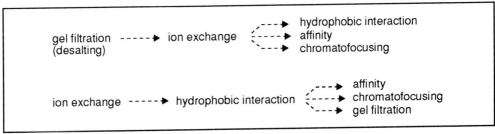

Figure 2.12 Some possible sequences of chromatographic purification steps.

optimisation of each chromatographic step

Each of the chromatography steps needs to be optimised, in terms of factors such as volume load, mass load, flow rate, resolution and cost. The first three of these factors determine the throughput, which can be shown to be dependent on flow resistance and binding capacity under working conditions.

Theories of chromatographic separations are rather well understood and will not be elaborated upon here. If you wish to pursue this aspect, we recommend the BIOTOL text, 'Product Recovery in Bioprocess Technology'.

Of key importance are, of course, throughput and resolution (quality of separation). It is no good having a high throughput with very poor resolution nor having superb resolution but extremely slow throughputs. Neither are industrially satisfactory.

Π Throughput and resolution are not the only important factors. Can you think of another important consideration? (Think for a moment about high performance liquid chromatography (HPLC) systems which use small particles which give excellent resolution and high pressures which may give reasonable flow rates).

The factor we hoped you would identify is cost.

The particles in HPLC systems are small (10 μm in diameter) and very carefully manufactured to high specifications. For large-scale separations, the costs of such particles may be prohibitive. Also, the pressure in the column may be very high, the equipment necessary for operating under such conditions expensive, and the load capacity too low to achieve a satisfactory throughput.

resolution might have to be sacrificed to reduce costs

Thus, although theory and practice show us that smaller particles give better resolution, at an industrial scale we may need to compromise. Particles of larger diameter (say 30-50 µm) may be more appropriate for use in industrial systems, provided the required separation can be obtained. Overall, it may be better to seek improvements by modifying the characteristics of the chromatographic process itself (for example in gel filtration, by trying a new gel with a different range of pore sizes), rather than in attempting to use smaller particles.

In some ways you may have found this section rather disappointing in so far as we have not given you any absolute schemes. It is impossible to do so because of the many variables (size, stability, concentration, contaminants, source) involved. Each enzyme purification scheme has to be more-or-less tailor-made for the particular situation. This is why developing a purification scheme on a pilot plant scale first is very important.

In the next section, we will consider issues which arise from scaling-up from a pilot scheme to full-scale operation.

SAQ 2.8	In a pilot scheme for the purification of an enzyme the steps indicated in the table below are used. Also reported in this table are the total volume of the enzyme preparation recovered at each stage, the total enzyme activity and the total amount of protein. Identify major weakness of this scheme, if any, and make suggestions for alternative approaches.

		volume (l)	total enzyme activity (U)	total protein
	initial extract	10.0	5.0×10^9	1 kg
step 1	ion exchange chromatography	0.5	4.8×10^9	200 g
step 2	gel filtration	1.0	4.6×10^9	50 g
step 3	affinity chromatography	0.2	4.3×10^9	1 g

2.7.7 Scale-up from pilot scheme to industrial process

scale-up

When a suitable pilot system has been designed and its components optimised, it may then be scaled-up for industrial use. The first scale-up is usually of the order of 100-fold, and this may, or may not, be followed by further increases in scale. In general, for a 100-fold scale-up, sample load, volumetric flow rate and column volume will all be increased by a factor of 100, whilst column bed height, linear flow rate and sample concentration will remain unchanged. If the results do not correspond to those obtained on a laboratory-scale, then of course some adjustments will be necessary.

components of a large-scale enzyme purification facility

The components of a typical scaled-up process are shown in Figure 2.13. The first tank module consists of containers storing sample, elution buffers and regeneration buffers. The delivery module, consisting of tubes, control valves and pumps, links these to the separation module, which contains the chromatography columns. Air traps and filters are included in the delivery module, to protect the columns. The monitoring module contains devices to monitor various characteristics of the column eluate, such as UV absorption, conductivity, pH and flow rate. This, together with continuous monitoring of pre-column factors such as pressure, air in the system, feedstock concentration and tank levels, is essential to ensure that the process is reliable and reproducible. After the monitoring module comes the fractionation module, which consists of control valves and tubes leading to the containers of the second tank module, where product is

Industrial production and purification of enzymes

collected. Not shown in Figure 2.13 is the important control module, which is programmed to open and close valves, adjust the flow rate, etc in response to signals from the monitoring module. Thus, for instance, only fractions containing protein, as indicated by the UV monitor, are collected.

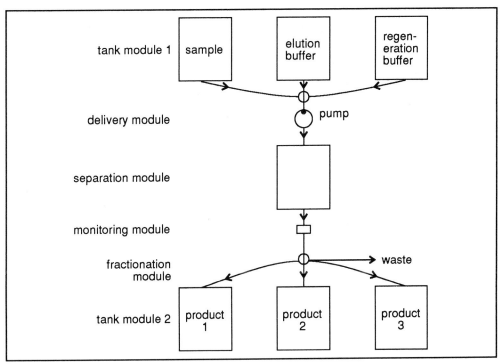

Figure 2.13 Simplified diagrammatic representation of the modules of a typical industrial chromatographic system for enzyme purification. The automatic control module is not shown.

2.7.8 An example of the purification of an enzyme on an industrial scale

As an example of enzyme purification on an industrial scale, let us consider a system for the production of human superoxide dismutase developed at Pharmacia LKB Biotechnology AB, Uppsala, Sweden. The enzyme was obtained by cultivation of recombinant yeast, and 3.5 l of the clarified yeast homogenate was used as the starting point for the purification process, approximately half of the total protein being superoxide dismutase. Electrophoretic pH-gradient investigations revealed that the best separations were obtained above pH 7, when all the proteins must have had a net negative charge. From this it was deduced that anion exchange chromatography at a pH above pH 7 should give a good separation of the product. For the product to be absorbed onto an anion exchange gel (Q Sepharose Fast Flow), the pH of the cell homogenate had to be adjusted upwards from 5.6, and an initial gel filtration step was carried out for the purposes of desalting. The anion exchange chromatography resulted in considerable purification being achieved, along with a 7-fold increase in concentration. SDS-PAGE (a form of polyacrylamide gel electrophoresis in which molecules separate according to size) was carried out on this partially-purified solution, and it was found that the remaining contaminants which were visualised differed significantly in size from superoxide dismutase.

superoxide dismutase

anion exchange chromatography

∏ What do you think this suggested about the type of chromatography which might be tried next?

gel filtration

It suggested that gel filtration might be effective. Note that the earlier gel filtration was solely for desalting and would not have separated proteins from each other. Indeed, gel filtration using Sephacryl S-200 (particle diameter 50 μm) resulted in 99% pure superoxide dismutase being obtained. The overall recovery for the process was 67%. This system was then successfully scaled-up 25-fold, 100-fold and finally 500-fold.

SAQ 2.9

Which of the following techniques are widely used in large-scale enzyme extraction and purification? (Do not include ones which are used solely for analytical purposes).

Ultracentrifugation; affinity chromatography; PAGE; sonication; hydrophobic interaction chromatography; freeze-pressing; ammonium sulphate precipitation; high-pressure homogenisation; tubular bowl centrifugation; detergent extraction.

2.8 Production of artificial enzymes

So far in this chapter we have considered, in some detail, the extraction of enzymes from living cells, and their subsequent purification. The enzymes in question might be naturally-occurring or produced as a result of the application of recombinant DNA techniques but, in each case, synthesis takes place within a cell via the normal processes of transcription and translation. In recent years, however, attempts have been made to produce artificial enzymes by quite different processes. These artificial enzymes have sometimes been given the name 'synzymes'. Here we can consider them only briefly.

synzymes

2.8.1 Synzymes

modification of myoglobin

One way to produce artificial enzymes is by modification of existing non-enzymatic molecules. For example, myoglobin, a carrier-protein which binds oxygen, can be given ascorbate oxidase activity by attaching $(Ru(NH_3)_5)^{3+}$ to three histidine residues.

The starting material need not be a protein. One such non-protein polymer is polyethyleneimine.

Polyethyleneimine, as might be guessed, is obtained by the polymerisation of ethyleneimine, which has the structure:

$$\underset{CH_2 - CH_2}{\overset{H}{\underset{|}{N}}}$$

The polymer is hydrophilic, and has a highly-branched three-dimensional structure containing primary, secondary and tertiary amines. If some of the primary amines are alkylated with 1-iododecane to give hydrophobic binding-sites, and others are alkylated with 4(5)-chloromethylimidazole, producing sites which can act as acid-base catalysts, the result is a synzyme which can act as an esterase, somewhat like chymotrypsin, but with less substrate specificity.

chymotrypsin-like synzymes

cyclodextrin-based synzymes

Synzymes with chymotrypsin-like characteristics have also been obtained based on cyclodextrins, which consist of 6-10 D-glucose units linked head-to-tail in a ring. Glycosidic oxygen atoms and -C-H linkages point inwards, creating a hydrophobic environment which can act as a binding-pocket. Catalytic activity is provided by the attachment of imidazole, hydroxyl and carboxyl groups. The resulting synzyme resembles chymotrypsin in its esterase activity, and is more stable. Alternatively, cyclodextrins may be linked to pyridoxal coenzymes, producing synzymes with transaminase activity.

2.8.2 Abzymes

abzymes

Yet another approach involves making antibodies with the characteristics of enzymes. These are sometimes called 'abzymes'. If a carefully-selected substance is injected into, say, a mouse, the immune system will make antibodies which will recognise and bind the substance, or very similar substances. With luck, some of these antibodies (which are proteins) may have the ability to catalyse a particular reaction.

∏ From your general knowledge of enzymology, suggest what type of substances might be injected to give a reasonable chance of obtaining an appropriate abzyme.

An analogue of the transition-state of the relevant reaction might be most effective, because they could stimulate the production of an antibody which would bind the substrate at a site suitable for a reaction to take place, provided suitable catalytic groups were present.

Antibodies with the required arrangement of catalytic groups cannot be made to order, so the formation of an abzyme with significant activity must be a matter of good fortune. Nevertheless, many antibodies, with a variety of different structures, will be produced, so a few might meet the requirements, at least to a certain extent. For example, abzymes with some (albeit low) esterase activity have been obtained by immunising with proteins to which phosphonate esters have been attached, phosphonate esters resembling the transition-state formed in carboxylic ester hydrolysis.

productions of monoclonal antibodies

Abzymes, like any other antibodies, may be proliferated by the techniques of monoclonal antibody production. So, for example, cells from a mouse's spleen may be removed several weeks after immunisation and fused with mouse myeloma cells, which can be grown in culture. Hybrid cells producing appropriate abzymes may then be selected and cultured to give rise to a large number of identical copies. Alternatively, abzymes may be produced in bacterial culture, following the selection and cloning of suitable antibody genes by means of recombinant DNA techniques.

The catalytic activity of abzymes may be enhanced by chemical modification. For example, metal ions have been added to the active sites of some abzymes to enable them to hydrolyse peptide bonds.

The first synzymes (including abzymes) to be produced were ones which catalysed simple, hydrolytic reactions, but the range is now expanding. Although they have not yet found significant use in industrial processes, that situation could soon change.

Summary and objectives

In this chapter we have examined the extraction and purification of enzymes in industry. We began by considering the sources of such enzymes. We learnt that micro-organisms are generally used as the source of enzymes for large-scale production, mainly because of the ease with which they can be grown in culture. Recombinant DNA techniques may be used to increase the yield of microbial enzymes, or to insert eukaryotic genes into micro-organisms. We also learnt that extracellular enzymes are the most convenient ones to process, but periplasmic enzymes and soluble intracellular enzymes are also extracted on an industrial-scale. Membrane-bound enzymes, particularly those which are integral proteins, have not generally been extracted for industrial purposes. When microbial cells need to be disrupted to extract soluble intracellular enzymes, the large-scale techniques most commonly used are bead mill grinding and high-pressure homogenisation. Liquid/solid separations are brought about mainly by centrifugation, filtration or aqueous biphasic partition. Downstream processing is carried out mainly by carefully-chosen sequences of chromatographic separations. Typical industrial systems include modules for delivery, separation, monitoring and fractionation. Amongst the factors which require monitoring are pressure, flow-rate, pH, UV absorption and conductivity. Linking everything, and responding to signals from the monitoring module, is a suitable-programmed automatic control module.

At the end of the chapter we discussed artificial enzymes, called synzymes or abzymes. They are of little current industrial significance, but may become much more important in future.

Now that you have completed this chapter you should be able to:

- discuss why micro-organisms are currently the main source for the industrial production of enzymes;
- indicate how recombinant DNA procedures may be applied in this context;
- discuss the relative merits of various extraction and purification procedures for the industrial production of enzymes;
- explain how the priorities for enzyme purification in an industrial process may be different from those in a scientific investigation;
- discuss how chromatographic systems may be linked together;
- outline the principles of scale-up for chromatographic systems;
- describe a typical industrial chromatographic system for enzyme purification;
- describe how the purity of enzyme preparations may be assessed;
- apply specific activity and yield calculations to an assessment of purification procedures;
- explain what is meant by artificial enzymes, giving examples, and discuss why artificial enzymes are of potential industrial importance.

The use of soluble enzymes in industrial processes

3.1 Introduction	52
3.2 Modes of use of enzymes in industry	52
3.3 The use of soluble enzymes	54
3.4 Advantages and disadvantages of soluble enzyme batch reactors	72
Summary and objectives	74

The use of soluble enzymes in industrial processes

3.1 Introduction

In earlier chapters, we briefly reviewed the major groups of enzymes used on a large scale in industry, the criteria we may apply to the selection of suitable enzymes for industrial use and the preparation of these enzymes. In this chapter, we will consider in outline, how enzymes may be used on a large scale. We will then discuss, in more detail, the design of batch processes using soluble enzymes and the limitations that solubility places on enzyme recovery and re-use. We will also examine the preparation of substrate and the influence of incubation conditions on enzyme stability.

3.2 Modes of use of enzymes in industry

Once we have identified a suitable enzyme, we are faced with the problem of how we will use it. There are three main options open to us:

- as a soluble enzyme;
- as an immobilised enzyme;
- as a non-purified preparation usually in the form of non-proliferating whole cells.

Soluble and immobilised enzymes can be either of intracellular or extracellular origin, but are usually used in a purified or semi-purified form. In 'non-proliferating whole cells', the enzyme remains in the intact cell. In some instances, the cells are made porous by treatment with, for example, a mild detergent. Non-proliferating whole cells may be used as suspensions or may be immobilised.

Some examples of the modes of use of different enzymes with industrial applications are given in Table 3.1. This table gives only a few examples of the many industrial processes that are now in operation.

The use of soluble enzymes in industrial processes

Mode of use of enzyme	Enzyme or micro-organism	Industrial application
soluble enzyme	*Bacillus* proteinases	used in washing powders
soluble enzyme	rennet (chymosin)	cheese manufacture
soluble enzyme	fungal α-amylase	baking processes
immobilised enzyme	glucose isomerase	high fructose corn syrup production
immobilised enzyme	penicillin G acylase	6-amino penicillanic acid production
immobilised enzyme	nitrilase	acrylamide production
non-proliferating whole cells	methanotrophs	epoxidation of gaseous alkanes
non-proliferating whole cells	*Pseudomonas putida*	benzene cis-glycol production
non-proliferating whole cells	*Xanthobacter sp.*	adipic acid production

Table 3.1 Some examples of modes of use of different enzymes with industrial applications.

Soluble and immobilised purified enzymes are normally used to catalyse a single reaction.

∏ What advantages and disadvantages may be cited for using 'non-proliferating whole cells'?

Several potential advantages may be cited. These include:

- the lower cost of producing the biocatalyst since there are no purification costs;

- whole cells may catalyse a sequence of reactions. Use may be made of the cellular organisation of enzymes into metabolic pathways leading to the sequential conversion of substrate to product;

- using whole cells may overcome problems associated with the requirements of cofactors. Many enzymes require cofactors (for example, NAD^+) and co-reactants (for example ATP). These are difficult to supply to purified enzymes used on a large scale but may be more readily supplied within cells;

- if the whole cell has an active transport mechanism, it may accumulate substrates supplied at low concentration thereby enhancing reaction rates;

- the ability to use an enzyme which is unstable after extraction, but stable within the cell.

The main disadvantages of using non-proliferating cells are that:

- since cells contain a multitude of enzymes, the substrate may be involved in a diversity of reactions leading to a multiplicity of products rather than the single, desired product. This obviously reduces yield and increases the complexity of downstream processing to isolate the desired product;

- the substrate has to gain access to the enzyme through the plasma membrane. Thus there may be a barrier between exogenously supplied substrate and the enzyme(s). For example, phosphorylated and other highly polar compounds may not be transported across the membrane. If this is the case, the membrane may be made more porous by, for example, treatment with detergent. (This may, of course, also allow essential cofactors to leak out of the cell!). In many cases, the transport of the substrate into the cell represents a rate-limiting step;

- the enzyme(s) that carry out the desired catalysis may represent only a small fraction of the total cellular material. Thus, the total amount of the enzyme we wish to use may be very limited.

We will not compare the advantages and disadvantages of using soluble and immobilised enzymes at this stage but will examine these in the appropriate section.

3.3 The use of soluble enzymes

Soluble enzymes are usually sacrificed after the reaction they catalyse is complete as the expense of recovering the enzyme is not economically justified. Good examples of this are the proteolytic enzymes used to produce meat extracts and the enzymes used in washing powder. The main advantages of using soluble enzymes are:

advantages

- homogeneous systems (providing the substrate is soluble) may be generated. This makes the predictions concerning reaction times and conversion efficiency much simpler;

- no additional costs other than producing the enzyme are generally incurred in presenting the enzyme to the reaction mixture. As we will see later, immobilising enzymes incurs costs relating to the use of physical supports and producing the immobilised system.

disadvantages The main disadvantages with using soluble enzymes are:

- the enzyme is difficult to recover from the reaction mixture and is usually lost;

- the product is contaminated by the enzyme which may continue to act after the reaction mixture has left the reactor. Thus in some instances deliberate steps may have to be taken to inactivate the enzyme;

- the enzyme (active or inactivated) contaminating the product may have to be removed to achieve the required product specification.

3.3.1 The kinetics of enzyme and catalysed reactions in solution

The key question in the design of an industrial process using a soluble enzyme is, 'how much enzyme do we need to use to achieve a desired product yield in a given amount of time?'

Fortunately, the kinetics of enzyme-catalysed reactions are well understood. Many, but not all, enzymes display Michaelis-Menten kinetics and here we will restrict ourselves to such enzymes. However, the principles we will describe may be applied to enzymes

The use of soluble enzymes in industrial processes

displaying different kinetics providing the relationships between parameters such as substrate concentration and reaction velocity are known.

single substrate and pseudo-single substrate reactions

We will predominantly focus on the single substrate and pseudo-single substrate reactions. Note that although many of the enzymes used in industry are hydrolases catalysing the following type of reaction:

$$\text{substrate} + H_2O \rightleftarrows \text{product}_1 + \text{product}_2$$

the fact that they are used in aqueous media means that H_2O is always present in non-limiting amounts. Such reactions are therefore pseudo-single substrate reactions and the rate of reaction depends upon the concentration of the substrate.

We remind you that according to the Michaelis-Menten equation, the rate of reaction is dependent upon the substrate concentration in the following way:

$$v_0 = \frac{V_{max}[S_0]}{K_M + [S_0]}$$

where v_0 = initial reaction rate; $V_{max} = k_2[E_0]$ = maximum rate of the reaction; $[S_0]$ and $[E_0]$ are the initial substrate and enzyme concentration; K_M is the Michaelis constant for the substrate.

k_2 requires further explanation.

Consider the interaction of substrate (S) with the enzyme (E) and the formation of product (P) in the following scheme:

$$E + S \underset{k_{-1}}{\overset{k_1}{\rightleftarrows}} ES \underset{k_{-2}}{\overset{k_2}{\rightleftarrows}} P + E$$

k_{cat} and the specificity constant (k_{cat}/K_M)

Note that we have included 4 reaction rate constants. If no product is present then $k_{-2}[P][E] = 0$ and the rate of formation of $P = k_2[ES]$. The maximum rate of formation of P is given when all of the enzyme is in the form of an ES complex. In other words $[ES] = [E_0]$ and thus $V_{max} = k_2[E_0]$. It is sometimes preferred to replace k_2 by k_{cat}. k_{cat} is known as the turnover number of the enzyme and is the maximum number of substrate molecules that the enzyme can 'turn over' to produce product per unit of time. Some examples of k_{cat} are $k_{cat} = 3$ s^{-1} for glucose isomerase and $k_{cat} = 500$ s^{-1} for α-amylase. The ratio of k_{cat}/K_M is known as the specificity constant of the enzyme and it reflects the relative rate of reaction at low substrate concentration. In evolutionary terms, enzymes have evolved to maximise their specificity constants (k_{cat}/K_M ratios).

∏ This maximisation of specificity constant is limited by two main factors. Can you identify what they are?

It is necessary for an enzyme to keep K_M in the approximate range in which the substrate concentration falls in nature. Also the specificity constant is limited by the rate at which substrates and enzymes encounter each other. This is also dependent upon diffusion. It is not surprising, therefore, that the specificity constants of enzymes are greatly influenced by naturally-occurring substrate concentrations and the diffusion rates of the natural substrates; the range of specificity constants observed is quite considerable (for example, catalase has a specificity constant of 4×10^7 l mol^{-1} s^{-1} and glucose isomerase has a specificity constant of 25 l mol^{-1} s^{-1}).

Π The k_{cat} of chymotrypsin using N acetyl-L-tyrosine has been determined as 184 s^{-1}. If, under the same incubation conditions, the K_M of this enzyme is 0.74 mmol l^{-1}, what is its specificity constant?

You should have calculated it to be 2.49×10^5 l mol^{-1} s^{-1} (from $\dfrac{184}{0.74 \times 10^{-3}}$).

We will not consider these kinetic relationships in greater detail at this stage since we have assumed that you have previously encountered these kinetics. (A full account is given in the BIOTOL text, 'Principles of Enzymology for Technological Application'). However, we will draw your attention to the significance of K_M to the industrial enzymologist.

Π What is K_M equal to?

K_M is equal to the substrate concentration which gives half the maximum velocity ($\tfrac{1}{2}V_{max}$). It is, in effect, a measure of the affinity of the enzyme for its substrate. A feature of Michaelis-Menten kinetics is that enzymes only attain maximum velocity at very high substrate concentrations.

Π What is v_0 in terms of V_{max} when $[S_0] = 9 \times K_M$?

By substituting in the Michaelis-Menten equation, you should have calculated that $v_0 = 0.9\, V_{max}$.

Π What is v_0 in terms of V_{max} when $[S_0] = 99 \times K_M$?

In this case $v_0 = 0.99\, V_{max}$.

We can represent the relationship between v_0 and substrate concentration (in terms of K_M) graphically as shown in Figure 3.1.

The use of soluble enzymes in industrial processes

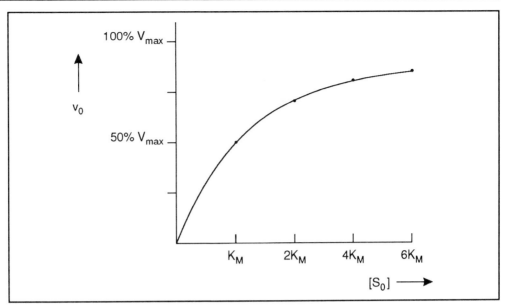

Figure 3.1 Relationship between v_0 and $[S_0]$.

We can conclude that high substrate concentrations will be needed to make efficient use of an enzyme. Is this likely to be a real problem in many industrial processes?

Let us do an intext activity to find out.

Π K_M values for enzymes are typically in the range of 10^{-2}-10^{-5} mol l^{-1}. Let us assume that the enzyme we propose to use has a $K_M = 10^{-3}$ mol l^{-1}. To be economically viable, 80% of the substrate has to be converted to product. Assume that the enzyme is not inhibited by high substrate concentration and that product is removed as it is formed so that no product inhibition occurs. We propose to apply the substrate at a concentration of 1 mol l^{-1}. Is the enzyme working as efficiently at the end of the reaction (ie when 80% of the substrate has been converted) as it was at the beginning? Attempt to answer this before reading on.

The rate of reaction at the beginning of the reaction would be:

$$v_0 = \frac{V_{max}(1)}{10^{-3} + (1)} = 0.999\ V_{max}$$

(that is about 99.9% of the maximum velocity).

At the end of the reaction we can still apply the Michaelis-Menten relationship because we have removed the product (ie $k_{-2}[P][E] = 0$). The substrate concentration has fallen to 0.2 mol l^{-1}. So the velocity of the reaction will now be:

$$v = \frac{V_{max}(0.2)}{10^{-3} + 0.2} = 0.995\ V_{max}$$

(that is, about 99.5% of the maximum velocity).

From this we might conclude that in industrial processes in which we use high substrate concentrations and enzymes with low K_M values, we can effectively use the enzyme throughout the conversion.

Our example, however, has a number of features which make it an over-simplification of the situation encountered in industry. Firstly, the product is not usually removed during the incubation and the product concentration builds up. Thus, simply from consideration of the laws of mass action, the reaction slows down. We will examine the consequences of this in the next section. Secondly, we do not always have to use substrates at high concentrations. For example, we might want to use an enzyme to remove a pollutant. If the concentration of the pollutant is low, then we will only use the enzyme inefficiently (that is, at only a small fraction of V_{max}).

consequences of product build up and of using low substrate concentrations

3.3.2 Calculating reaction times in soluble enzyme reactors

At the heart of the design of a batch process is the decision that has to be taken concerning how much enzyme to use. This decision will, of course, be influenced by such conditions as:

- the size of the available vessel in which the reaction is to take place;

- the quantity of product to be made;

- the relative cost of the enzyme and the operational costs of the reactor;

- the availability and operational capability of downstream processing.

We are particularly concerned here with calculating how much enzyme is required to achieve a particular product yield in a specified time.

It follows from the Michaelis-Menten equation that the rate of reaction is proportional to the concentration of enzyme. Thus, in principle, twice as much enzyme should halve the reaction time needed to achieve a particular amount of substrate conversion. In practice, however, the Michaelis-Menten equation is not entirely helpful in aiding us to calculate the amount of enzyme required to achieve the target conversion in a specified time.

Remember that the Michaelis-Menten equation relates to the initial velocity (v_0) and assumes the concentration of product (P) is zero.

We need to consider the situation in which the product accumulates. Thus the back reaction E+P $\to$ ES begins to increase and, at equilibrium E+P $\to$ ES will equal the forward reaction ES $\to$ E+P.

In other words $k_2[ES] = k_{-2}[E][P]$ (see Section 3.3.1 for symbols).

In practice, the time taken to convert a known amount of substrate to product is related to the amount of enzyme and the K_M of the enzyme in the following way:

$$V_{max}t = K_M \ln \frac{[S_0]}{[S_t]} + ([S_0] - [S_t])$$

or:
$$V_{max}t = 2.303\, K_M \log \frac{[S_0]}{[S_t]} + ([S_0] - [S_t])$$

where t is the time during which the reaction has taken place; $[S_0]$ = initial substrate concentration; $[S_t]$ = substrate concentration at time t.

This equation can be derived from integrating the change in [S], over the time $t_0 \to t_t$ in the following way:

The rate of reaction per unit volume may be expressed in terms of the change in substrate concentration in the reactor and time:

$$v = -\frac{d[S]}{dt} = \frac{V_{max}[S]}{K_M + [S]}$$

(Note the minus sign, substrate is being consumed).

Therefore:

$$\int_0^t V_{max}\, dt = -\int_{[S_0]}^{[S_t]} \left(1 + \frac{K_M}{[S]}\right) d[S]$$

Integration using the boundary conditions that $[S] = [S_0]$ at $t = 0$.

$$V_{max}t = \left(K_M \ln \frac{[S_0]}{[S_t]} + ([S_0] - [S_t])\right)$$

(This is sometimes written as: $V_{max}t = ([S_0] - [S_t]) - K_M \ln \frac{[S_t]}{[S_0]}$)

If the fractional conversion is F where:

$$F = \frac{[S_0] - [S_t]}{[S_0]}$$

then:

$$V_{max}t = K_M \ln \frac{1}{1-F} + [S_0]F$$

This is quite a useful form of this equation, as it is generally applicable.

Effectively $([S_0]-[S_t])$ is the change in concentration at time t. Remember that $V_{max} = k_2[E_0] = k_{cat}[E_0]$.

Let us do some sample calculations using this relationship.

SAQ 3.1

1) It is proposed to use an enzyme to hydrolyse a soluble ester. The ester is to be supplied to the reactor at a concentration of 1 mol l^{-1}. The enzyme has a K_M of 1 mmol l^{-1}. The enzyme supplied to the reactor has a specific activity of 100 µmoles min^{-1} mg^{-1} protein. If the reactor is supplied with 100 mg protein l^{-1}, how long will it take for:

   a) 50% of the ester to be hydrolysed?

   b) 75% of the ester to be hydrolysed?

   c) 90% of the ester to be hydrolysed?

2) Assuming the same conditions as in 1, calculate the incubation times required to achieve the same amount of hydrolyses as in 1a), 1b) and 1c), if the enzyme has a much lower affinity for its substrate, with a K_M of 100 mmol l^{-1}.

3) Assuming the same conditions as in 2), how much enzyme would be required to halve the times needed to achieve the same amount of hydrolysis?

We have already indicated that as the reaction approaches equilibrium, the net forward reaction rate slows because of the build-up of product. This effect is even more pronounced if the enzyme is subject to product inhibition. In circumstances where strong product inhibition occurs, batch processes are very ineffective unless some way can be found to remove product as it is formed.

3.3.3 Heterogeneous soluble enzyme reactors

One of the attractions of using soluble enzymes is that it enables us to produce a homogeneous reaction mixture. This facilitates the calculation of reaction times and substrate conversion. However soluble enzymes are often used in heterogeneous systems.

Π Identify the properties of substrates which may frustrate our attempts to produce a homogeneous system.

solubility and viscosity effects

The key properties of these substrates are their solubility and their effects on the viscosity of the reaction mixture. Both low solubility and high viscosity will reduce the ease with which adequate mixing can be achieved. Inadequate mixing will reduce enzyme-substrate interaction and reduce the efficiency with which we use the enzyme.

Substrates which tend to be insoluble or to produce highly viscous solutions are usually pre-treated in order to aid dispersion or reduce viscosity. For example starch suspensions are often preheated to 55-80°C (depending on the source). Although this does little to solubilise the starch, it causes gelatinisation and makes the substrate more accessible.

Obvious examples of insoluble substrates commonly used in industrial processes, especially in the food industry are proteins and lipids. Meat, for example, is insoluble and lipids form immiscible droplets in aqueous suspensions. In these cases, it is important to increase the water-substrate interfacial area. This is achieved by homogenisation (meat) or by producing emulsions (lipids). There is good evidence that in these cases Michaelis-Menten kinetics can be applied except that the concentration of substrate should be replaced by interfacial area per unit volume.

∏ If we do this, what will be the units of K_M?

K_M will be the interfacial area per unit volume at which $½V_{max}$ is achieved. The units could be, therefore, $m^2 l^{-1}$.

As you might have anticipated, matters are not quite as simple as that. With particulate material (for example homogenised meat), there is often an initial reaction rate which is proportional to the surface area followed by a new, slower, rate. Presumably, the enzyme carries out the reaction with loosely-bound, surface substrates followed by a slower reaction with more compacted substrate.

Because of the structural complexity of such substrates (meat is, after all, chemically very complex and contains many molecules not utilised by the enzyme), it is difficult to build mathematical models of these systems. Reaction kinetics in such systems depend as much on the pre-treatment and nature of the substrate as it does on the amount of enzyme added. In these systems much effort is directed towards substrate pre-treatment in order to use the enzyme most effectively.

∏ What other factors might you consider in preparing a complex substrate for an enzymatic reaction?

presence of inhibitors or inactivators

Apart from obvious considerations such as choosing the concentration to use, the other major consideration is whether or not inhibitors or enzyme-inactivators are likely to be present. For example, $CuSO_4$ is a potent inhibitor of α-amylase and ethanol inhibits pectinase. In addition, some biologically-derived substrates may contain natural inhibitors. For example, the pulses (beans, peas etc) contain potent inhibitors of some proteases. Also many biologically-derived substrates contain proteases which may degrade the enzyme you are using.

Although the presence of inhibitors in the substrate are undesirable in the reaction mixture, we can also make use of knowledge of the effects of inhibitors. For example, α-amylase can be stopped by the addition of about 1 g l^{-1} $CuSO_4$ to the reaction mixture and this is easier and cheaper to apply than inactivating the enzyme by heat. Clearly, the subsequent presence of Cu^{++} ions must be acceptable, or the Cu^{++} ions themselves must be removed.

SAQ 3.2

1) Pectinases are used to remove pectin from wine to reduce haze. Should the enzyme be used before or after fermentation? (Give reasons for your choice).

2) Glucose isomerase catalyses the reversible isomerisation of glucose to fructose. The enzyme is inhibited by oxygen. The enzyme glucose oxidase which produces gluconic acid from glucose is inhibited by D-arabinose.

We are attempting to use glucose isomerase to convert glucose to fructose. We discover that our enzyme is contaminated by a small amount of glucose oxidase. Is this an advantage or disadvantage? (Give reasons for your choice). If you selected disadvantage, explain how you might overcome this disadvantage.

3) α-amylase is known to be inhibited by ascorbic acid, oxalate, phosphate, alcohols, maltose and sucrose. The enzyme catalyses the hydrolysis of α-1-4 links in starch to release maltose and short oligosaccharides.

In a process designed to produce low molecular weight sugars from starch using α-amylase, should one of the inhibitors cited above be used to stop the reaction at the end of the incubation? The low molecular weight products are to be used in the manufacture of food.

SAQ 3.3

We are proposing to use a proteolytic enzyme to produce soluble peptides from beef offal. For this purpose, the offal is first homogenised. Two different homogenisers have been tested and some of their features are recorded below:

	Homogeniser A	Homogeniser B
capital cost	5020$	4800$
running cost (kg^{-1} meat)	0.14$	0.12$
capacity (kg h^{-1})	105	101
average particle size produced (mm^3)	25	85

On the basis of this data, which of the two homogenisers appears to be the most suitable to use.

3.3.4 Media formulation for soluble enzyme systems

Π From your previous knowledge, cite at least two factors which must also be considered in the formulation of the reaction mixture in addition to the presence of inhibitors and the presentation of substrate.

pH, ionic strength and -SH protectants

The two factors we anticipate you might cite are the pH of the reaction mixture and its ionic strength. You might also have suggested the inclusion of additives which might increase the stability of the enzymes (for example antioxidants to prevent the oxidation of essential -SH groups on the enzyme) and cofactors.

activity vs stability

Clearly we need to design a medium which has an appropriate pH. You will be well aware that each enzyme has its own pH optimum for activity. What might surprise you is that the activity:pH profile of an enzyme may not be identical to its stability:pH profile. For example, the proteolytic enzyme, pepsin, is most active at about pH2 but is most stable at pH6. In most cases, however, there is close coincidence between stability and activity. In batch systems in which we use the enzyme only once, provided the enzyme is not speedily inactivated at the operational pH, we can select the pH which optimises activity. It is only in long incubations or when we want to re-use the enzyme (for example if we are operating a continuous process using an immobilised enzyme) that choosing a pH for stability rather than optimum activity becomes an important issue.

Π There are two main ways in which we can maintain the pH at a chosen value. Can you identify them?

The most common way is to use a buffer but in more sophisticated systems an automatic pH titration system may be used. We will discuss each in turn.

Π List the features you would look for in a buffer which would make a buffer suitable for an industrial process using an enzyme.

The features we hope you would cite are:

- capability of buffering pH in the range needed by the enzyme;

- non-inhibitory;

- easy to separate from the product and not impairing the use of the product (for example non-toxic);

- cheap.

self-buffering substrates

In some cases, buffers are not required as the substrate itself has sufficient buffering capacity. For example, proteins, with their many ionisable sidechains, are strong buffers. In such cases, the pH of the reaction mixture is adjusted by the addition of acid or alkali before the addition of the enzyme.

In automatic pH titration devices, the pH of the reaction mixture is monitored by a pH probe. This is connected to a controller. If the pH falls below a set value, the controller activates a pump which pumps in alkali. If the pH rises above a set value, the controller switches on a pump coupled to an acid reservoir.

pH and equilibrium position

One final point, we would like to make about the effects of pH on enzyme reactions is that pH may also affect the equilibrium point of the reaction. This is especially true in those reactions in which substrates and products are ionisable. Changes in pH may alter the molecular species present and thus the equilibrium of the reaction. There are, however, other examples. For example, in the interconversion of glucose and fructose by the enzyme glucose isomerase, alkaline conditions favour the enolisation of fructose which, in turn, favours the formation of glucose. Under acidic conditions, the formation of fructose is favoured.

influence of ionic strength

The ionic strength also influences the activity of many enzymes. This is especially so when the reaction involves charged substrates and/or the movement of charged groups within the active site. If the reaction involves two charged groups approaching each other, then a relationship has been established that:

$$\log k = \log k_0 + Z_A Z_B \sqrt{I}$$

where k = actual rate constant, k_0 = rate constant at zero ionic strength and Z_A and Z_B are the electrostatic charges on the reacting species and I is the ionic strength of the solution.

Thus if Z_A and Z_B are opposite to each other (one positive, the other negative), there is a decrease in reaction rate with increasing ionic strength. If the charges are the same (both positive or both negative) then there is an increase in reaction rate with increasing ionic strength. Thus it is important to control the ionic strength of the reaction mixture. We will do an SAQ on this later.

cysteine 2-mercaptoethanol

The addition of enzyme protectants is a common practice especially in circumstances where enzymes are particularly unstable. Common additives are -SH protectants such as cysteine or 2-mercaptoethanol. Commercial judgements need to be made as to whether the cost of the protectant and its subsequent removal from the product is less than the cost arising from enzyme inactivation.

Cofactor requirement, especially organic cofactors such as NAD^+, raises particular difficulties. These cofactors are expensive and a system is needed to recover them. As we will see in later chapters, enzymes requiring such complex cofactors have not been successfully used in soluble enzyme reactions for large-scale processes. However, many enzymes require simple inorganic ions (for example Mg^{2+}, Mn^{2+}, Ca^{2+}) for activity. Although adding to the cost of the process, addition of these ions presents little technical difficulty in batch operations.

Finally, before we leave this section on reaction mixture formulation, let us present another problem.

∏ We are intending to use trypsin to hydrolyse a suspension of homogenised meat. We are using an incubation mixture with a pH close to neutral and a temperature of 37°C. Our chosen incubation time is 24 hours. Can you identify a problem that might arise in our reaction mixture?

The problem we hope you identified was the probability that the reaction mixture would become heavily contaminated by micro-organisms. The substrate we are using is nutritionally rich and the incubation conditions, if suitable for an enzyme, are almost certainly suitable for some micro-organism or other. This is a general problem that arises in many enzyme-catalysed processes with extended incubation times.

∏ How may we prevent the build up of microbial contamination?

antibiotics not really suitable

It is not really sufficient to simply say add an inhibitor of microbial growth. Antibiotics would do this but they are expensive, difficult to remove from the product and often only inhibit specific organisms. What we would ideally like is a non-specific, cheap and easy to remove anti-microbial agent. A common solution to this problem is to use a few drops of chloroform or toluene. These cause damage to the plasma membranes of most

The use of soluble enzymes in industrial processes

micro-organisms effectively killing them. On completion of the incubation, these reagents can be easily removed by evaporation.

∏ It would be a helpful form of revision to make yourself a check list of the items you should consider in preparing a reaction mixture for incubation with an enzyme. Do this by re-reading through the previous sections and compare your check list with our list displayed in Table 3.2. You should realise that we not only have to consider the requirements of the enzyme, but also cost and the subsequent use of the product.

substrate	Is it soluble? If not, how can we maximise contact with the enzyme? Does it contain enzyme inhibitors/inactivators? If so, how can we remove these?
pH	What pH do we need to operate the process at (kinetic and equilibrium effects)? How is the correct pH best achieved (buffer, titration device)? Will the buffer affect the activity of the enzyme or cause difficulties in downstream processing?
enzyme protectants	Do we need to protect the enzyme from, for example, oxidation? if so what is the best protectant to use (cost, ease of removal)?
cofactor requirements	Does the enzyme need cofactors? If so, how can we best add these?
ionic strength	Will this influence reaction rates?
anti-microbial agents	Is our reaction mixture likely to become contaminated and/or overrun by micro-organisms? If so, how can we prevent this?

Table 3.2 A check list of items to be considered in the formulation of a reaction mixture for a process using a soluble enzyme.

SAQ 3.4

1) It is discovered that a protein extract prepared from beans contains a potent inhibitor of the enzyme trypsin. This inhibitor is itself a protein. It had been proposed to use trypsin to produce peptides from the protein extract from beans. What options are open to us, to overcome this problem?

2) Pectinase is to be used to increase the yield of juice from crushed fruit. It has been proposed to use a small quantity of pectinase and a long (60 hour) incubation time. What problems do you perceive in this proposal?

3) An enzyme to be used in a soluble-enzyme reactor has a requirement for Fe^{3+} ions for activity. It binds these ions quite avidly and a concentration of 1 mmol l^{-1} is sufficient to achieve maximal activity. If the product of the enzyme is to be used as a food additive, select the most appropriate source(s) of ferric ions to be added to the incubation mixture from those listed below:

a) ferric oxalate; b) ferric oxide; c) ferric acetate; d) haemoglobin; e) ferric chloride.

SAQ 3.5

1) Below is a graph showing the pH:activity and the pH:stability profiles of an enzyme. From this data select the most appropriate pH range in which the enzyme should be used. (Note that the pH:stability profile is reported in terms of a rate constant k_i of the inactivation of the enzyme, thus high k_1 values indicate high levels of instability.

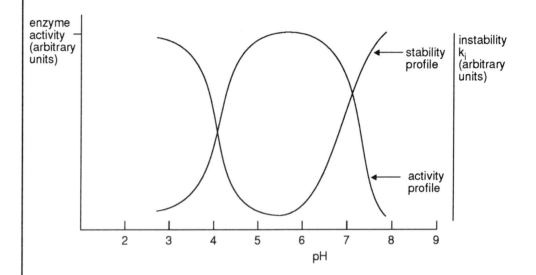

2) In the figure below we have plotted both the activity of an enzyme (in terms of initial reaction velocity) and log equilibrium constant (K_{eq}) for the reaction $A \rightarrow B$ catalyse by this enzyme, against pH. Note $K_{eq} = \frac{[B]}{[A]}$. From this data, what pH should we use to incubate the enzyme with A in order to produce B at maximum rates?

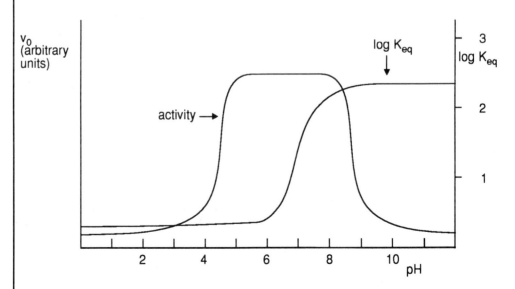

The use of soluble enzymes in industrial processes

SAQ 3.6

1) The catalytic mechanism of chymotrypsin is well understood and the rate-limiting step is known to involve the approach of two positively charged groups on histidine and arginine. Would doubling the ionic strength of the incubation mixture used with this enzyme increase or decrease the catalytic rate (k_{cat}) of this enzyme and if so by how much? (Assume that the ionic strength is X and 2X in each case).

2) In the above incubation, which one of the following would have the greatest effect on the reaction rate? (Note ionic strength is given by $I = 0.5 \Sigma(cZ^2)$ where c = concentration of an ion, Z = charge of an ion. To calculate I we summate the ions present in solution).

   a) inclusion of 0.1 mol l^{-1} CaCl$_2$;

   b) inclusion of 0.2 mol l^{-1} NaCl;

   c) inclusion of 0.1 mol l^{-1} FeCl$_3$.

   (Assume that the action of these salts is non-specific).

3.3.5 Incubation conditions and enzyme stability

Earlier in this chapter (Section 3.3.2), we discussed how we might calculate how much enzyme is needed to achieve a particular amount of substrate conversion in a specified time. You should recall that providing the enzyme is not sensitive to substrate or product inhibition, the key relationship is

$$V_{max}t = K_M \ln \frac{[S_0]}{[S_t]} + ([S_0] - [S_t])$$

V_{max} is related to the amount of enzyme present in the system. In our calculations, we have thus far assumed V_{max} to be constant. In other words, we have assumed that the enzyme is stable. Experience has shown us that this does not reflect reality. Most enzymes are labile and become inactive. Thus V_{max} will decrease during the course of incubation. In this section, we will consider the factors which mainly affect enzyme stability and how we can adjust our reactor parameters to allow for this loss in enzyme activity.

Previously we have discussed the importance of controlling pH to optimise activity and stability of an enzyme so in this section we will predominantly focus on the influence of temperature.

Reactions are influenced by temperature according to the Arrhenuis relationship.

$$k = Ae^{-\Delta G^*/RT}$$

Arrhenius relationship

where k is the kinetic rate constant for the reaction, A is the Arrhenius constant, ΔG^* is the standard change in Gibbs function of activation, R is the gas constant and T the absolute temperature. It should not be surprising that the rate of an enzyme-catalysed reactions increases with temperature. But whilst increase in temperature accelerates the reaction, it can also speed up denaturation of the enzyme.

Examine the data shown in Figure 3.2. This shows how the amount of product formed increases with time at various temperatures.

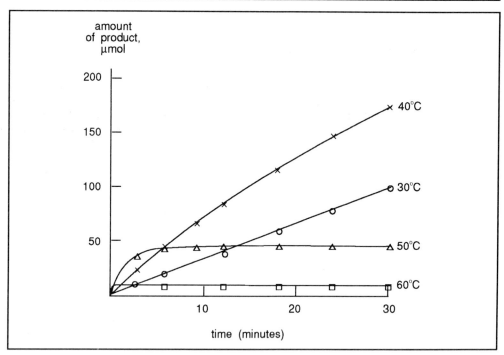

Figure 3.2 Effect of temperature on the amount of product formed in an enzyme-catalysed reaction.

At 30°C the rate is low, at 40°C it is about double that observed at 30°C. (This is sometimes expressed as the temperature quotient Q_{10}, which is the increase in rate over a 10°C rise in temperature. In this case $Q_{10} = 2$).

Denaturation begins to occur at this temperature.

∏ What evidence is there for the assertion that denaturation occurs at 40°C?

Did you notice that the reaction rate is no longer linear with time? At higher temperatures, denaturation is even faster. At 50°C, the reaction is faster still, but complete denaturation has occurred in less that 10 minutes. At 60°C, the reaction is extremely rapid but very short lived! The most suitable operating temperature will, therefore, depend upon the total reaction time and the stability of the enzyme.

The thermal denaturation of enzymes usually follows first order kinetics and can be described by:

$$[E_t] = [E_0].e^{-k_d t}$$

where $[E_t]$ = concentration of enzyme at time t; $[E_0]$ = concentration of enzyme at time t = 0; k_d = deactivation rate constant; t = time. We could also write this in the form: $\log[E_t] = \log[E_0] - 2.303\, k_d t$.

From this arises an important quantitative definition of stability - the half-life ($t_{\frac{1}{2}}$), which is the time needed for the activity of the enzyme to fall to half its original value.

The use of soluble enzymes in industrial processes

Thus since $[E_t] = 0.5[E_0]$

$\ln 0.5 = -k_d t_{\frac{1}{2}}$ and $t_{\frac{1}{2}} = 0.693/k_d$.

Thus with enzymes which display first order deactivation kinetics, the half-life of the enzyme is inversely proportional to the rate of denaturation. Such deactivation kinetics are typical of many soluble enzymes in solution but with some soluble enzymes and with immobilised enzymes, more complex denaturation kinetics are displayed.

complex deactivation kinetics

These more complex kinetics may usually be modelled by a serial deactivation scheme in which the enzyme may exist in a number of distinct forms. We can represent the scheme by:

$$E_1 \to E_2 \to E_3$$

where E_1 is the native enzyme and E_2 and E_3 are alternative species. The various forms of the enzyme have different activities. E_2 might be more, or less, active than E_1. Usually E_3 has little or no activity.

In this situation E_1 and E_2 have their own first order deactivation rates (k_{d1} and k_{d2}). Thus in these cases, the decay curves are complex and depend upon the relative proportions of E_1, E_2 and E_3, the rates of interconversion of E_1, E_2 and E_3, the specific activities of E_1, E_2 and E_3 and the deactivation constants (k_{d1} and k_{d2}).

Such systems can be modelled and experimental data can be made to fit such models. However, because reliable and reproducible data are difficult to obtain, deactivation data should be regarded as rather error-prone. In most cases, it is better to rely on experimental data rather than on hypothetical models. We will return to this when we discuss immobilised enzymes in later chapters. It is with the long-term incubations used with immobilised enzymes that enzyme stability is particularly important.

Let us, for now, return to the simpler first-order deactivation observed with soluble enzymes. This deactivation has two important consequences:

- on the conditions used for long-term storage;

- on the amount of enzyme we require to achieve a particular conversion/time regime, in a batch incubation.

We will deal with each in turn.

Denaturation and storage

cold storage and the use of anti-freezing agents and other additives

Since deactivation is temperature dependent (k_d increases with temperature), storage of enzymes is done at low temperatures. Most enzymes are stable for months at refrigeration temperatures (0-4°C). Cooling to below 0°C can increase stability even further provided the enzyme solution is not frozen. This is achieved by adding anti-freezing agents (for example glycerol). k_d values may be further reduced by forming enzyme-ligand complexes. Usually the enzyme is incubated with substrate, product or an inhibitor. Also, inclusion of thiol anti-oxidants may improve thermal stability. One final point to bear in mind is that thermal protein denaturation is primarily due to the interaction of proteins with the aqueous environments. k_d values are higher in dilute solutions. Thus concentrated enzyme solutions are more stable than dilute solutions. Enzymes dried under non-denaturing conditions may be much more stable than enzymes in solution.

Denaturation and enzyme use

The total product made by enzyme catalysis in a fixed time is dependent upon the activity of the enzyme and the stability of the enzyme. We have seen that increases in temperatures may increase the initial activity of the enzyme but may also increases its rate of deactivation.

We need to be able to determine the most suitable operating temperatures. This will depend upon the total reaction time and the stability of the enzyme.

Let us consider a real example. Pectinase is more-or-less stable at 40°C but about 90% of its activity is lost by incubation at 55°C for 10 minutes (that is, k_d is strongly temperature dependent). However, the initial rate of reaction is more than twice as fast at 55°C than it is at 40°C. Thus, if we were to plot a graph of amount of pectin hydrolysed against time at these two temperatures they would cross as shown in Figure 3.3.

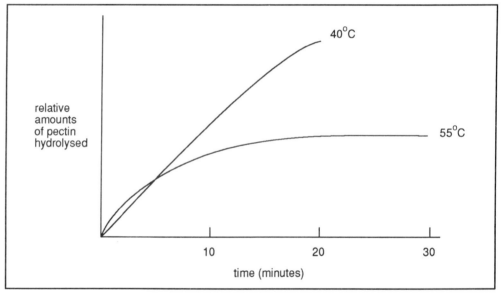

Figure 3.3 Stylised representation of the hydrolysis of pectin by pectinase at 40°C and 55°C (see text for discussion).

Thus if we want to use pectinase on a batch system from which we do not intend to recover the enzyme after incubation the choice of temperature depends on our chosen incubation time. If we intend to have a short incubation time, we could incubate the reaction at 55°C as this would require less enzyme to achieve the same amount of hydrolysis. If, on the other hand, we intended to use longer incubation periods (for example, 20 minutes or longer), we would make better use of the enzyme if we incubated it at 40°C.

Now returning to our relationship relating the amount of substrate converted (S_0]-[S_t]) to the amount of enzyme (V_{max}):

$$V_{max}t = 2.303 \, K_M \ln \frac{[S_0]}{[S_t]} + ([S_0] - [S_t])$$

it will be evident that V_{max} will change with time as a result of deactivation. If we assume that deactivation occurs by first order kinetics then:

$$\frac{d[V_{max}]}{dt} = -k_d[V_{max}]$$

and V_{max} at time t (V_{max}^t) is given by $V_{max}^t = [V_{max}^0] e^{-k_d t}$ where V_{max}^0 is the initial amount of activity.

V_{max} and k_d are both temperature dependent. Thus by determining V_{max} and k_d at different temperatures and by combining the reaction kinetics with the deactivation kinetics, it becomes possible to establish the optimum temperature to achieve maximum conversion in a fixed time.

In practice, however, we usually adopt a compromise temperature. This is because the reaction rate usually has only limited temperature dependence (for example Q_{10} values are usually only of the order of 2) whereas k_d values show marked temperature dependence. This situation is illustrated in Figure 3.4.

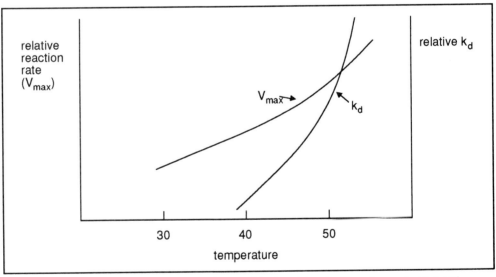

Figure 3.4 Stylised relationship between reaction rate (V_{max}) and denaturation rate constant (k_d) and temperature.

Π Given the data presented in Figure 3.4, at what temperature would you incubate the reaction mixture?

It would be best to operate at temperatures in the order of 35-38°C. Although this would give us a slightly lower V_{max}, we are unlikely to have significant denaturation of the enzyme. In practice, it is usually impossible to ensure localised overheating does not occur (for example close to the heat exchangers). If we operated close to 40°C, then any localised overheating would result in significant enzyme denaturation and this would more than counter the advantages gained by the increase in V_{max}.

SAQ 3.7

1) Aliquots of an enzyme solution are incubated in buffer at various temperatures. At time intervals, samples are removed, immediately cooled and the activity of the enzyme in each sample assayed at 30°C. The data from these assays are recorded below. Use these data to calculate the deactivation constant for the enzyme at the chosen temperatures (assume a first order reaction).

	Activity of the enzyme (μmol mg^{-1} protein min^{-1})			
Time (Min)	40°	50°	53°	58°
0	100	100	100	100
20	98.9	94.8	90	59
40	97.2	90.0	81	34.5
60	96.8	83.5	73	20.4
80	96.0	81.0	65.5	10.2

2) Use the values of k_d you calculated in 1) to determine the half-life of the enzyme at 50°C and 58°C.

3) It is proposed to run a reaction vessel using this enzyme in a variety of temperature:time regimes (a-e). In each case, calculate the anticipated % of the original enzyme activity that will be present in the reaction vessels at the end of incubation.

a) 2 hours at 40°C;

b) 2 hours at 50°C;

c) 2 hours at 58°C;

d) 30 hours at 40°C;

e) 30 hours at 53°C.

3.4 Advantages and disadvantages of soluble enzyme batch reactors

The advantages of using soluble enzymes are principally that we can generate homogeneous systems and there is little pre-preparation of the enzyme we wish to use other than its initial isolation. The use of enzymes in this way, has, however, some serious disadvantages. The first of these are economic. Commercially, the price of enzymes covers a wide range but typically a few milligrams may cost 1-10$. In other words, enzymes are not cheap. Since enzymes are not consumed in a process (they are catalysts), this high cost should not be a problem unless, of course, they lose activity by denaturation. With most batch systems, however, the enzyme is usually used only once. Despite its high cost, it is even more costly to recover it from the reaction mixture. This does mean that invariably processes using soluble enzymes have high costs for the enzymes they use.

A second disadvantage also arises from the difficulty of recovering the enzyme from the reaction mixture. It means that the enzyme contaminates the product. In many product specifications, the presence of this residual enzyme is undesirable since it may continue to modify the product or reduce its acceptability for its proposed use.

Both these types of problems would be reduced if an easy way was devised to separate the enzyme from the reaction mixture. This is a principal motivation behind immobilising enzymes.

By immobilising an enzyme by attaching it to a support or entrapping it in a matrix, we are able to recover and re-use the enzyme or, as we shall see, use it in a continuous process. At the same time by entrapping or attaching the enzyme to a support we are able to remove it from the reaction mixture thereby simplifying downstream processing. Enzyme immobilisation is the topic of the next chapter.

Summary and objectives

In this chapter we have examined some important aspects of using soluble enzymes in industry. We began by briefly describing the various modes of enzyme use and explained some of the advantages of using whole cells instead of isolated enzymes. We also explained that the use of soluble enzymes allowed us to create homogeneous systems providing the substrate was compatible with an aqueous system. We then examined the kinetics of enzymes in solution and provided relationships between substrate conversion (product yield), enzyme activity and incubation time. We also examined aspects of substrate preparation and media formulation. In particular we discussed the preparation of particulate and insoluble substrates and examined the need to carefully select buffers and other additives to be included in the incubation mixture. In the final part of the chapter, we discussed the need to control incubation conditions, particularly temperature, in order to maximise product yield. A key element in this discussion was the thermal denaturation of the enzyme.

Now that you have completed this chapter you should be able to:

- calculate reaction times from supplied data relating to enzyme activity, K_M and substrate conversion;

- explain the advantages of using enzymes with low K_M values;

- explain how knowledge of enzyme inhibition is important in industrial enzymology;

- explain the importance of surface:liquid interfacial area when insoluble substrates are used and to select appropriate substrate preparation strategies;

- identify potential problems arising from substrate impurities and from long incubation periods and suggest strategies for overcoming these problems.

Enzyme immobilisation

4.1 Introduction	76
4.2 The major forms of enzyme immobilisation	77
4.3 Techniques for immobilising enzymes	78
4.4 The consequences of immobilisation	97
4.5 The effects of diffusion on the kinetics of immobilised enzymes	103
Summary and objectives	117

Enzyme immobilisation

4.1 Introduction

At the end of Chapter 3, we pointed out the main disadvantages with using soluble enzymes and made a case for using an enzyme in an immobilised form. So far we have used the term immobilised enzyme in a rather generalised way. By immobilised we do not mean that the enzyme is totally immobilised. If it was so rigidly held that no movement was possible, then it would cease to have catalytic activity. By immobilisation, we mean that its movement is physically restricted or localised in a defined space but its vibrational and more complex movements related to catalytic activity are able to take place without hindrance. This is achieved by fixing the enzyme to, or within, some other material. As we will learn later in this chapter, this may be achieved using a variety of insoluble support materials.

meaning of immobilisation

In this chapter, we will describe the processes by which enzymes may be immobilised and explain the consequences the process of immobilisation has on the activities and properties of enzymes. We will also examine the consequences of the physical separation of the enzyme from the bulk solutions. This will enable us to examine the design and performance of reactors using these immobilised systems in Chapter 5.

∏ Before reading on see if you can list the potential advantages and disadvantages of immobilising enzymes.

There are many potential advantages. The ones we anticipated you will have identified are:

- because the enzyme is held on, or in, an insoluble support, it may be readily separated from soluble substrates and products. This means that we can recover the enzyme and re-use it. It also makes possible the design of continuous processes in which substrate is continually supplied to the enzyme and the product is removed;

- since the enzyme can be readily removed from the reaction mixture, downstream processing is easier and contamination of the product by the enzyme is reduced;

- since we can readily remove the enzyme from the reaction mixture, it is relatively easy to design processes where the contact time between substrate and enzyme is controlled. In other words, we can simply terminate the reaction at a desired point by separating substrate and enzyme;

- immobilisation may lead to increased enzyme stability.

The main disadvantages are that:

- the immobilisation process incurs processing costs (cost of supports and reagents);

- the immobilisation process may lead to some inactivation of the enzyme;

- since immobilised enzymes are physically separated from the bulk solution, substrate access to the enzyme may be restricted by diffusion or other factors. We will discuss this in detail later.

choosing between a soluble and on immobilised system

Whether or not we elect to use a soluble or an immobilised enzyme depends upon the relative advantages and disadvantages of the two systems for each particular process. For example, if a process uses a cheap enzyme and contamination of the product with the enzyme does not present a problem, we would probably elect to use a soluble enzyme system. If, on the other hand, our enzyme is expensive, the ability to re-use the enzyme would be of prime importance and we would choose to use an immobilised system. By the time you have completed this chapter you should be in a stronger position to make judgements about whether or not to use an immobilised system.

4.2 The major forms of enzyme immobilisation

Immobilised enzymes can be broadly divided into three types:

- enzymes attached to a support;
- crosslinked enzyme molecules;
- enzymes contained within a support.

These different types are illustrated in Figure 4.1. Note that there are two basic mechanisms for attaching an enzyme to a support, either by covalent binding or by physical adsorption. Similarly, we can also identify two groups of methods for producing an enzyme contained within a support; these are micro-encapsulation and entrapment in a polymeric matrix.

Π Examine Figure 4.1 carefully. Do any of the forms of immobilised enzymes mimic the situation found in nature?

The two basic types of enzyme immobilisation involving a physical support may be considered as equivalent to the natural occurrence of immobilised enzymes. For example, enzymes attached to a support may be considered to be analogous to enzymes attached to membranes or to cell walls. Similarly, enzymes contained within a support especially those that are encapsulated may be regarded as being analogous to soluble enzymes being restricted to micro-compartments within cells.

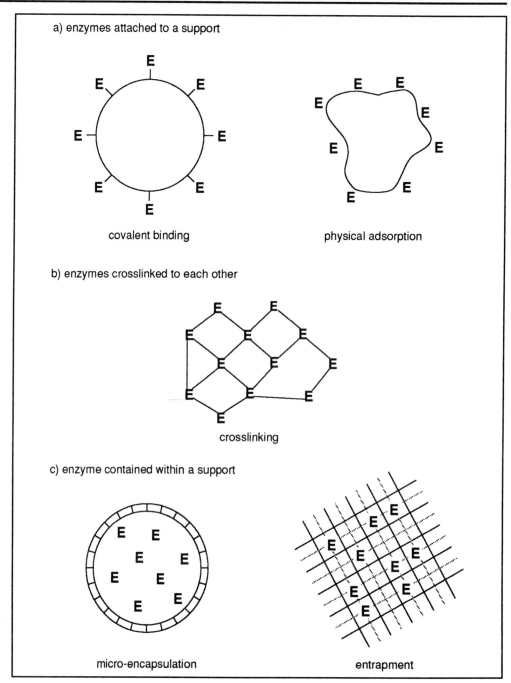

Figure 4.1 Examples of the different types of enzyme immobilisation.

4.3 Techniques for immobilising enzymes

In this section, we will first examine the key issues involved in enzyme immobilisation. Then we will examine each of these groups of techniques.

4.3.1 Key issues in the immobilisation of enzymes

The key issues involved in enzyme immobilisation are:

- to immobilise enzymes in such a way that they retain their activity;

- to achieve immobilisation of enzymes without impairing substrate access to the enzymes or product diffusion away from them;

- to produce immobilised enzyme systems which are stable under the conditions in which they are to be used.

Obviously, it is senseless to carry out an immobilisation of an enzyme if its activity is destroyed by the process of immobilisation. It is also fruitless to immobilise an enzyme in a form in which the substrate can no longer gain access to the enzyme.

Π Is micro-encapsulation likely to be successful if we wish to use α-amylase to hydrolyse starch?

You should have concluded that it is not likely to be successful. We could quite readily encapsulate α-amylase but the polymeric starch molecules would not penetrate the capsule.

Micro-encapsulation and entrapment of enzymes means that there are physical barriers between the enzyme and the surrounding medium containing the substrate. This is likely to restrict substrate access to the enzyme. We may also visualise the fact that large particles of crosslinked enzyme may restrict access of substrates to some enzyme molecules as well.

Π Which of the processes described in Figure 4.1 are most likely to lead to inactivation of an enzyme?

Covalent binding of an enzyme to a support and chemically crosslinking enzymes means that the enzyme is chemically reacted with a linking agent or an activated support. The reaction between the linker (or the support) and the enzyme raises the probability that the structure of the enzyme (and thus its activity) is altered. Covalent binding and crosslinking are, therefore, most likely to lead to inactivation of enzymes.

These two major problems with enzyme immobilisation are of major concern in enzyme immobilisation and much effort needs to be undertaken to ensure that they are overcome.

diverse protocols are used

Currently there are no universally accepted protocols for determining which method to use when contemplating enzyme immobilisation. The vast number of publications on this topic reflects this fact. Often a method which works well with one enzyme is less successful with another. Here we will give a generalised description of each of the strategies of immobilisation. We will not give a detailed list of protocols. It is more important that you understand the principles involved. The optimisation of the conditions for immobilisation of an enzyme you are working with can only be determined by experimentation. We will return to this point later.

4.3.2 Covalent binding

Π From what we have described so far, explain how covalent binding of an enzyme to its support can affect the enzyme.

Immobilisation of an enzyme by covalent binding to its support is normally irreversible and also brings about chemical modification of the enzyme. This chemical modification may have little or no effect upon the enzyme with regard to substrate-binding and enzyme activity.

effects of immobilisation on substrate-binding and catalytic activity

Conversely, covalent binding may have drastic effects upon an enzyme, altering its ability to bind the substrate or reducing its catalytic activity thus rendering the immobilised form of the enzyme unsuitable for industrial application. These alterations to substrate-binding ability and catalytic activity of an enzyme due to covalent binding may, however, in some circumstances be advantageous and so make the enzyme more industrially attractive following immobilisation. Covalent binding may also increase enzyme stability.

stabilisation of α-amylase by immobilisation

Let us give a specific example of this potential advantage. The soluble enzyme α-amylase has been reported to lose its activity in one day if incubated at 45°C. If, however, it is covalently attached to cellulose by a diazonium derivative (see later), the enzyme retains 66% of its activity if maintained at 45°C for a day. In fact, about 20% of its activity is retained after 7 days at this elevated temperature. It should be noted, however, that about 45% of the initial activity of the soluble enzyme is lost during the initial coupling reaction.

supports are called 'carriers'

This method of enzyme immobilisation involves the covalent binding of an enzyme to activated support material. These support matrices are also referred to as carriers and are generally water-insoluble. There is a variety of suitable supports including natural carriers (agar, agarose, chitin, cellulose, collagen, etc), organic polymers (polyurethanes, polyoxirane, polystyrene, nylon, etc) and inorganic carriers (glass, alumina, silica gel, magnetite, nickel oxide, etc). There are, however, advantages and disadvantages with each carrier and this must be borne in mind when selecting a carrier for use. The enzyme will be bound to an appropriate group on the support by a suitable amino acid residue. Whether an amino acid is suitable to form a covalent bond with a carrier depends upon it possessing the appropriate functional group.

Π List the functional groups found on the amino acid residues which may be used as candidates to crosslink enzymes with carriers via a linking agent.

reactive functional groups used in linking to carriers

Such groups include the carboxyl group ($-COO^-$) on the side chains of aspartic acid and glutamic acid as well as the C-terminal carboxyl group, the phenol group of tyrosine, the mercapto group ($-SH$) of cysteine, the ε-amino group ($-NH_2$) of lysine and the N-terminal amino group, the hydroxyl group ($-OH$) of threonine and serine and sometimes the imidazole group of histidine. Hydrophobic amino acids are not usually involved in forming covalent bonds during enzyme immobilisation as they are often 'hidden' within the enzyme.

Enzyme immobilisation by covalent binding essentially involves two stages. It begins with the activation of the water-insoluble carrier by a reagent, followed by a coupling reaction involving the addition of the enzyme which bonds covalently to the carrier. The bond is formed between the appropriate functional group on the enzyme and the functional group on the carrier.

Enzyme immobilisation

cyanogen bromide activation

The activation of the functional groups on the support can be brought about by different reagents. A common method which is used for the immobilisation of enzymes to polysaccharides involves the activation of the polysaccharide with cyanogen bromide (CNBr) to give an inert carbamate and a reactive imidocarbamate as shown in Figure 4.2. In this example the enzyme is immobilised by an isourea bonding.

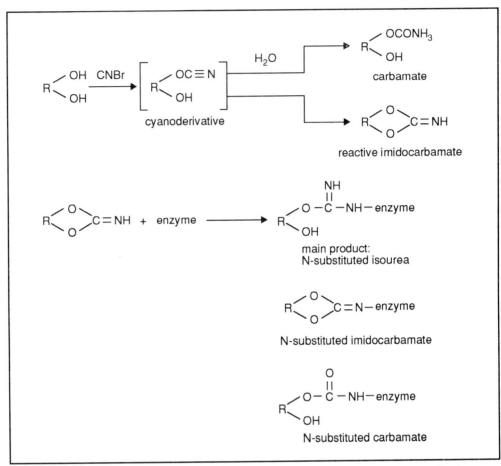

Figure 4.2 Activation of a polysaccharide to enable it to be used as a carrier in enzyme immobilisation.

We can discriminate a number of methods for covalent bonding of enzymes to insoluble supports into a number of subgroups. These are:

- via a diazonium intermediate;
- by formation of peptide links;
- by alkylation;
- by using a polyfunctional intermediate;
- other methods.

Some examples of these are illustrated in Figure 4.3. Examine this figure carefully.

a) via a diazonium intermediate

reaction:

$$R-X-NH_2 \xrightarrow[\text{HCl}]{\text{NaNO}_2} R-X-\overset{+}{N} \equiv NCl^- \xrightarrow{\text{enzyme}} R-X-N=N-\text{enzyme}$$
(support)   diazonium

functional groups involved:

tyrosine

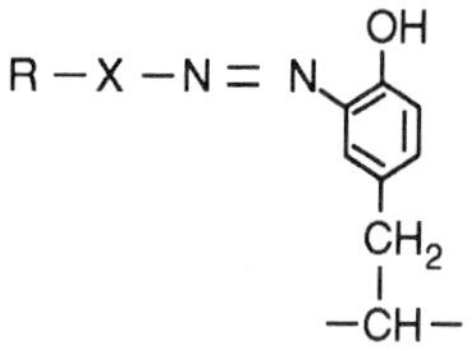

histidine

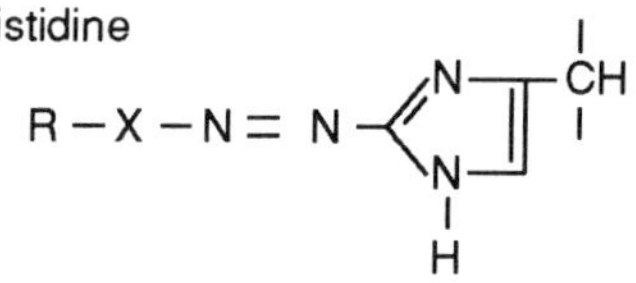

examples of support used

derivatives of polysaccharides:
- p-aminobenzyl cellulose 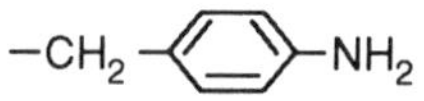
- p-aminobenzoyl cellulose 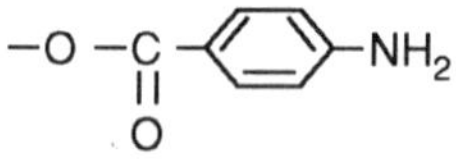
- aminobenzoyl derivative of Sephadex®

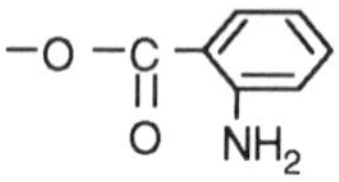

amino copolymers:
- poly L-leucine, p-amino D-L phenylalanine

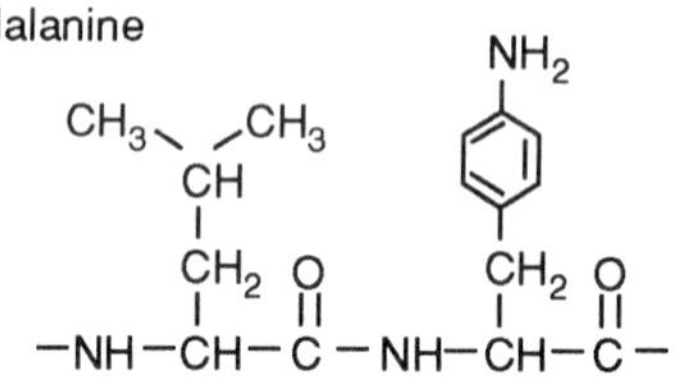

polystyrene derivatives

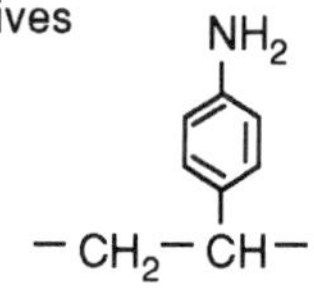

polyacrylamide derivatives (marketed as BIOGEL® and Enzacryl®)

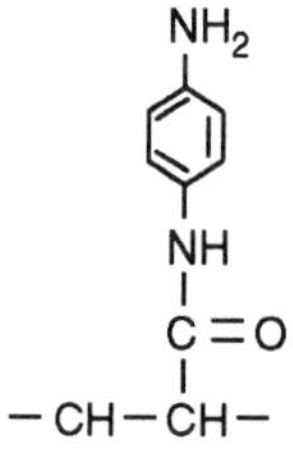

silane derivative, eg α-aminopropyltriethoxysilane

Figure 4.3 a) Summary of the main reactions and supports used to immobilise enzymes by covalent bonding. Sephadex®, BIOGEL® and Enzacryl® are registered trademarks.

Enzyme immobilisation

b) by formation of peptide links

i) via azide formation

reaction:

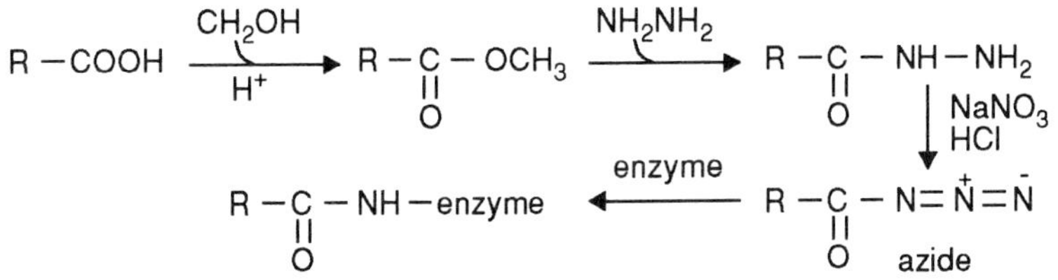

$$R-COOH \xrightarrow{CH_3OH, H^+} R-\underset{\underset{O}{\|}}{C}-OCH_3 \xrightarrow{NH_2NH_2} R-\underset{\underset{O}{\|}}{C}-NH-NH_2 \xrightarrow{NaNO_3, HCl} R-\underset{\underset{O}{\|}}{C}-N=\overset{+}{N}=\overset{-}{N} \text{ (azide)} \xrightarrow{enzyme} R-\underset{\underset{O}{\|}}{C}-NH-\text{enzyme}$$

functional groups involved
reacts with: COOH groups of glutamate and aspartate residues
supports used: acidic or hydrazine derivatives of cellulose

ii) via polymeric derivatives of maleic anhydride

reaction:

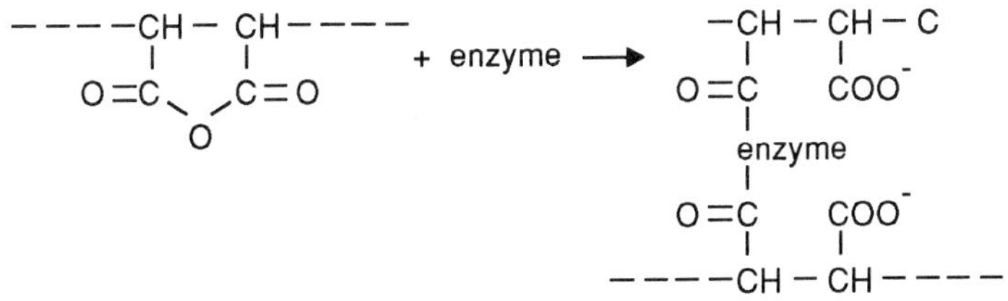

```
----CH—CH----              -CH—CH—C
    |   |                   |   |
    O=C C=O   + enzyme →    O=C COO-
     \ /                     |
      O                    enzyme
                             |
                           O=C  COO-
                             |   |
                          ---CH—CH---
```

functional groups involved: amino groups of, for example, lysine
supports: co-polymers of ethylene and maleic anhydride

iii) via isocyanate derivative

reaction:

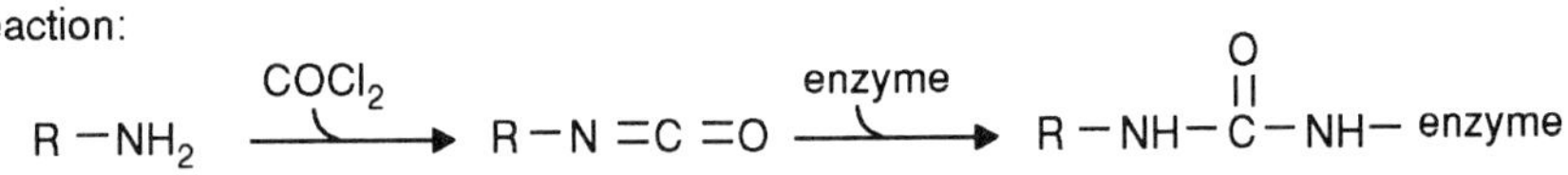

$$R-NH_2 \xrightarrow{COCl_2} R-N=C=O \xrightarrow{enzyme} R-NH-\underset{\underset{}{\overset{O}{\|}}}{C}-NH-\text{enzyme}$$

functional groups involved: amines
supports: typically amine derivatives (see a), usually amine derivative of styrene

iv) via cyanogen bromide

reaction:

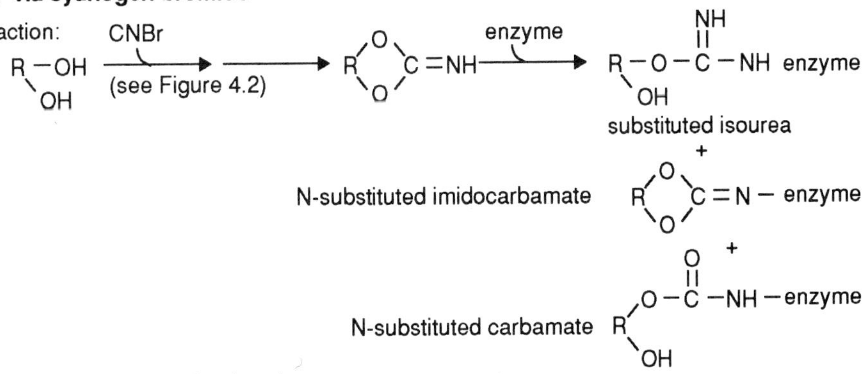

$$R\underset{OH}{\overset{-OH}{\diagdown}} \xrightarrow[\text{(see Figure 4.2)}]{CNBr} R\underset{O}{\overset{O}{\diagdown}}C=NH \xrightarrow{enzyme}$$

$R-O-\underset{OH}{\overset{NH}{\underset{\|}{C}}}-NH\text{ enzyme} \quad$ substituted isourea

N-substituted imidocarbamate $\quad R\underset{O}{\overset{O}{\diagdown}}C=\overset{+}{N}-\text{enzyme}$

N-substituted carbamate $\quad R\underset{OH}{\overset{O-\underset{\|}{\overset{O}{C}}-NH-\text{enzyme}}{\diagdown}}$

functional groups involved: amines
supports: various but each with α-β dihydroxy groups
$\left(-\underset{OH}{\overset{}{C}}H-\underset{OH}{\overset{}{C}}H-\right)$ typically, cellulose, Sephadex® or Sepharose®

Figure 4.3 b) Summary of the main reactions and supports used to immobilise enzymes by covalent bonding. Sephadex® and Sepharose® are trademarks.

c) by alkylation of phenolic, amine and sulphydryl groups on the enzyme

via halogen derivative

reaction:

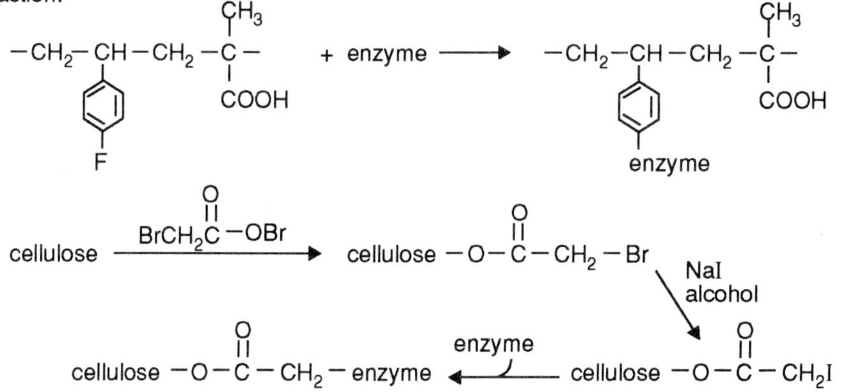

$-CH_2-CH-CH_2-\underset{COOH}{\underset{|}{C}}-\underset{F}{\underset{|}{C_6H_4}}-CH_3$ + enzyme ⟶ $-CH_2-CH-CH_2-\underset{COOH}{\underset{|}{C}}-\underset{enzyme}{\underset{|}{C_6H_4}}-CH_3$

cellulose $\xrightarrow{BrCH_2\overset{O}{\overset{\|}{C}}-OBr}$ cellulose $-O-\overset{O}{\overset{\|}{C}}-CH_2-Br$ $\xrightarrow[\text{alcohol}]{NaI}$

cellulose $-O-\overset{O}{\overset{\|}{C}}-CH_2-$ enzyme $\xleftarrow{\text{enzyme}}$ cellulose $-O-\overset{O}{\overset{\|}{C}}-CH_2I$

functional groups include: hydroxyl amine and sulphydryl groups on the enzyme

supports: homo- and co-polymers of halogenated derivatives of methacrylate, cellulose derivatives

d) by using polyfunctional intermediates*

reaction:

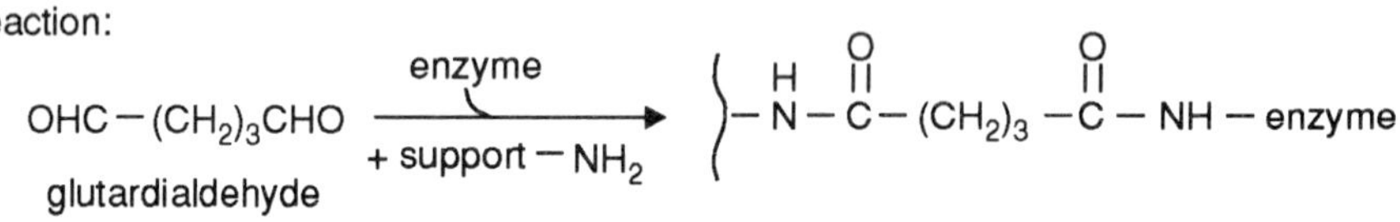

$OHC-(CH_2)_3CHO$ $\xrightarrow[\text{+ support}-NH_2]{\text{enzyme}}$ $\left\{\begin{matrix}\\\end{matrix}\right.$ $-\overset{H}{\overset{|}{N}}-\overset{O}{\overset{\|}{C}}-(CH_2)_3-\overset{O}{\overset{\|}{C}}-NH-$ enzyme

glutardialdehyde

supports: typically amine derivatives of cellulose (AE-cellulose, DEAE- cellulose) and Sepharose®, collagen, chitin, mineral supports coated with alkylamines, nylon derivatives

Figure 4.3 c) Summary of the main reactions and supports used to immobilise enzymes by covalent bonding. * polyfunctional intermediates are discussed in the section on crosslinking.

In Figure 4.3 we have given rather a lot of information concerning covalent bonding of enzymes to supports. We have given some details of the chemistry of the reactions involved, the functional groups on enzymes involved in covalent bonding and the structure of a variety of supports. Examine this figure carefully. We would not expect you to remember all of the details of the structures shown but you should understand the principles involved in each reaction type. You must realise, however, that the information given in Figure 4.3 is only a rather brief summary of the multitude of reaction types and supports that are available. It might be helpful for you to redraw this figure to aid your understanding.

Π Faced with this rather daunting array of possibilities, how do you set about choosing the appropriate activated support for an enzyme?

use of published information

A key point is to examine the information available in the literature (we have provided some examples at the end of this text). Manufacturers of supports also provide useful data concerning reaction mechanisms, protocols and binding capacities of supports. It is worthwhile researching the literature thoroughly before embarking on laboratory experiments. Even then you cannot be guaranteed success the first time. Since the behaviour of enzymes in the coupling process is largely unpredictable, we cannot be certain of success with any particular protocol and activated support. It all depends upon the reactivity and distribution of side chains and their contribution to the catalytic function of the enzyme. Reaction with the 'wrong' side chain could lead to binding accompanied by inactivation.

∏ In examining published protocols, what criteria will you place on the method(s) you select?

Apart from obvious issues such as; is it easy to carry out, how expensive are the reagents, is the method inherently safe or dangerous and so on, the important things to look for are:

- has the method been applied to the same (or a similar) enzyme?
- what is the binding capacity of the support?
- are data indicating the level of enzyme denaturation during coupling available?
- are data available on the stability of the enzyme coupled to the support?

To give you an idea of the types of data that are available in the literature, we have given an example of a summary of one immobilisation protocol (see Panel 4.1). This one was taken from the first edition of, 'Handbook of Enzyme Biotechnology', edited by Wiseman and published by Ellis Horwood, 1975. Although quite an old reference, this summary shows the sort of elements you should look for in examining the available techniques.

Panel 4.1

Example of a summary of a protocol for immobilising an enzyme by covalent binding (derived from 'Handbook of Enzyme Biotechnology', Ed Wiseman, Ellis Horwood, 1975).

Linkage of enzymes by 3(4 amino phenoxy) 2-hydroxypropyl ether of cellulose.

diazonium derivative

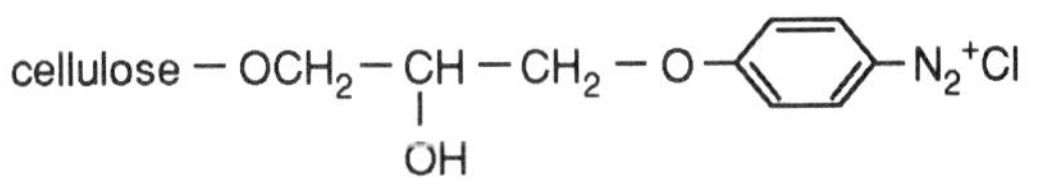

cellulose — OCH_2 — CH — CH_2 — O — ⟨phenyl⟩ — N_2^+Cl
                              |
                              OH

Enzymes coupled:

α-amylase, β-amylase, glucamylase, glycosidase mixture (note that in the original, specific references to the original reports were given).

Preparation. Cellulose (10 g) was reacted under nitrogen at 50°C with a 10% solution of p-nitrophenylglycidyl ether (20 ml) in the presence of 10% aqueous sodium hydroxide (10 ml) for 24 hours. The product was ground in 2 mol l^{-1} acetic acid and then, after washing with 2 mol l^{-1} acetic acid, dried with acetone.

Specimen result. A sample (100 mg) containing 20.7 μmol active functional group/g cellulose reacted, after diazotization, with 1 ml of 0.5% α-amylase solution at pH 7.6-7.7 and 0.5°C for 18 hours, retained 53% activity after coupling and 1.09 mg protein was attached/100 mg cellulose. Another aliquot of the same sample reacted at pH 6.3-6.4 bound 0.83 mg protein/100 mg and retained 35% activity. Whereas soluble α-amylase lost all activity in one day at 45°C, the insoluble α-amylase retained 66% after 1 day and 21% after 7 days.

SAQ 4.1

1) Which of the following amino acid side chains react with diazonium intermediates on an activated support?

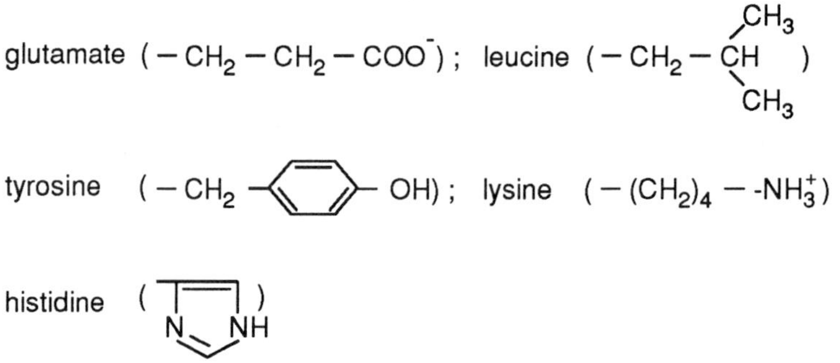

glutamate $(-CH_2-CH_2-COO^-)$; leucine $(-CH_2-CH(CH_3)_2)$

tyrosine $(-CH_2-C_6H_4-OH)$; lysine $(-(CH_2)_4-NH_3^+)$

histidine (imidazole ring with N, NH)

2) Which of the following could be used to produce diazonium intermediates suitable for immobilisation of an enzyme?

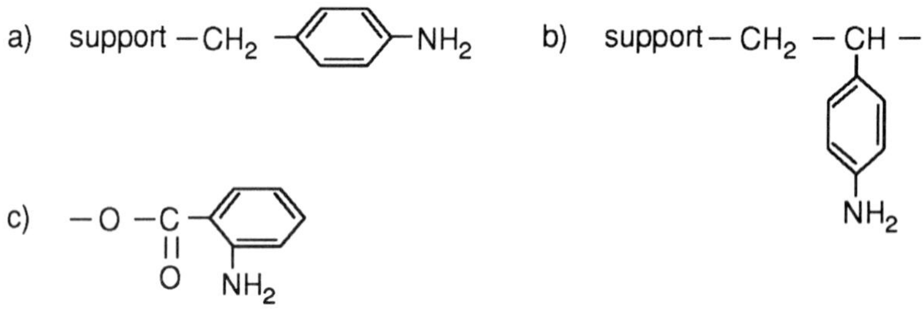

a) support $-CH_2-C_6H_4-NH_2$

b) support $-CH_2-CH-$ (with phenyl-NH_2 substituent)

c) $-O-C(=O)-C_6H_4-NH_2$

3) An enzyme is immobilised to a co-polymer of ethylene and maleic anhydride by the following reaction:

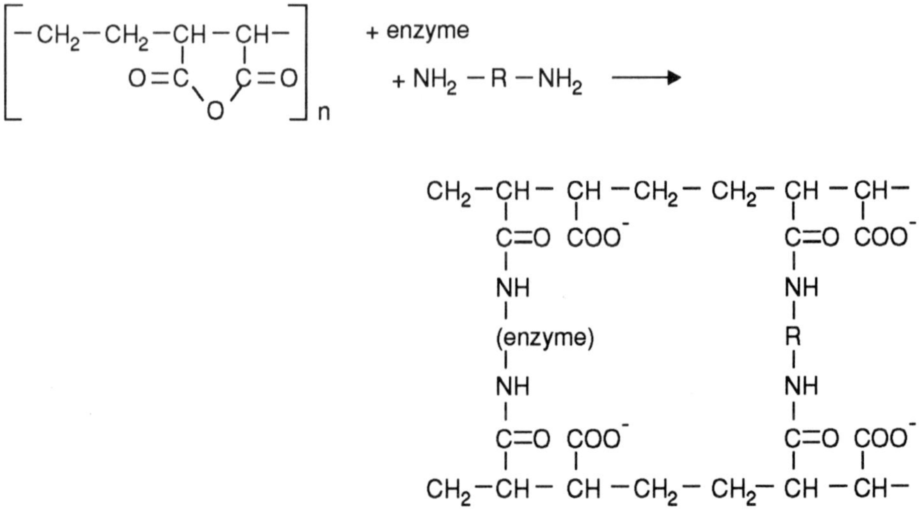

$$\left[-CH_2-CH_2-CH-CH- \atop \quad\quad O=C\diagdown_{O}\diagup C=O \right]_n + \text{enzyme} + NH_2-R-NH_2 \longrightarrow$$

$$\begin{array}{c}
CH_2-CH-CH-CH_2-CH_2-CH-CH- \\
\quad\quad C=O \quad COO^- \quad\quad\quad\quad C=O \quad COO^- \\
\quad\quad NH \quad\quad\quad\quad\quad\quad\quad\quad NH \\
\quad\quad (\text{enzyme}) \quad\quad\quad\quad\quad\quad R \\
\quad\quad NH \quad\quad\quad\quad\quad\quad\quad\quad NH \\
\quad\quad C=O \quad COO^- \quad\quad\quad\quad C=O \quad COO^- \\
CH_2-CH-CH-CH_2-CH_2-CH-CH-
\end{array}$$

a) Explain why NH_2-R-NH_2 is added to the reaction mixture.
b) What charge will be present on the immobilised enzyme support at neutral pH?

Enzyme immobilisation

SAQ 4.2a

1) An enzyme is being linked to cellulose via diazonium groups. A solution containing the enzyme trypsin (20 gl^{-1}) was mixed with the diazonium derivative of cellulose. At time intervals, samples were removed and unreacted diazo groups were quenched using β-naphthol solution at pH8.

The amount of protein attached to the cellulose and the activity of the enzyme were determined. From these data, the specific activity of the enzyme attached to the cellulose was determined and reported as a % of the specific activity of the soluble enzyme. The data are reported graphically below:

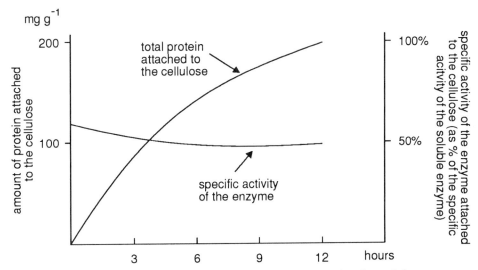

a) From these data select the most appropriate incubation time for the production of immobilised trypsin.

b) Can it be concluded from these data that linkage via a diazonium intermediate of the enzyme to the support is denaturing the enzyme? If not, how could you test whether or not the enzyme is denatured by diazonium linking?

SAQ 4.2b

2) In an experiment similar to that described in 1), ribonuclease was coupled to a copolymer of p-aminophenylalanine and leucine via diazonium groups:

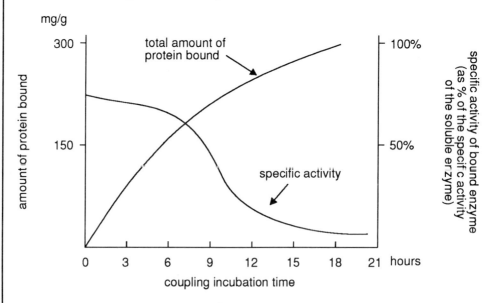

Data analogous to those reported in 1) was collected:

a) Explain how you would use these data to determine the optimum period for the coupling reaction.

b) Explain what the likely consequences would be if the leucine moieties of the co-polymer were replaced by glutamate residues. We remind you that the structure of glutamate is:

$$\begin{array}{c} COO^- \\ | \\ CH_2 \\ | \\ CH_2 \\ | \\ {}^-OOC-CH-NH_3^+ \end{array}$$

Read our response carefully as it provides useful re-inforcement of some of the concepts we have introduced.

Despite the complexities and uncertainties of covalently binding enzymes to carriers, this procedure offers many advantages over some alternative methods.

Enzyme immobilisation

∏ See if you can list some of these advantages.

The sort of advantages we anticipated that you might cite are that the enzyme is firmly attached to the support and will not readily be washed away and that the enzyme is held on the surface and should be readily accessible to the substrate.

A few of the many examples of enzymes that have been covalently linked to carriers are reported in Table 4.1.

Since covalent binding to a support has features in common with the crosslinking of enzymes (both involve the formation of covalent bonds), we will discuss crosslinking next.

Method of covalent bond formation	Enzyme immobilised	Support matrix
diazotisation	trypsin	cellulose
diazotisation	β-amylase	cellulose
diazotisation	protein kinase	agar
alkylation/arylation	α-amylase	polystyrene
alkylation/arylation	β-D-galactosidase	polyacrylamide
alkylation/arylation	chymotrypsin	agarose
(peptide) bond formation	glucose oxidase	collagen
amide bond formation	xanthine oxidase	cellulose
amide bond formation	trypsin	glass
Schiff base formation	choline oxidase	polyacrylonitrile
Schiff base formation	urate oxidase	nylon
Schiff base formation	papain	cellulose

Table 4.1 Enzymes immobilised by different types of covalent bond formation. Table adapted from 'Handbook of Enzyme Biotechnology', (1985), Wiseman, A. (Ed) Second Edition, Ellis Horwood Limited, Chichester, England. Tables 4.12, 4.13, 4.14, 4.15 Pages 171-179.

4.3.3 Crosslinking

Glutardialdehyde is one of the most common reagents used to crosslink enzyme molecules. It has the structure:

$$\begin{array}{c} CHO \\ | \\ (CH_2)_3 \\ | \\ CHO \end{array}$$

∏ From this structure, see if you can predict how glutardialdehyde forms crosslinks between enzyme molecules.

An enzyme immobilisation method which results in the insolubilisation of the enzyme following the formation of chemical bonds (intermolecular crosslinkages) between the enzyme molecules relies on the use of bifunctional or multifunctional reagents. Typical reagents used include glutardialdehyde, which forms a Schiff base between enzyme and reagent, and isocyanate derivatives, which give rise to the formation of peptide bonds. Figure 4.4 illustrates a typical crosslinking of enzyme molecules using glutardialdehyde as an example. Because enzymes contain more than one surface NH_2-groups, a complex network of enzymes crosslinked by glutardialdehyde can be developed.

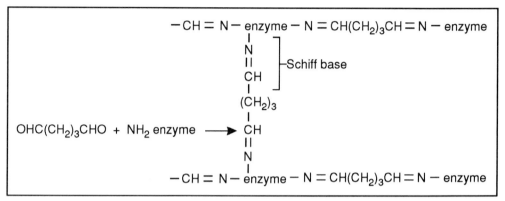

Figure 4.4 Enzyme immobilisation by the crosslinking method using glutardialdehyde as the bifunctional reagent.

There is a wide variety of bifunctional reagents that may be used showing different specifications. Some examples are shown in Table 4.2.

∏ From bifunctional reagents shown in this table, which is the most common functional group in enzymes for crosslinking?

You should have identified amino (-NH_2) groups. We must, however, distinguish between bifunctional and multifunctional reagents.

Bifunctional reagents possess two identical functional groups whilst multifunctional reagents, as their name suggests, possess two or more different functional groups. In some instances an insoluble support matrix is used to aid crosslinking and in such cases multifunctional reagents are usually employed, whilst crosslinking between enzyme molecules themselves without the use of a support is normally brought about by bifunctional reagents.

The crosslinked protein aggregates that form following treatment with the bi- or multifunctional reagents are themselves water-insoluble and can, therefore, be retained in a reaction vessel while substrate and products can be added to, or removed from, the reaction vessel.

Enzyme immobilisation

Some bifunctional reagents	Preferred coupling groups of the protein
glutardialdehyde $OHC-(CH_2)_3-CHO$	$-NH_2$
hexamethylene bisisocyanate $OCN-(CH_2)_6-NCO$	$-NH_2$
N,N-hexamethylene bismaleimide	$-SH$
bisoxirane $CH_2-CH-(CH_2)_n-CH-CH_2$ with epoxide oxygens	$-NH_2, -OH$
bisimidate $H_3CO-C(=^+NH_2)-(CH_2)_n-C(=^+NH_2)-OCH_3$	$-NH_2$
bisdiazobenzidine-2, 2-disulphonic acid $N\equiv N^+-\text{(biphenyl with R)}-N^+\equiv N$ (R = SO_3^-)	HO–⟨phenyl⟩–
divinylsulphone $H_2C=CH-SO_2-CH=CH_2$	$-NH_2$
1,5-difluoro-2, 4-dinitrobenzene	$-NH_2$, HO–⟨phenyl⟩–
2,4,6-trichloro-s-triazine	$-NH_2, -OH$

Table 4.2 Some bifunctional reagents used to couple proteins. Table redrawn from 'Characterisation of Immobilised Biocatalysts', (1979), Buchholz, K (Ed) Dechema Monograph. Vol 84, Verlag Chemie, Weinheim/New York. Table 4.1.1, page 150.

This method of enzyme immobilisation is perhaps not as satisfactory as covalent bonding to a water-insoluble matrix as sometimes the active site is involved in the formation of crosslinks and so activity is lost. Also, due to the crosslinking, the overall shape of the protein aggregate can inhibit access, by the substrate, to the active site. The method of crosslinking of pure enzymes has not as yet been an economically interesting technique. Crosslinking with the aid of macroporous matrices is closely related to the coupling of (purified) enzymes to chemically activated support materials, albeit the crosslinking technique is simpler and less specific.

crosslinking often results in the loss of activity

4.3.4 Physical adsorption onto a support

Π What are the binding forces involved in physical adsorption of an enzyme?

many different supports can be used

Physical adsorption is probably the most straightforward method of enzyme immobilisation. In this process, the enzyme is actually adsorbed onto a solid matrix or carrier without the formation of covalent bonds. Again there is a variety of solid support matrices including activated charcoal, kaolinite, porous glass, silica gel, anion and cation exchange resins, hydroxylapatite and calcium phosphate gel. The binding of the enzyme to the carrier may be mediated by ionic forces, hydrogen bonding, hydrophobic interactions, or even van der Waal's forces. The preparation of an immobilised enzyme by physical adsorption to a support matrix basically involves mixing the enzyme in solution with the carrier for a reasonable time period, and then washing away any unbound enzyme. For instance, the immobilisation of α-amylase requires the enzyme to be incubated with activated carbon for 1 hour. The mixture is then filtered, after which the immobilised α-amylase-carbon complex can be packed into a plug flow reactor, fed with starch substrate and the product (glucose) can be collected in the eluent. The most commonly used supports for enzyme immobilisation by physical adsorption are the ion-exchange resins as these readily adsorb most proteins.

ionic bonds, hydrogen bonds and hydrophobic forces are important

From your knowledge of the purification of enzymes using ion exchange chromatography, you should be able to make some predictions about how the immobilisation of enzymes onto ion exchange resins is carried out.

SAQ 4.3

1) An enzyme has a pH optimum of 6 and a pI of 7.3. Which of the following two resins would be suitable for immobilising this enzyme? (We have given the pKa values of the ionisable groups on the resins).

$$R - COOH \quad pKa\ 4$$
$$R - {}^+NH_3 \quad pKa\ 9$$

2) We have an ion exchange resin of the type R-COOH which has a pKa value of 4. If our enzyme has an isoelectric point (pI) of 7 and shows maximum activity at pH 8, would this ion exchange resin be suitable for immobilising this enzyme?

Π From your knowledge of ion exchange chromatography, what factor other than pH might affect the binding of an enzyme to an ion-exchange resin?

The factor we hoped you identified was ionic strength. High salt concentrations will tend to elute enzymes (see Chapter 2 if you have forgotten how this occurs).

Enzyme immobilisation

leaking can be a problem

Physical adsorption is reversible and some loss of enzyme will be unavoidable during use as the fores of attraction are not as great as with covalent bonding. This means that it may be necessary to remove desorbed enzyme from the product as is the case with soluble enzymes. A slight alteration in the ionic strength of the media or a change in pH can cause leaching of the enzyme. It is, therefore, required that all variable parameters are carefully monitored so as to avoid unnecessary leakage of enzyme from the support. This is particularly important in the food and health care industries where it is essential that products are not contaminated with the biocatalyst. Physical adsorption methods also have the problem of binding and concentrating unwanted substances that may be present in the substrate posing an added difficulty in reactor maintenance and control.

physical adsorption mimics some enzymes in vivo

Despite the drawbacks of physical adsorption as a method for enzyme immobilisation there are also advantageous factors, for example the process is relatively inexpensive and easy to perform. Another benefit of this method is that it allows easy renewal of the immobilised enzyme, thus permitting semi-continuous operation. Physical adsorption is the method of enzyme immobilisation which most greatly mimics *in vivo* membrane-bound enzymes and is, therefore, important to scientists investigating such systems.

combinations of physical and chemical binding may be used

To prevent the problem of desorption, a combination of physical and chemical bonding is sometimes applied. In such a procedure, the physically-obtained bonding is followed by chemical crosslinking of the enzyme. Many immobilised enzymes in commercial use are mixtures of different immobilisation techniques, for example: a glucose isomerase preparation (Maxazyme GI Immob) is prepared by entrapment of a crude enzyme preparation within gelatin (physical adsorption of the enzyme to the gelatin prevents leakage). This is followed by polymerisation of the gelatin *and* crosslinking of the enzyme to the gelatin and its impurities with glutardialdehyde as a bifunctional reagent.

4.3.5 Entrapment and micro-encapsulation

Π Before reading on see if you can outline the problems occurring with covalent binding and physical adsorption of an enzyme to its support that are not encountered with entrapment and encapsulation techniques.

Immobilisation of an enzyme by entrapment or micro-encapsulation is in principle the 'isolation' of the enzyme molecule together with a number of solvent molecules from the bulk phase. Because the enzyme is enclosed in, rather than attached to, a support matrix, problems such as loss of activity owing to the covalent attachment of an important amino acid do not occur. Similarly steric problems are usually not encountered.

With both entrapment and micro-encapsulation one important factor is the pore size of the gel matrix or membrane which need to be small enough to retain the enzyme in order to prevent leaching, but large enough to allow substrate and product to pass through unhindered. This method of enzyme immobilisation therefore is not suitable for use with substrates or products with high molecular weights owing to diffusion limitations.

Examples of some enzymes that have been immobilised within a support matrix are shown in Table 4.3.

Method of immobilisation	Enzyme	Support material
gel entrapment	urease	polyacrylamide
gel entrapment	chymotrypsin	polyacrylamide
gel entrapment	α-amylase	2-hydroxyethyl methacrylate
fibre entrapment	glucoamylase	cellulose acetate
fibre entrapment	aminoacylase	cellulose acetate
fibre entrapment	penicillin amidase	cellulose acetate
micro-encapsulation	β-D-galactosidase	nitrocellulose
micro-encapsulation	asparaginase	nylon
micro-encapsulation	catalase	polystyrene

Table 4.3 Examples of some enzymes that have been immobilised within a support matrix. Adapted from 'Handbook of Enzyme Biotechnology' (1985) Wiseman, A (Ed), Second Edition, Ellis Horwood Ltd, Chichester, UK, pp150-152.

Examine this table carefully as it includes examples of the support materials that are used to either entrap or encapsulate enzymes.

One of the simplest ways of entrapping enzymes is to use alginate. All that is required is that the enzyme is made up in a solution of sodium alginate which is dropped into a solution containing divalent ions (usually Ca^{2+}). The divalent Ca^{2+} ions crosslink the alginate to form beads. The enzyme is trapped within the beads. Unfortunately we must maintain the level of Ca^{2+} ions in all subsequent operations using these beads, otherwise the alginate gel solubilises. It is for this reason that alginate gels do not find use in commercial operations. The process of forming alginate gels for entrapping enzymes is illustrated in Figure 4.5.

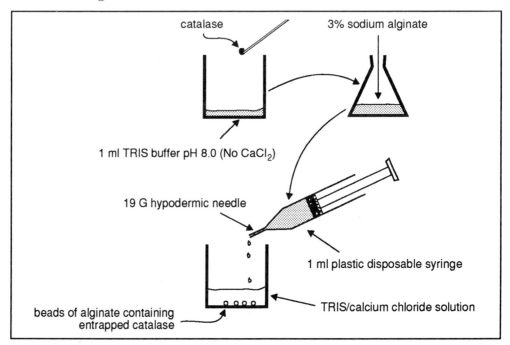

Figure 4.5 A simple laboratory demonstration of enzyme immobilisation using an alginate gel.

Enzyme immobilisation

Entrapment

As we have seen, entrapment is achieved by inclusion of the enzyme molecule within the gel matrix of a crosslinked polymer. Immobilisation is brought about by mixing the enzyme with the monomer or soluble polymer followed by polymerisation or crosslinking to form a gel. The result of this is that the enzyme can move freely in solution but movement is restricted within the interstitial spaces of the gel.

Typical polymeric matrices that have been used for enzyme immobilisation include polyacrylamide, Ca-alginate, chitosan and polyvinyl alcohol. Leakages may occur, however, due to the broad distribution of pore sizes in the gel. Of critical importance, therefore, is the control of the pore sizes in the entrapping gel. With polyacrylamide gels, for example, the gel pore size is predominantly controlled by the concentration of the monomers and the ratio acrylamide to N,N^1-methylene bisacrylamide used in the polymerisation procedure (see Figure 4.6).

gel pore size is critical

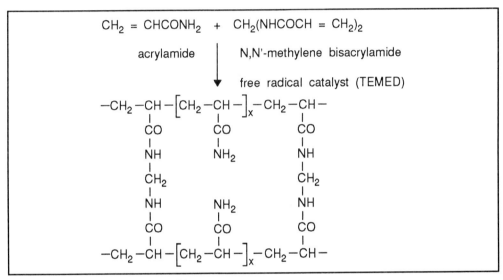

Figure 4.6 Polymerisation of acrylamide.

The crosslinking is normally done using a chemical polymerisation procedure using an initiator such as ammonium persulphate and the catalyst N,N,N^1,N^1-tetramethylenediamine (TEMED).

The final polymerised gel should be considered as a crosslinked three-dimensional matrix containing 'pores' through which substrates and products may pass.

SAQ 4.4 The enzyme trypsin is to be used to hydrolyse soluble proteins. Is this enzyme a suitable candidate to be entrapped in a polyacrylamide gel?

A special form of entrapment is the inclusion of enzymes inside porous membranes or hollow fibres. We have illustrated an example of a membrane reactor in Figure 4.7.

membrane reactors

You will notice that in this reactor, the enzyme is circulated on one side of the membrane, whilst substrate passes around the other. During incubation, the small substrate molecules pass through the membrane and interact with the enzyme. Products diffuse in the opposite direction. There are many different configurations of these types of systems. They are also usable in circumstances in which the biocatalysts are whole cells and the product is quite a large molecule. The main criteria is, of course, that the membrane allows passage of the substrate(s) and products(s) - but not of the biocatalyst. For example, such systems can be used to produce monoclonal antibodies using cultured animal cells.

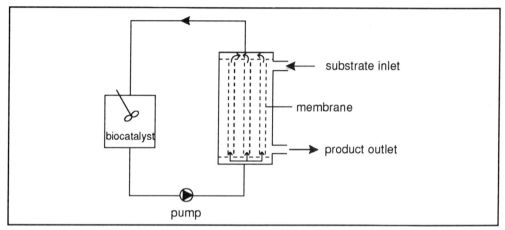

Figure 4.7 Membrane reactor with circulating enzyme. Low molecular weight components diffuse through the membrane and are removed from the system.

Instead of flat-bed membranes, the system may be arranged in the form of hollow fibres. Also, it is possible to attach (covalently or by physical adsorption) the enzyme onto the surface of the membrane as shown in Figure 4.8.

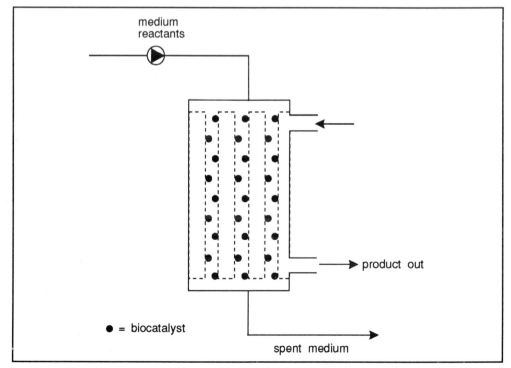

Figure 4.8 Membrane reactor incorporating an immobilised biocatalyst.

The enzymes can be incorporated during the preparation of the membranes. Such systems can be used to function as a reaction surface between two different liquid phases. This avoids the need to form emulsions. This technique is very useful when two non-miscible liquids, such as an organic solvent and water, are used. For example, the enzyme lipase converts lipids, carried by an organic solvent, to glycerol and fatty acids soluble in the aqueous phase.

∏ Can you think of another possible advantage of using a membrane reactor? (Think about enzymes that are inhibited by their own products).

Using a membrane reactor potentially offers us a way of overcoming product inhibition. By allowing products to diffuse out of the enzyme compartment, they can be removed thereby reducing the inhibition of the enzyme. Even when product inhibition is not encountered, the removal of product will favour the forward (S → P) reaction simply as a consequence of the law of mass action.

clogging of membranes

The commercial applications of the membrane reactors are, however, limited. This is due to the unpredictability of the membranes. Membranes tend to clog quite often and have a limited life span. They are also relatively expensive.

Micro-encapsulation

membranes comprised of nylon, nitrocellulose, ethylcellulose, polystyrene or phospholipids

As its name suggests this method of enzyme immobilisation involves the encapsulation of enzymes within a semi-permeable membrane. The various types of membrane employed include nylon, nitrocellulose, ethyl cellulose, polystyrene and phospholipids and the pore size of the different membranes can vary in diameter from 1μm to 100μm. The use of membranes means that more than one enzyme can be immobilised at one time, while several steps may be required to covalently bind multiple enzymes to a carrier. The use of semi-permeable microcapsules or liposomal vesicles for the encapsulation of enzymes can result in a 100% retainment of enzyme activity and enzyme loss is minimal due to the uniform pore size in the membrane. The practical application of this technique, however, is limited because of the limited life span of the micro-capsules and their susceptibility to clogging.

4.4 The consequences of immobilisation

Before we move on to consider the reactor configurations that use immobilised enzymes in more detail, we will examine, in general terms, the effects of immobilisation on enzyme activity.

In order to determine the effects of immobilisation, an examination of the differences between the enzyme characteristics before and after immobilisation is required. Both differences in the enzyme structure and in the direct surroundings or micro-environment of the enzyme molecule can be involved. In the environment of a soluble enzyme the physical conditions (pH, ionic strength, dielectric constant, substrate concentration) are essentially identical to those in the bulk of the solution. However, with an immobilised enzyme the polymer matrix can provide the enzyme with a micro-environment that is quite different to the bulk phase of the solution (see Figure 4.9). Partition effects can occur due to the repulsion or attraction of molecules in the micro-environment which can be attributed to the charge of the support. Also diffusion of molecules in and out of the micro-environment can be restricted compared to that which occurs in the bulk phase.

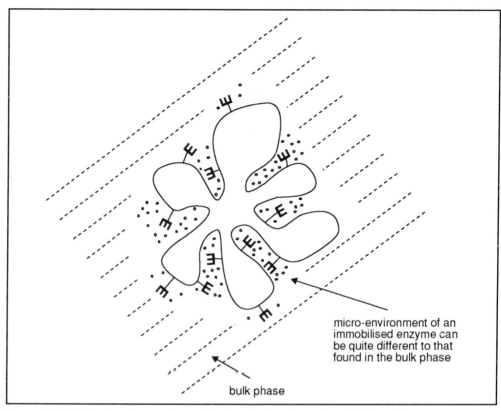

Figure 4.9 Illustrating the micro-environment that can be formed upon immobilisation of an enzyme.

enzyme molecules may be in diferent micro-environments

You should realise, however, that not all of the enzyme molecules in an immobilised system are identical. They may be linked in different ways. This means that they may be distorted differently. Furthermore, since the enzyme molecules find themselves in different micro-environments within the support matrix, this may have an important effect on, for example, accessibility of the active site. Some molecules may be on the surface of the matrix, others may be deeply embedded in the support.

Here we will divide the discussion of the consequences of immobilisation into effects brought about by changes to the enzyme and phenomena which arise through the heterogeneity of the immobilised system.

4.4.1 Effects caused by changes to enzyme structure

changes in structure and movement can modify catalytic activity

One obvious change that may occur to an enzyme following immobilisation is an alteration in its structure. Because of the inherent strong interaction between the enzyme molecule and the support, restrictions to the conformational structure of the enzyme may occur. The function of an enzyme in catalysing a specific reaction is closely related to the conformational degrees of freedom. The lowering of the activation energy of a reaction originates from the ability of a protein structure to convert all kinds of binding and conformational energies and entropies into the one conformational change that will bring about catalytic activity. Therefore, it is not only the active site of the enzyme that plays an important role in catalysis, also more distant parts of the enzyme molecule may be essential in the conformational effects that take place prior to the catalytic step. Covalent binding of an 'essential' amino acid residue may cause chemical modifications, steric hindrance and an inaccessible active site. Conversely, binding or

Enzyme immobilisation

chemical modification of an enzyme molecule can sometimes give extra protection against inactivation or denaturation of the protein, resulting in enhanced stability.

Also the kinetics of an enzyme are affected by immobilisation. Upon immobilisation V_{max}, but particularly K_M often do change. Industrial enzymes are preferentially used at high initial substrate concentrations thus ensuring that a change in K_M will not affect the enzyme activity. However, industrial applications also intend to rely on a high substrate/product conversion ratio, and an increase of K_M due to immobilisation will thus cause an increase in the contact time between biocatalyst and substrate needed to achieve the target conversion.

K_M influenced by support

In practice, we should describe the K_M of an enzyme in an immobilised system as the apparent K_M (written K_M'). Not only may the binding of substrate to an enzyme be modified by the structural changes brought about by immobilisation, but also access of the substrate to the binding sites may be changed. Remember that substrate will have to diffuse into the matrix and thus the concentration of substrate in the matrix may be substantially lower than that in the bulk liquid. Even when the enzyme is on the surface of the support, the stagnant (unmixed) layer of liquid around the support means that the substrate concentration in the micro-environment of the enzyme may be lower than in the bulk liquid. We can represent this in the following way:

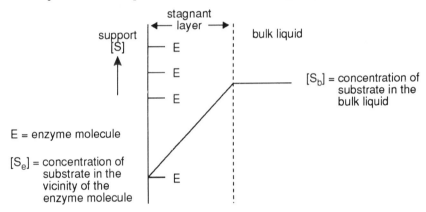

Thus in experiments in which we determine v_0 at various [S] values, we usually measure $[S_b]$ not $[S_e]$. Thus K_M values determined from such data give only the apparent (not necessarily the real) K_M value.

V_{max} may be changed

We can also give analogous arguments for changes in V_{max} for immobilised enzymes. Since conformational changes occur during immobilisation and movements of the enzyme are constrained, this may influence the catalytic efficiency of the enzyme and thus the kinetics of the enzyme-catalysed process. This will be reflected both in v_0 and V_{max} values.

4.4.2 Phenomena arising through the heterogeneity of immobilised systems, pH and ionic strength

Consider a molecule X in the bulk solution. This will partition between the solution and the support with a characteristic partition coefficient K_x.

$$K_x = \frac{[X_s]}{[X_b]}$$

where [X$_s$] = concentration of X in the support (ie the micro-environment of the enzyme) and [X$_b$] = concentration of X in the bulk solution.

If our support is negatively charged, it will attract positively charged ions, in particular it may attract protons. These would reach an equilibrium position with a characteristic partition coefficient:

$$K_{H^+} = \frac{[H_s^+]}{[H_b^+]}$$

where [H$_s^+$] = proton concentration in the vicinity of the enzyme;

[H$_b^+$] = proton concentration in the bulk solution.

Note we can calculate that an electropotential is set up at the interface of the support and solution using the Donnan equation:

$$E = \frac{RT}{F} \ln K_H^+ = \frac{RT}{F} \ln \frac{[H_s^+]}{[H_b^+]}$$

where F = Faraday constant; T = temperature; R = gas constant.

Because protons have been attracted to the support [H$_s^+$] > [H$_b^+$].

In other words the pH in the support (pH$_s$) is lower than the pH in the bulk liquid (pH$_b$). Thus if we determine the activity of the enzyme at various pH values for the bulk liquid, the pH optimum of the enzyme appears to have shifted.

We have represented this situation in Figure 4.10.

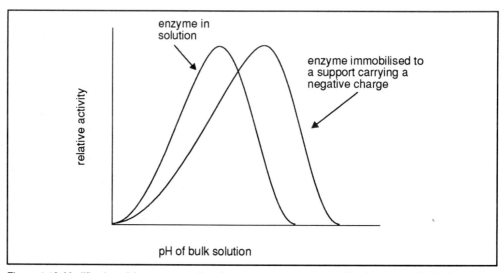

Figure 4.10 Modification of the apparent pH optimum of an enzyme immobilised on a negatively charged support.

charge distribution of ions may influence K_M and V_{max}

We have learnt that the charge on a support can influence the pH of the micro-environment of an enzyme held by that support. We might also anticipate that we may experience similar disturbances in the concentration of ions other than H$^+$ ions by a charge on the support. Since ionic strength influences the interaction between ionic species (see Section 3.3.4) and this may affect both K_M and V_{max}, then we might anticipate that an electrostatically charged support may also change these parameters.

In SAQ 3.6, we used the example of chymotrypsin. We remind you that the rate-limiting step in the catalysis carried out by this enzyme involved the interaction of two positively charged groups (on histidine and arginine). In a study on this enzyme using N-acetyl-L tyrosine, Goldstein has examined the effects of ionic strength on the catalytic constant (k_{cat}) and K_M of this enzyme. The results are shown in Table 4.4.

	Ionic strength (mol l^{-1})	k_{cat} (s^{-1})	K_M (mmol l^{-1})
free enzyme	0.05	184	0.74
	1.0	230	0.55
enzyme attached to a positively charged support	0.05	119	7.10
	1.0	165	5.82
enzyme attached to a negatively charged support	0.05	300	2.50
	1.0	280	1.93

Table 4.4 The effects of ionic strength on k_{cat} and K_M of soluble and immobilised chymotrypsin (Goldstein, 1972) (see text for details).

Note that the positively charged support increases the apparent K_M quite significantly reduces k_{cat} whilst the negatively charged support has less effect on the apparent K_M and greatly enhances the rate of catalysis. It should be noted that the partitioning effects resulting from the charge on an immobilised support is most noticeable when using solutes of weak buffering capacity. In industry, weak buffering capacity is the 'norm' rather than the exception because adding buffering capacity to the system raises costs.

Before we leave the effects of supports on the pH and ionic strength in the micro-environment of the enzyme, we need to consider the effects of using hydrophobic supports.

The creation of a hydrophobic environment around the enzyme molecule may greatly influence the ionisation of side groups, the ionic and hydrophobic interactions within the enzyme molecule and the partitioning of substrates from the bulk liquid. Generally the more hydrophobic an environment, the stronger the ionic interactions (Na^+ and Cl^- ions will not dissociate in petrol but they readily dissociate in water). Thus we might expect that the pKa's of ionisable groups on the enzyme will be modified by a hydrophobic support. Similarly hydrophobic substrate will partition into the support from the bulk solution whilst hydrophilic molecules will partition out. Thus we may anticipate that hydrophobic supports may influence both the K_M and the catalytic activity of an enzyme. An example of this effect is observed with the enzyme alcohol dehydrogenase. This enzyme will catalyse the oxidation of butanol but, attached to a polar support (eg, polyacrylamide), it has a high apparent K_M (K_M = 0.1 mol l^{-1}). On a more hydrophobic support (a co-polymer of methacrylate and acrylamide) the apparent K_M is reduced to 0.025 mmol l^{-1}.

influence of hydrophobic supports on K_M and catalytic function

4.4.3 Substrates and products

We have already demonstrated that the support used to immobilise an enzyme may influence the properties of an enzyme by changing the micro-environment of the enzyme. We also pointed out that the K_M' and velocity of catalysis (k_{cat} or V_{max}) may also be changed because the enzyme is structurally modified by its attachment to the support. We now turn our attention to examining the effects of immobilisation on substrate and product concentrations in the vicinity of the enzyme.

Π Using the description of the effects of electronic charge on the support on H^+ ion distribution, explain how a negatively charged support may influence the K_M' of an enzyme if the substrate is positively charged (assume that substrate diffusion into the support is not rate-limiting).

[S_s] may be different from [S_b]

We would expect the substrate to be attracted by the support. In other words $[S_s] > [S_b]$, where $[S_s]$ = substrate concentration in the vicinity of the support and $[S_b]$ = substrate concentration in the bulk liquid. Thus at a relatively low $[S_b]$ value we may get a faster reaction than we might anticipate from studies with the free enzyme. In other words K_M' would appear to be lower. We have represented this situation in Figure 4.11.

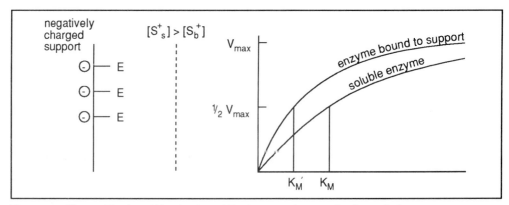

Figure 4.11 The modification of K_M of an immobilised enzyme (see text for discussion).

Thus there is an apparent shift in K_M because of the partitioning of substrate between the bulk liquid and the support. We may also get analogous partition of the product. Thus we might anticipate that a positively charged product would be retained more by a negatively charged support.

Since the rate of the forward reaction S → P (substrate to product) depends on the relative concentrations of S and P, we might anticipate that the overall reaction kinetics of an enzyme may be quite different in an immobilised system from that observed with the same enzyme in bulk solution where product is free to diffuse away from the enzyme. However, the overall reaction kinetics are not only influenced by such partitioning effects, they are also influenced by the rates of diffusion of both substrate and product. We will examine these effects in the next section.

You will also have anticipated that immobilisation may influence the distribution of inhibitors in a system. We will examine this aspect in Section 4.5.2

SAQ 4.5

Assume that the binding of an enzyme to a support does not influence its structure nor its catalytic activity. Also assume that the system we have developed is not limited by diffusion rates. Which of the following situations are likely to lead to faster rates of catalysis than would be observed with the enzyme in its soluble form. (Give reasons for your choice).

1) A negatively charged support with a negatively charged substrate and a positively charged product.

2) A hydrophobic support with a negatively charged substrate and an uncharged product.

3) A negatively charged support with a positively charged substrate and a positively charged product.

4) A negatively charged support with a positively charged substrate and a negatively charged product.

4.5 The effects of diffusion on the kinetics of immobilised enzymes

In the previous section, we examined changes to the kinetics of enzyme-catalysed reactions arising through the structural changes to the enzyme and/or as a result of the partitioning of substrate and product between the bulk solution and the immobilised system. Of critical importance in the design of immobilised enzyme systems are the effects of diffusion of substrates and products on the kinetics of the reaction catalysed by the immobilised enzyme. In this section, we will examine these effects, firstly by providing a generalised description of the diffusion of reactants and products and secondly by a more detailed discussion of the effects such diffusions have on the kinetics of the reaction. A detailed discussion of the process engineering aspects of these systems is given in the BIOTOL test 'Bioreactor Design and Product Yield'.

4.5.1 Diffusion of substrates and products on immobilised enzyme systems

The net movement of substrates and products can be divided into two stages:

- external diffusion in which substrates diffuse from the bulk liquid, through the stagnant liquid layer surrounding the immobilised enzyme to the surface of the immobilised system. Products diffuse in the reverse direction;

- internal diffusion in which substrates and products diffuse within the immobilised enzyme matrix.

We can anticipate that a series of concentration gradients will be set up within the system. These are illustrated in generalised form in Figure 4.12.

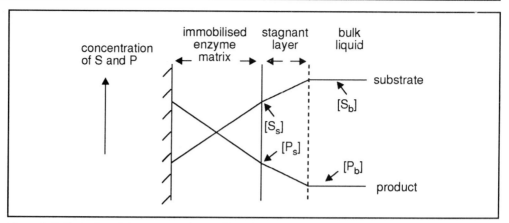

Figure 4.12 Illustration of the concentration gradients in an immobilised system (see text for discussion). [S_b] and [P_b] = concentrations of substrate and product in the bulk liquid, [S_s] and [P_s] = concentration of substrate and product at the liquid:enzyme matrix interface.

Examine Figure 4.12 carefully. You will notice that the concentration of substrate in the immobilised enzyme matrix is significantly lower than in the bulk liquid. On the other hand, product concentration is higher in the immobilised enzyme matrix than in the bulk liquid. We might anticipate, therefore, that the velocity (v) of the reaction taking place in the matrix is likely to be smaller than that which would be observed in a system in which the same amount of enzyme was used in free solution (v_{sol}). The ratio of v/v_{sol} is referred to as the effectiveness factor (η) of the system. Thus:

$$\eta = \frac{v}{v_{sol}}$$

Π If substrate and product only diffuse slowly through the stagnant layer and the immobilised enzyme matrix, will the effectiveness factor be higher or lower than if the diffusion rates were high?

You should anticipate that η would be lower. The slower the diffusion rates, the greater the concentration gradients and thus [S] would be low and [P] high in the immobilised enzyme matrix. Therefore, v would be low and η would also be low.

If diffusion was fast the concentration gradients would be shallow and [S_s] and [P_s] would be similar to the values of [S_b] and [P_b]. Thus the value of v would be similar to v_{sol} and η would approach 1.

This is a somewhat simplified picture of the concentration gradients which are observed in practice. Remember that substrates and products may also partition between the liquid and the immobilised system. In Figure 4.13 we have drawn concentration profiles for a system in which the substrate and product preferentially partition into the immobilised system.

Enzyme immobilisation

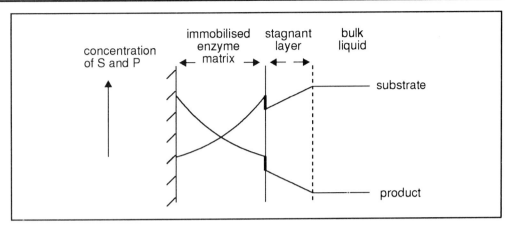

Figure 4.13 Concentration profiles in which substrates and products partition into the immobilised system.

Note the discontinuous nature of the concentration profiles at the liquid:immobilised enzyme interface. We could draw analogous concentration profiles for substrates and products which preferentially partition out of the enzyme matrix. In this case the steps in the concentration gradients would be as shown in Figure 4.14.

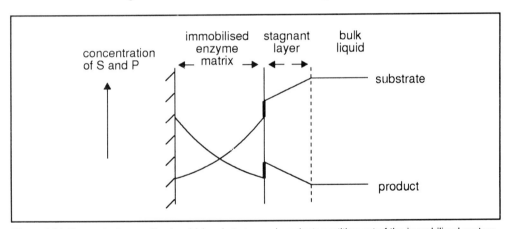

Figure 4.14 Concentration profiles in which substrates and products partition out of the immobilised system.

Thus we might anticipate that the concentrations of substrates and products in the micro-environment of the enzyme would reflect the rate of diffusion and the partitioning properties of these solutes. These in turn will be reflected in the velocity (rate) of the reaction.

To simplify matters, in the following sections we will mainly concern ourselves with the effects of diffusion rates in systems which do not display partitioning effects. We will, however, divide our discussion into two parts. In the first part, we will discuss the effects of diffusion when the immobilised enzyme is presented on a flat impervious support. In the second part, we will discuss the effects of diffusion when the immobilised enzyme is presented in the form of porous matrix.

4.5.2 Diffusion and impervious immobilised enzyme systems

We can visualise these systems in the following way:

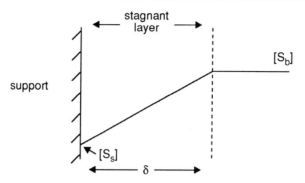

The flow (flux) of substrate to the surface of the support can be calculated from Fick's law.

Fick's law can be written as:

$$\phi_s = D_s \frac{\Delta S}{\delta}$$

where $\Delta S = [S_b] - [S_s]$; ϕ_s = flow rate of substrate; D_s = diffusion coefficient ($m^2 s^{-1}$); δ = thickness of the stagnant layer.

The rate of change in substrate concentration at the surface of the support as substrate diffuses in, is given by:

$$\left\{\frac{d[S_s]}{dt}\right\}_{diffusion} = \phi_s = \frac{D_s}{\delta}([S_b] - [S_s])$$

But substrate is being used at the surface by enzyme converting it to product. We can write this in the form:

$$\left\{\frac{d[S_s]}{dt}\right\}_{reaction} = V_{max}\, f[S_s]$$

Since the reaction is related to the amount of enzyme present (V_{max} = maximum rate of reaction catalysed per unit area of surface) and a function (f) of the substrate concentration. Thus:

$$\frac{d[S_s]}{dt} = \frac{D_s}{\delta}([S_b] - [S_s]) - V_{max}\, f[S_s]$$

When the reaction has progressed for some time a steady state will be reached. Thus:

$$\frac{d[S_s]}{dt} = 0$$

and:

$$\frac{D_s}{\delta}([S_b] - [S_s]) = V_{max}\, f[S_s]$$

In other words, the rate of the enzyme catalysed reaction will equal the rate of substrate diffusion to the support surface. If we assume that the enzyme behaves according to Michaelis-Menten kinetics.

Enzyme immobilisation

$$f = \frac{[S_s]}{K_M' + [S_s]}$$

and, therefore:

$$\frac{D_s}{\delta}([S_b] - [S_s]) = \frac{V_{max}[S_s]}{K_M' + [S_s]}$$

But from Fick's law $\frac{D_s}{\delta} = k_L$ where k_L is the mass transfer coefficient (ms^{-1}).

and so we can write:

$$k_L([S_b] - [S_s]) = \frac{V_{max}[S_s]}{K_M' + [S_s]}$$

Clearly we would like to maximise the velocity of the reaction.

Π From the relationships we have described so far, how might we try to maximise the rate of the reaction?

Since the rate of the reaction is equal to the rate of diffusion, we can try to maximise diffusion.

changing the rate of diffusion

The rate of diffusion can be written as $\frac{D_s}{\delta}([S_b] - [S_s])$ or as $k_L([S_b]-[S_s])$. We could, therefore, try to increase diffusion by using high $[S_b]$ values or by reducing δ, the stagnant layer around the support. δ depends on hydrodynamic conditions and is reduced by increasing the rate of stirring. There is, however, a limit on how much δ can be reduced by physical agitation and may not be possible at all in, for example, packed bed reactors (we will discuss these later).

You might have thought that you could have increased V_{max} (by increasing the amount of enzyme per unit area of support) to increase the rate of reaction. Let us think what will happen when we put more enzyme onto the surface of the support. With every increase in V_{max}, we will consume substrate at the surface faster. At low V_{max}, substrate diffusion to the surface will not be rate-limiting. However, when we increase V_{max}, substrate diffusion becomes progressively more rate-limiting and $[S_s]$ will decline. Thus since:

$$v = \frac{V_{max}[S_s]}{K_M' + [S_s]}$$

doubling V_{max} will not necessarily double v. When V_{max} is extremely high, increasing it even further will have no effect on v because the increase in V_{max} will be compensated for by a further decline in $[S_s]$. We can represent this graphically (see Figure 4.15, read the legend carefully).

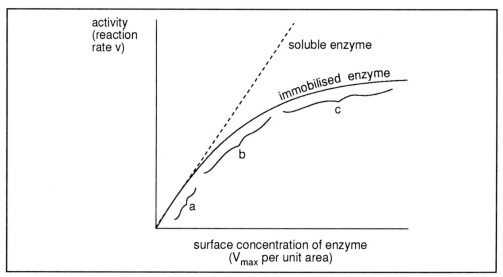

Figure 4.15 Stylised relationship between reaction rate (v) and the concentration of enzyme (V_{max} per unit area) on a support surface: a) the kinetics are controlled by the amount of enzyme; b) kinetics influenced by diffusion and by the amount of enzyme on the surface; c) kinetics dominated by the rate of transport of substrate across the stagnant layer.

Let us convert this into a mathematical form.

If $[S_s]$ is much greater than K_M' then $\dfrac{V_{max}[S_s]}{K_M' + [S_s]}$ approaches V_{max}. Thus under these conditions:

$$v_A = V_{max} = k_L([S_b]-[S_s])$$

where v_A is the rate of reaction catalysed per unit area of the immobilised enzyme.

In many instances, for example when we have a large amount of enzyme or slow diffusion rates, $[S_s]$ is much lower than K_M.

In this case the rate of the reaction becomes:

$$v_A = \frac{V_{max}[S_s]}{K_M'}$$

Thus:

$$[S_s](V_{max}/K_M') = k_L([S_b] - [S_s]).$$

and:

$$[S_s] = \frac{k_L[S_b]}{k_L + (V_{max}/K_M')} = \frac{[S_b]}{1 + (V_{max}/K_M')(1/k_L)}.$$

Thus substituting for $[S_s]$ in:

$$v_A = [S_s]\left(\frac{V_{max}}{K_M'}\right)$$

$$v_A = \frac{S_s(V_{max}/K_M')}{1 + (V_{max}/K_M')(1/k_L)}$$

and:

$$v_A = \frac{[S_b]}{1/k_L + 1/(V_{max}/K_M')}$$ (E 4-1)

Thus the rate of the reaction (v_A) is dependent upon the mass transfer coefficient of the substrate (k_L) and the catalytic activity (V_{max}/K_M') of the enzyme.

If k_L is much greater than V_{max}/K_M then the overall kinetics reduces to:

$$v_A = [S_b](V_{max}/K_M')$$ (E 4-2)

If, on the other hand, k_L is much smaller than V_{max}/K_M, then the reaction rate becomes:

$$v_A = k_L[S_b]$$ (E 4-3)

We can use Equations E 4-1, E 4-2 and E 4-3 to explain the shape of the curve shown in Figure 4.15.

At low V_{max} values $v_A = [S_b](V_{max}/K_M')$ and v_A is proportional to V_{max}. (Remember that (V_{max}/K_M') is the specificity constant of the enzyme). At high V_{max} values $v_A = k_L[S_b]$ and v_A is independent of V_{max}. At intermediate values:

$$v_A = \frac{[S_b]}{1/k_L + 1/(V_{max}/K_M')}$$

and is thus dependent on but not directly proportional to V_{max}.

SAQ 4.6

List as many factors as you can that will increase the likelihood that an immobilised system will be limited by external diffusion, rather than by the amount of enzyme attached to the support. When you have done this read our response carefully as it provides a summary.

Effects of inhibition

If the enzyme is inhibited by the product (P), then we would anticipate that this product inhibition would have a greater effect in an immobilised system than in a system using a soluble enzyme. The product, made at the surface of the support has to diffuse through the stagnant layer. The actual profile would be dependent upon the rate of formation of P, D_P and δ (where D_P = diffusion coefficient of P).

∏ If a competitive inhibitor (I) was added to the bulk liquid, would this have more or less effect on an immobilised enzyme than on a soluble enzyme?

It depends on the relative diffusivity of I and S. If I diffused faster than S then we would anticipate the following concentration profiles:

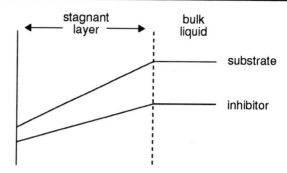

Thus the ratio [I] to [S] would be greater at the surface of the support than in the bulk liquid. Thus we would anticipate greater inhibition than would be the case with a soluble enzyme. If I diffuses more slowly than S, then we would expect less inhibition.

This is, however, a great simplification. As we have seen, in practice, immobilisation may change the binding characteristics of an enzyme. Thus not only may K_M be changed, but also K_I (the binding constant of an inhibitor). Since the reaction rate of an enzyme catalysed reaction in the presence of a competitive inhibitor is given by:

$$v = \frac{V_{max}[S]}{K_M\left(1+\frac{[I]}{K_I}\right) + [S]}$$

(see Section 1.3.4)

we must anticipate that since in an immobilised system $[I] = [I_s]$, $[S] = [S_s]$ and K_M and K_I may be changed (to K_M' and K_I'), predicting the outcome of the presence of the competitive inhibitor depends on being able to predict $[I_s]$, $[S_s]$ and K_M' and K_I'.

We can argue similarly for uncompetitive and non-competitive inhibitors using the appropriate kinetic equations (see Section 1.3.4).

Prediction of the effects of an inhibitor is further complicated by partitioning of the inhibitor between the bulk liquid and the immobilised enzyme. Thus although the inhibitor may be present in only low concentrations in the bulk liquid, its partitioning may cause its concentration in the micro-environment of the enzyme to become relatively much higher. This is particularly a problem in reactors in which immobilised enzymes are used continuously and the inhibitor, present in the bulk liquid, is strongly attracted to the immobilised enzyme support.

SAQ 4.7

Which of the following factors are likely to increase the rate of a reaction catalysed by an enzyme attached to the surface of a support?

1) Increased agitation of the bulk liquid containing the substrate.

2) Continued replacement of the bulk liquid containing the substrate.

3) Increased concentration of the substrate in the bulk liquid.

4) Increased viscosity of the substrate solution.

Enzyme immobilisation

SAQ 4.8

1) The diffusion coefficient of a substrate (D_s) is 10×10^{-10} m^{2}s^{-1} and the thickness of the stagnant layer adjacent to a flat immobilised enzyme support has been determined as 5 µm. Use these data to determine the maximum rate of reaction if we use a substrate concentration of 1 mol l^{-1}. The ratio V_{max}/K_M' for this system has a value of 20 ms^{-1}. Assume that product does not inhibit the enzyme.

2) What would be the rate of reaction if we reduce of the amount of enzyme on the support used in 1) by a factor of 10^6?

3) What would be the rate of reaction in 2) if the ratio V_{max}/K_M' was 1×10^{-4} ms^{-1}?

4.5.3 Diffusion and porous immobilised enzyme systems

In our discussion of the changes in the properties of enzymes attached to a support, we have described how the rate of substrate diffusion to the enzyme may limit the rate of substrate conversion to product. We learnt that simply increasing the amount of enzyme attached to the support may not greatly enhance the rate of substrate conversion (see for example, Figure 4.15). In such circumstances, improvement of reactor performance may be achieved by increasing agitation of the bulk fluid (decreasing the thickness of the stagnant layer) or by increasing the concentration of substrate in the bulk fluid ($[S_b]$). The treatment of this problem in Section 4.5.2 depicted the enzyme as simply being on the surface of the support. In practice, however, the enzyme may be attached to or entrapped within the whole matrix of the support. It is this situation which we will examine here.

Imagine that we have an entrapped or attached enzyme in the form of a thin film coating a solid support. Substrate will diffuse into this film from the bulk liquid. Inside of the film, the enzyme present will begin to convert substrate (S) to product (P). The product will of course diffuse back into the bulk liquid. What we need to be able to do is to predict substrate concentration throughout the film in order to determine whether or not the reaction rate is limited by the availability of substrate or enzyme.

Using Fick's law of diffusion, the rate of diffusion of the substrate across the stagnant layer and in the film can be predicted. Inside the film, it may be used by the enzyme by zeroth order kinetics (if [S] is much greater than K_M') or first order kinetics (if [S] is much lower than K_M'). At [S] values closer to K_M', we would have to use the Michaelis-Menten relationship to predict substrate conversion rates (providing, of course, the enzyme behaved according to the Michaelis-Menten equation! If not we would have to apply the appropriate relation between v and [S]).

If we know how much enzyme is present per unit volume of the immobilised film (V_{max} per unit volume) and the K_M' of the enzyme, we could, in principle, calculate the rate of consumption of substrate in each layer of the film and the predicted substrate concentration in each layer.

Thus, we could draw substrate concentration profiles across the film. In Figure 4.16, we show two examples.

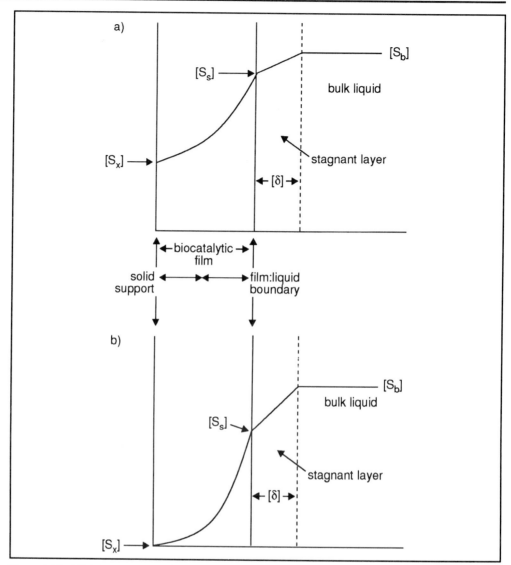

Figure 4.16 a) Substrate concentration profile of a biocatalytic film with complete substrate penetration. b) Substrate concentration profile of a biocatalytic film with substrate depletion (see text for description).

Π Which of the examples shown in Figure 4.16 indicates that the reaction is limited by diffusion of substrate into the film?

You should have identified b). Notice that part of the film in b) contains very little substrate. Often we can again use the term 'effectiveness factor' (η) to describe this situation. If an immobilised catalyst has V_{max} activity per unit volume, then the maximum rate of substrate converted per unit area of the biocatalyst is given by $V_{max}X$ where X is the thickness of the film. If [S] throughout the film is much greater than K_M', then the rate of conversion of substrate approaches $V_{max}X$ per unit area. The effectiveness factor is said to approach 1. If substrate levels fall to values around K_M' or below, then the rate of conversion of substrate will be lower than $V_{max}X$. In this case we can define the effective factor η as:

Enzyme immobilisation

$$\eta = \frac{\text{actual reaction rate}}{\text{maximum reaction rate}}$$

Obviously we would like to design a system in such a way that the effectiveness factor is close to 1.

∏ What could we change in situation b) in Figure 4.16 to increase η?

There are several factors we could change. The ones we hope you would suggest are:

- use a thinner film;
- use less enzyme per volume of film;
- increase $[S_b]$ in the bulk liquid or decrease the thickness of the stagnant layer.

The calculation of the effectiveness factor is quite complex and depends upon the geometry of the biocatalyst (for example whether it is in the form of a thin film or in the form of spheres), the order of the reaction (for example zeroth, first order, Michaelis-Menten), the thickness of the biocatalyst and the diffusion coefficient within the biocatalyst. Equations relating these parameters are given in Table 4.5. We will not derive these equations here but they are derived in the BIOTOL text, 'Operational Modes of Bioreactors' if you wish to follow this up.

Here we confine ourselves to some general observations.

Biocatalyst geometry	Reaction order	Effectiveness factor and remarks	
slab	0th	$\eta = \sqrt{(2[S_b]D)/R^2 k_0}$ $= 1 - \left(\frac{R-r}{R}\right)$	$\eta = 1$ providing $R\sqrt{(k_0/D[S_b])} = < \sqrt{2}$
sphere	0th	$\eta = 1 - \left(\frac{R-r}{R}\right)^3$	$\eta = 1$ providing $R\sqrt{(k_0/D[S_b])} = < \sqrt{6}$
slab	1st	$\eta = \frac{\tanh(R\sqrt{(k_1/D)})}{R\sqrt{(k_1/D)}}$	
sphere	1st	$\eta = \frac{3}{\sqrt{(k_1/D)}.R}$	$\cdot \left\{ \frac{\cosh(k_1/D).R}{\sinh(k_1/D).R} \right\} - \frac{1}{\sqrt{(k_1/D)}.R}$
slab	Michaelis-Menten	$\eta = \frac{3D\frac{d[S]}{d_r}}{R\,v(S_b)}$	

Table 4.5 Expressions for the effectiveness factor for zeroth, first order and Michaelis-Menten kinetics in flat plate (slab) and spherical geometry biocatalysts.
KEY: $[S_b]$ = substrate concentration in the bulk liquid; D = diffusion coefficient in the biocatalytic matrix; k_0 = zeroth order rate constant; k_1 = first order rate constant; R = thickness of the biocatalyst (slab) or radius of the spherical biocatalysts; r = penetration depth of substrate; $\frac{d[S]}{d_r}$ = substrate concentration gradient at the liquid/biocatalyst interface; $v(S_b)$ = velocity of the reaction at the bulk liquid substrate concentration; *tanh, cosh, sinh* = hyperbolic functions where $\cosh x = \frac{1}{2}(e^x + e^{-x})$, $\sinh x = \frac{1}{2}(e^x - e^{-x})$ and $\tanh x = \frac{\sinh x}{\cosh x}$.

η = 1 only if the reaction follows zeroth order kinetics

First you should note that it is only possible to achieve an effectiveness factor of 1, if the reaction is zeroth order (ie independent of substrate concentration). With higher order reactions in which the reaction rate is dependent upon substrate concentration then it is impossible to achieve a reaction rate within the biocatalyst as high as that at its surface.

With enzyme catalysed reactors, if [S] is maintained at high values compared to K_M' then the reaction approaches zeroth order kinetics. If [S] is low compared to K_M' throughout the biocatalyst, the reaction approaches first order kinetics. If [S] within the biocatalyst is close to K_M', then the kinetics of the reaction is a part way between zeroth and first order.

We have represented these situations in Figure 4.17. Note that in this figure we have assumed that the stagnant layer does not act as a barrier to diffusion and that $[S_s] = [S_b]$.

Π Examine Figure 4.17 carefully. In Figure 4.17 a), is the effectiveness factor of the biocatalyst = 1?

The answer is no. Note that the substrate has only penetrated to a depth of r. A portion of the biocatalyst (R-r) is not being used.

Π If R = 2 mm and r = 1.5 mm, what is the effectiveness factor of the biocatalyst shown in Figure 4.17 a).

The answer is $\eta = \dfrac{1.5}{2} = 0.75$. Another way of thinking about this is that 75% of the biocatalyst is being used.

Π If the substrate profile shown in Figure 4.17 a), was in a spherical rather than in a slab form, what would the effectiveness factor be? (You may need to refer to Table 4.4 to obtain the appropriate relationship).

The answer is $\eta = 1 - \left(\dfrac{0.5}{2}\right)^3 = 0.984$. In other words 98.4% of the biocatalyst is used.

What your calculation has shown you is that, all other things being equal (ie D, k_0, R and $[S_b]$), then spherical catalysts give higher effectiveness factors.

This observation is true whether or not the reaction follows zeroth order, first order or Michaelis-Menten kinetics. It is purely a product of the geometry of the system.

spherical biocatalysts give higher effectiveness factors than films

The relationships relating effectiveness factors to the biocatalyst geometry, substrate diffusion coefficient, reaction rate and biocatalyst thickness described in this section can be applied to immobilised biocatalysts be they immobilised enzymes or immobilised non-growing cells. If growing cells act as the biocatalyst, obviously the relationship becomes much more complex as the reaction rate constant (k) becomes time dependent. We will not, however, tackle this more complex situation here.

You should note that in systems subject to inhibition by product or by other components in the bulk liquid, the kinetics of conversion becomes very complex to model. Although many attempts have been made to do this, such models tend to be unreliable. In practise, it is often safer to build pilot (experimental) systems and to determine η experimentally rather than attempt to build theoretical models. For this reason we will not extend our discussion of the effects of inhibition here. If you wish to follow this aspect further, you will find some of the texts in 'Suggestions for Further Reading' at the end of this text helpful.

Enzyme immobilisation

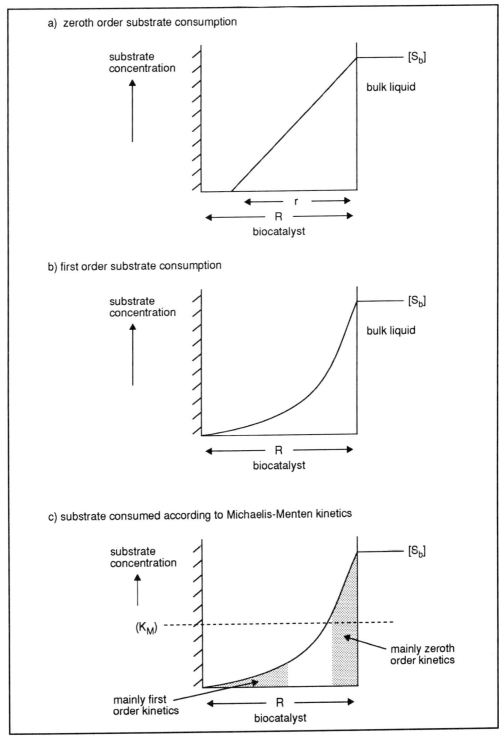

Figure 4.17 Simplified substrate profiles in slabs with a) zeroth, b) first order and c) Michaelis-Menten kinetics (symbols as in Table 4.1). Note that we have simplified these substrate profiles by assuming that there is no diffusion barrier in the stagnant layer around the biocatalyst and that the substrate concentration at the biocatalyst:liquid interface is the same as in the bulk liquid.

SAQ 4.9

The thickness of an immobilised biocatalyst in the form of a slab is 1 mm. The zeroth order reaction constant for substrate consumption in the biocatalyst is 1 mol m^3 h^{-1} and its surface area is 100 m^2. The rate of consumption of substrate by this biocatalyst has been measured as 0.08 mol h^{-1}.

1) Does the substrate penetrate the whole biocatalyst?

2) If not, how far does the substrate penetrate the biocatalyst?

SAQ 4.10

We provide the following data:

A biocatalyst is in the form of a 2 mm thick slab supported by a packing material. Substrate is consumed by the biocatalyst according to zeroth order kinetics with a zeroth order reaction rate constant of 1×10^{-3} mol s^{-1} kg^{-1}. 100 kg of the biocatalyst is immobilised with a concentration of 100 kg m^{-3}. The effective diffusion coefficient of the substrate is 2.2×10^{-9} m^2 s^{-1}. The substrate is consumed at a rate of 0.06 mol s^{-1}.

1) Calculate the volume of the biocatalyst and its area.

2) Determine whether the substrate has penetrated the whole depth of the biofilm and, if not, what depth it has penetrated.

3) Calculate the effectiveness factor for the biocatalyst.

4) Calculate the substrate concentration in the bulk liquid. (You will need the relationship shown in Table 4.4).

SAQ 4.11

Identify which of the following are true:

1) If an immobilised enzyme is sensitive to product inhibition, this will tend to lower the value of η.

2) If a positively charged inhibitor is included in the bulk liquid, then an enzyme entrapped in a negatively charged support will be more inhibited than the same enzyme in free solution.

3) A particular substrate at high concentrations is known to inhibit its enzyme. A high substrate concentration in the bulk liquid will tend to lead to complete inhibition of the enzyme if the system is not diffusion limited.

4) In an immobilised system which is diffusion limited, the enzyme will appear to be less sensitive to pH than will the same enzyme in solution.

5) In an immobilised system which is diffusion limited, the enzyme will appear to be less sensitive to inhibitors than will the same enzyme in solution.

Summary and objectives

In this chapter we began by examining the potential advantages and disadvantages of using immobilised enzymes. We then moved on to explain how enzymes may be immobilised and described the techniques available to achieve immobilisation. We then examined the consequences of immobilisation on the properties of enzymes. The apparent changes to major enzyme parameters such as K_M and V_{max} arise through structural changes to the enzyme molecules and as a result of changes in the micro-environment of these molecules. We also examined the consequence of partitioning the enzymes and bulk reaction mixtures with particular emphasis on the effects of external and internal diffusion of substrate on the kinetics of reactions catalysed by immobilised enzymes.

Now that you have completed this chapter you should be able to:

- explain the advantages and disadvantages of using immobilised enzymes;

- describe and give examples of the different methods of enzyme immobilisation;

- interpret data relating to the covalent binding of enzymes to carriers (supports) to optimise the conditions used during the binding process;

- identify appropriate ion exchange resins to use as supports for immobilising enzymes by physical adsorption;

- explain how immobilisation may bring about apparent and real changes in the properties of enzymes;

- list the factors which will tend to lead to diffusion limitation of the rates of substrate conversion catalysed by immobilised enzymes;

- calculate reaction rates of processes catalysed by immobilised enzymes from data relating to mass transfer (K_L), $K_M{'}$, V_{max} and substrate concentrations.

Immobilised enzyme reactors

5.1 Introduction	120
5.2 Reactor configuration	120
5.3 Stirred tank batch reactors	122
5.4 Plug flow reactors (PFRs)	126
5.5 Continuous flow stirred tank reactors (CSTRs)	130
5.6 Fluidised bed reactors (FBRs)	139
5.7 Membrane and hollow fibre reactors	140
5.8 Satisfying cofactor and cosubstrate requirements	142
5.9 Reactor development is a team activity	146
Summary and objectives	148

Immobilised enzyme reactors

5.1 Introduction

In Chapter 4, we explained the potential advantages of using immobilised enzymes and described the techniques that may be employed to achieve immobilisation. We also examined the consequences of immobilisation on the properties of enzymes and the velocity of reactions catalysed by immobilised enzymes. In this chapter, we will examine the design and performance of reactors employing the use of immobilised enzymes. To do this, we have predominantly adopted the perspective of the enzymologist rather than that of the process engineer. The details of process technology associated with bioreactor design and operation are dealt with elsewhere in the BIOTOL series. We particularly recommend the BIOTOL text, 'Bioprocess Technology: Modelling and Transport Phenomena', 'Operational Modes of Bioreactors', and 'Bioreactor Design and Product Yield', if you wish to follow up this aspect of industrial enzymology.

In this chapter, we will first give a brief overview of the various designs of bioreactors before comparing the performances of these reactors. In the final part of the chapter, we will briefly examine the problems associated with the requirement for cofactors and indicate how these problems may be overcome.

5.2 Reactor configuration

Enzyme reactors consist of vessels used to achieve the conversion of substrates to products using enzymes as catalysts. There are many different configurations available to us. Here we will focus on the major options. These are:

- stirred tank batch reactors (STRs);

- plug flow reactors (PFRs), sometimes referred to as packed bed reactors;

- continuous flow stirred tank reactors (CSTRs);

- membrane/hollow fibre reactors (MRs);

- fluidised bed reactors (FBRs).

We have illustrated these in Figure 5.1.

Immobilised enzyme reactors

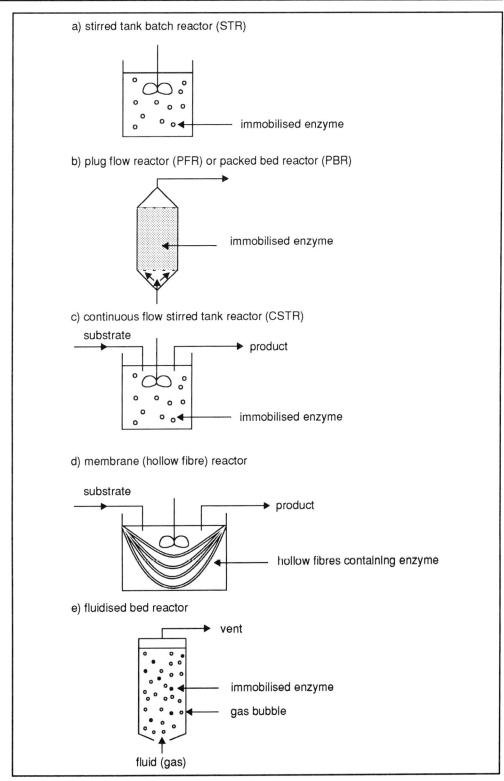

Figure 5.1 Reactor configurations (see text in Sections 5.2-5.5 for description).

In selecting a reactor, there are a number of factors which influence the choice. These include:

- the amount of product that has to be made;
- the cost of operation, including the cost of substrate and downstream processing;
- running costs including depreciation;
- capital cost of equipment;
- ease of operation with special attention to maintaining physical (pH, temperature) and chemical conditions (addition of substrate, removal of product);
- the stability of the enzyme.

We will briefly examine these aspects for each type of reactor.

5.3 Stirred tank batch reactors

batch and batch-fed systems

These types of reactors normally consist of a tank fitted with a stirrer and with fixed baffles which improve mixing. In these reactors, the enzyme, usually entrapped or encapsulated in beads or adsorbed onto particulate supports, simply replaces the soluble enzyme in the batch systems described in Chapter 3. In these systems, the enzyme and substrate(s) have the same residence time in the reactor. In some cases, for example if the enzyme is relatively unstable, additional enzyme may be added during the reaction. Alternatively, especially in cases where substrate inhibition occurs, we might use low initial substrate concentrations and add additional substrate as substrate is consumed. These latter types of operations are described as batch-fed systems or fed-batch systems.

Π See if you can list some advantages and disadvantages of replacing soluble enzymes in stirred tanks by enzymes entrapped in beads or adsorbed onto particles.

limitations to using immobilised enzymes in STRs

The main advantage is that the beads or particles can be readily separated from the reaction mixture at the end of the reaction. Thus the enzyme is recovered for re-use and the product is free of contamination from the enzyme. The disadvantages are that such immobilised enzymes cannot be used effectively to convert particulate substrate (for example, homogenised meat, cellulose, starch) or for large molecular species which will not penetrate the beads. Also, as we learnt in Chapter 4, even with low molecular weight substrates, the diffusion of substrates to the enzyme may be rate limiting. Thus an encapsulated/entrapped enzyme process is likely to be slower than a similar process using the same amount of soluble enzyme. Even with covalent attachment or physical adsorption onto the surface of a support, the reaction is likely to be slower because of heterogeneity of the system. Further disadvantages of using an immobilised enzyme are the cost of the immobilisation and the loss of activity during the immobilisation process.

Thus, whether we choose to use a soluble or an immobilised enzyme in a stirred tank batch reactor depends on the balance between the savings achieved on enzyme re-use

Immobilised enzyme reactors

and on easier downstream processing and the additional costs incurred by increased reaction times and by the immobilisation process.

∏ Would there be any advantage of using an enzyme immobilised in beads if the desired conversion required two enzymes to sequentially convert the substrate to product?

In principle, the answer is yes. By entrapping both enzymes in the same matrix, the product of the first enzyme would be generated in close proximity to the second enzyme. This second enzyme would of course remove the product of the first enzyme, thereby enhancing the rate of the first step (by the laws of mass action or by reducing product inhibition) providing the process was not diffusion limited.

5.3.1 The importance of mixing in stirred tank reactors

In our discussion of immobilised enzymes in Chapter 4, we emphasised the importance of the stagnant (unmixed) layer surrounding an immobilised enzyme as a diffusion barrier. Ideally we would like this layer (we designated it as δ) to be as thin as possible. The thickness of this layer depends upon the fluid flow around the particles containing the enzyme. Fluid flow can be of two types, laminar or turbulent. We can represent these two flow types as:

laminar and turbulent flows

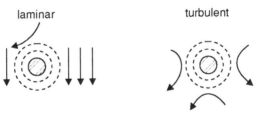

In laminar flow, each layer of fluid flows in parallel and thus solutes have to diffuse through each layer of the bulk liquid and the stagnant layer to reach the enzyme. In turbulent flow, solutes diffuse directly from the bulk liquid through the stagnant layer. We can represent the concentration gradients in the two systems in the following way:

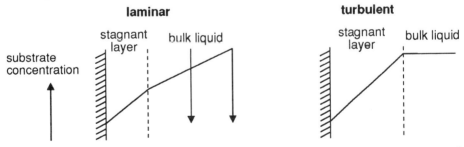

Thus we can anticipate that diffusion is more likely to be rate limiting in a laminar flow system than in a turbulent flow system.

Whether or not the flow of the system is laminar or turbulent depends upon the inertia of the liquid (its tendency to continue on its way) and the viscous forces (the friction between neighbouring layers) which resist the flow. The ratio of these two forces is called the Reynolds number (Re) and can be written as:

$$Re = \frac{\text{inertia pressure}}{\text{viscous stress}}$$

At low Re values (slow fluid flow rates and/or fluids of high viscosity) the flow is laminar.

In batch reactors we may attempt to ensure turbulent flow especially in a system that is likely to be diffusion limited. This is most easily achieved by increasing the power of the impeller to ensure that high fluid flow rates are achieved. There is, however, a cost balance to be taken into account. High power input costs money and this must be more than compensated by the reduced reaction times to be worthwhile. Furthermore, turbulence imposes physical stresses on the particulate immobilised system and can lead, in extreme cases, to its physical disruption.

achieving turbulent flow by high power input

(The theory behind mass transport and Reynolds numbers are described in detail in the BIOTOL text, 'Bioprocess Technology: Modelling and Transport Phenomena').

5.3.2 Productivity of stirred tank batch reactors

Of key importance in the design of a batch process is determining the reaction time to achieve a particular level of conversion of the substrate. This, of course, will depend upon the amount of enzyme, the substrate concentration and the effectiveness factor of the immobilised system.

For our purposes, let us assume that the enzyme follows the Michaelis-Menten equation:

$$v_0 = \frac{V_{max}[S]}{K_M + [S]}$$

The rate of reaction (v) may be expressed in terms of the change in substrate concentration with time:

thus:
$$v = \frac{-d[S]}{d_t} = \frac{V_{max}[S]}{K_M + [S]}$$

therefore:
$$\int_0^t V_{max} d_t = -\int_{[S_0]}^{[S_t]} \left(1 + \frac{K_M}{[S]} d[S]\right)$$

Using the boundary condition $[S] = [S_0]$ when $t = 0$, integration gives:

$$V_{max} t = K_M \ln \frac{[S_0]}{[S_t]} + ([S_0] - [S_t])$$

This equation is essentially the same as that derived for a soluble enzyme in a batch reactor (see Chapter 3).

For immobilised enzymes used in a batch system, we may have to modify this equation.

Π Examine this equation and see if you can identify the ways in which it may need to be modified.

The equation assumes that all of the enzyme molecules have access to the substrate. With immobilised systems, this is only true if the system is not diffusion limited and it has an effectiveness factor $\eta = 1$.

You will recall that in an immobilised system, the effectiveness factor is dependent upon the substrate concentration in the bulk liquid, the mass transport coefficient and the concentration of the enzyme. Since the substrate is being consumed during the reaction, η may also alter with time.

Let us take two extreme examples. In the first let us assume that the substrate concentration in the bulk liquid is high compared to K_M even at the end of the reaction and that the system is not diffusion limited. In this case the equation derived above holds. In the second, let us assume that the reaction velocity is diffusion controlled throughout the process. In this case, the rate of reaction (v) is given by: $v = K_L[S]$ (see Section 4.5.2).

In this case, the rate of substrate conversion is first order with respect to [S].

Thus in this case we can write:

$$-\frac{d[S]}{dt} = k_L[S]$$

Integrating between $t = 0$ and $t = t$, we obtain:

$$\int_{[S_0]}^{[S_t]} \frac{d[S]}{[S]} = -\int_0^t k_L dt$$

$$\ln \frac{[S_t]}{[S_0]} = -k_L t$$

or:

$$[S_t] = [S_0] e^{-k_L t}$$

In practice, industrial processes often use substrates at molar concentrations and the K_M values are often typically in the range 10^{-3}-10^{-5} mol l^{-1}. Thus, providing the biocatalyst is not too thick, the system is well mixed and substrate diffuses readily into the biocatalyst, then the effectiveness factor is close to 1.

Thus in these circumstances our equation:

$$V_{max} t = K_M \ln \frac{[S_0]}{[S_t]} + ([S_0] - [S_t])$$

holds.

An alternative way of expressing this equation is:

$$V_{max} t = K_M \ln \frac{1}{1-F} + F[S_0]$$

where F = fraction of substrate converted to product = $\frac{[S_0] - [S_t]}{[S_0]}$.

Π Can you think of a circumstance in which diffusion might be rate limiting throughout incubation?

The most obvious case is one in which the substrate is supplied at very low concentrations. A typical example would be in the conversion of low levels of pollutants.

5.3.3 The problems associated with reproducibility and scale-up.

Batch reactors often suffer from batch-to-batch variations. Small temperature fluctuations lead to accumulated differences. Small differences in the media and substrates used also lead to accumulated differences between different batches. A further difficulty that is encountered is in scaling-up the process. It is quite difficult to ensure efficient (turbulent) mixing in large vessels. The increased power required is quite substantial and leads to localised physical stresses in the vicinity of the impeller.

For these reasons, stirred tank batch reactors using immobilised enzyme are usually used for small-scale operations involving the production of high priced products. The simplicity of these devices makes them attractive in situations in which the same equipment is used for a variety of conversions. It is relatively straightforward to monitor and control physical (eg temperature, pH) conditions and chemical requirements (eg substrate, cofactors) on a small scale.

STRs used for small-scale operations

Since the real advantage of using immobilised enzymes is being able to design continuous processes, we will not extend the discussion of batch processes here.

SAQ 5.1

1) Let us assume we are using an immobilised enzyme in the form of small beads in a stirred batch reactor of total volume 10 m^3. Sufficient enzyme is added to give a maximum reaction rate of 10 mol min^{-1} m^{-3}. If the K_M of the enzyme is 1 mol m^{-3} and the substrate is added at an initial concentration of 1000 mol m^{-3}, how long will it take for 60% of this substrate to be converted to product? You should assume that:

   a) the enzyme is not inhibited by high substrate concentrations nor by the build up of product;

   b) substrate and product rapidly diffuse into and out of the biocatalyst and that the concentration of substrate in the centre of the beads never falls below 40 mol m^{-3}.

2) Assuming the same condition as in 1) except that the K_M value of the enzyme is 100 mol m^{-3}. How long will it take for 60% of the substrate to be converted to product in this case?

5.4 Plug flow reactors (PFRs)

In this type of reactor, the immobilised enzyme is usually packed into a column and the substrate is pumped through the system (see Figure 5.2).

Immobilised enzyme reactors

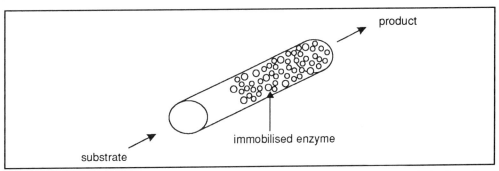

Figure 5.2 Stylised representation of a PFR. Note that since the enzyme is usually present in the form of a packed bed, these reactors are also referred to as packed bed reactors (PBRs).

Such reactors may be arranged horizontally or vertically and substrate can be pumped in either at the bottom or at the top. In the ideal case, the substrate stream flows evenly throughout the column and there is no back-mixing. In practise, this is difficult to achieve but there is some evidence that an upward flow is better than a downward flow. This usually produces less clogging and a more even flow.

In our discussion, we will mainly focus on the ideal case in which material present at any given cross section has an identical residence time in the reactor.

The critical calculation that needs to be conducted is the calculation to determine the size of the reactor that is required to achieve a particular conversion of substrate.

∏ List the factors which may influence the size of a column that is required to achieve a particular substrate conversion.

The factors which will influence the size of the column required are:

- the activity of the enzyme present in the column;

- the flow rate (slow flow rates give longer residence times);

- the geometry of the particles containing the enzyme;

- accessibility (diffusion rates) of the substrate to the enzyme.

In the next section we will provide details of these relationships.

5.4.1 The performance of PFRs (PBRs)

profile of reaction rates along the column

The arrangement of the reactor in the format shown in Figure 5.2 means that the enzyme molecules at the start of the column are exposed to substrate. Enzyme molecules at the end of the column are exposed to minimal levels of substrate but maximal concentrations of product. Thus we can anticipate that there will be a profile of reaction rates along the column resembling the reaction rate:time profile we observed in batch reactors.

In plug flow reactors, the distance the substrate solution has moved down the column is a function of the flow rate. If it takes θ minutes for the substrate to pass through the column (ie the maximum residence time = θ minutes), then in $\frac{1}{2}\theta$ minutes, it will have moved half way along the column.

We can, therefore, replace the t term in the batch equation by $\frac{\theta \times l}{L}$ where θ is the maximum residence time, l is the distance along the column and L is the total length of the column.

Thus since our equation for stirred batch reactors was:

$$V_{max}t = K_M \ln \frac{[S_0]}{[S_t]} + ([S_0] - [S_t])$$

For a PBR, the equation becomes:

$$V_{max}\frac{\phi \times l}{L} = K_M' \ln \frac{[S_0]}{[S_l]} + ([S_0] - [S_l])$$

where $t = \frac{\theta \times l}{L}$ and $[S_l]$ = substrate concentration at l. $[S_0]$ = substrate concentration at $l = 0$ (the input concentration).

This relationship holds fairly well, unless the reaction is limited by diffusion of the substrate to the enzyme (ie η becomes less than 1).

The total residence time θ of the material in the column is given by:

$$\frac{\text{liquid volume}}{\phi} = \frac{(\text{cross sectional area} \times L) - \text{volume of the support}}{\phi}$$

where ϕ = flow rate.

SAQ 5.2

A continuously fed PFR is to be set up. The data concerning this PFR are as follows. The substrate is supplied in solution at a concentration of 1 mol l^{-1}. The enzyme used has a K_M' of 100 mmol l^{-1}. The maximum rate of the reaction that can be achieved using this PFR is 10 mmols min^{-1} l^{-1} of the reactor. The reactor has a cross-sectional area of 0.1 m^2 and the volume taken up by the enzyme and its support is 50% of the total reactor volume.

If the flow rate (ϕ) is 1 litre min^{-1}, how long will the column need to be to:

1) achieve 50% conversion of the substrate?

2) achieve 75% conversion of the substrate?

(Assume that the system is not inhibited by product and that the process is not diffusion limited).

The sorts of calculation we have just carried out for a batch reactor (SAQ 5.1) and plug-flow (SAQ 5.2) are only true if the system is operating throughout with η values close to 1. In other words, the system is not diffusion limited. In these circumstances

Immobilised enzyme reactors

plug flow reactors are kinetically equivalent to batch reactors with soluble enzymes. Let us turn our attention to those circumstances in which η is not close to 1 (ie we are operating in a diffusion limited system).

η alters with time or distance in respectively batch and plug-flow reactors

If η remains constant throughout the reaction, the calculation is relatively straightforward. Since η represents the fraction of the biocatalyst which is used then we simply replace V_{max} in the equations we have used by ηV_{max}. In practice this is a purely hypothetical case since η is dependent upon substrate concentration in the bulk liquid ($[S_b]$). In batch systems and in plug-flow, [S] changes with time (or distance along the column); thus η must also change. Let us take an example to show the type of calculation that needs to be done.

Let us assume we have a biocatalyst which follows Michaelis-Menten kinetics and is in a slab geometry. In Chapter 4 (see Table 4.4) we mention that the effectiveness factor for such a system is given by:

$$\eta = \frac{3D\frac{d[S]}{dr}}{Rv_{[S_b]}}$$

where $\frac{d[S]}{dr}$ = substrate concentration gradient at the liquid: biocatalyst interface;

$v_{[S_b]}$ = the velocity of the reaction at the bulk liquid substrate concentration;

D = diffusion coefficient of the substrate;

R = thickness of the biocatalytic slab.

In batch systems and plug flow systems, the values of $\frac{d[S]}{dr}$ and $v_{[S_b]}$ both change with time (or distance along the column) as $[S_b]$ is changing. Since the actual value of the reaction rate = ηV_{max}, this also alters with time. We can write the rate of change of η as:

$$\frac{d\eta}{dt} = \frac{3D\frac{d^2[S]}{drdt}}{R\frac{dv_{[S_b]}}{dt}} \quad \text{(batch system)}$$

or:

$$\frac{d\eta}{dl} = \frac{3D\frac{d^2[S]}{drdl}}{R\frac{dv_{[S_b]}}{dl}} \quad \text{(plug flow)}$$

By integrating this over the total incubation time (for the batch system) or the length (L) of column (for the plug-flow system), we can, in principle determine the overall effectiveness factor for the system and this can be used to calculate substrate conversion over time (batch system) or length of the column (plug flow system). The mathematics are of course complex and we will not elaborate on them further here. If you wish to follow this up, appropriate references are given in 'Suggestions for Further Reading' at the end of this text.

Having pointed out that, kinetically, batch and plug-flow immobilised enzyme reactors behave identically, we might ask why we use PBR systems? The key point is that PBRs can be operated as a continuous process whereas a batch process has substantial 'down-time' while the reactor is being emptied and re-filled. If x is the proportion of the total time that a batch reactor is operational, using the same amount of enzyme in a batch and a PBR system will give product yields in the ratio of x:1 (batch:PBR).

SAQ 5.3 A batch reactor using an immobilised enzyme is operated with a standard incubation time of 20 hours. It then takes 5 hours to unload the reactor and refill it with enzyme and substrate. The same reaction can be carried out using the same amount of enzyme and operational conditions in a continuous packed bed reactor (PBR). What is the product yield of the batch system expressed as a % of the product yield of the PBR?

5.4.2 Some disadvantages of PBRs

In practice, the flow rate of liquid through the PBR depends upon the pressure drop across the PBR. This pressure drop depends on the bed height, flow rate and the viscosity of the substrate stream. If we increase the flow rate, this may distort the particles and this in turn may restrict the flow. This in turn increases the back pressure which may further deform the particles of the immobilised enzyme thereby restricting the flow even further. Thus there are limits to the bed heights and flow rates that can be used and these are dependent upon the deformability of the particles that make up the bed.

effects of pressure drops

It is also often necessary to cover the packed bed with a guard layer. The particles of the immobilised enzyme are usually small to maximise the surface area of the immobilised enzyme. This means that the liquid channels between particles are also small. In these circumstances beds may easily become blocked. To prevent this a filter (guard) layer may be placed on the entry face of the packed immobilised enzyme bed to prevent fouling by colloidal or particulate material.

filter (guard) layers

A further problem with PBRs is that it is difficult to control pH or temperature within the column. With wide reactors with diameters 20cm or more, localised heat output may lead to localised temperature changes.

localised temperature changes

Furthermore, uneven packing of the bed or uneven application of the substrate stream may result in uneven flow and, therefore, in uneven residence time. It is important, therefore, to use uniformly sized biocatalysts to reduce this type of problem.

5.5 Continuous flow stirred tank reactors (CSTRs)

With the production of immobilised enzymes, it is possible to produce CSTRs using enzymes. In these, some sort of filtering device enables withdrawal of the product without removal of the enzyme. Alternatively, the product and particulate enzyme may be withdrawn together from the tank after which the particulate enzyme is separated from the product and returned to the tank. We have illustrated both reactor types in Figure 5.3.

A third alternative is simply to remove product and enzyme from the reactor and to return only a portion of this directly to the reactor. (Note that, if the filter device shown in Figure 5.3a) has small enough pores, we could use it to keep a soluble enzyme in the

Immobilised enzyme reactors

system. This then becomes essentially a membrane reactor of the type we discussed in Chapter 3).

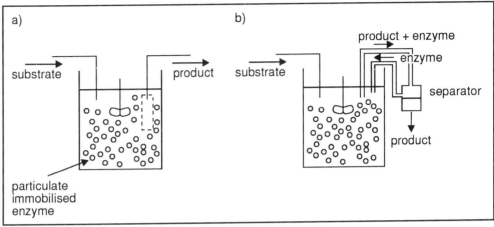

Figure 5.3 Simplified diagrams of CSTRs using immobilised enzymes: a) the enzyme is retained by an internal filter; b) the enzyme is returned to the tank via an external separator.

5.5.1 Performance of CSTRs

We can best describe the performance of CSTRs by drawing up a substrate balance for the reactor. We will use the reactor of the type shown in Figure 5.3a). We will assume that the reactor behaves in an ideal way such that the enzyme is homogenously suspended in the total reaction volume and that substrate solution entering the tank is mixed instantaneously.

The rate of change of substrate in the tank can be written as:

$$\text{Vol}_r \frac{d[S]}{dt} = \begin{bmatrix} \text{flow of substrate} \\ \text{into the tank} \end{bmatrix} - \begin{bmatrix} \text{flow of substrate} \\ \text{out of the tank} \end{bmatrix} - \begin{bmatrix} \text{amount of substrate} \\ \text{converted in the tank} \end{bmatrix}$$

Mathematically this can be written as:

$$\text{Vol}_r \frac{d[S]}{dt} = \phi[S_0] - \phi[S_r] - v_r$$

where $[S_0]$ = concentration of substrate in the feed; $[S_r]$ = concentration of substrate in the reactor output; Vol_r = volume of the reactor; v_r = velocity of the reaction in the reactor (for the total reactor). ϕ = flow rate.

If we divide by Vol_r, then this gives us a substrate balance per unit volume of the reactor. Thus:

$$\frac{d[S]}{dt} = \frac{\phi}{\text{Vol}_r}[S_0] - \frac{\phi}{\text{Vol}_r}[S_r] - \frac{v_r}{\text{Vol}_r}$$

$\frac{\phi}{\text{Vol}_r}$ is usually referred to as the dilution rate (D) ie $D = \frac{\phi}{\text{Vol}_r}$

thus:
$$\frac{d[S]}{dt} = D([S_0] - [S_r]) - v$$

where v = velocity (rate of reaction) per unit volume of the reactor.

But, if we assume Michaelis-Menten kinetics:

$$v = \frac{V_{max}[S_r]}{K_M' + [S_r]}$$

where V_{max} = maximum velocity per unit volume, K_M' = apparent K_M of the immobilised enzyme.

If we allow the device to run for some time, we will reach a steady state. In other words:

$$\frac{d[S]}{dt} = 0$$

Thus under these conditions:

$$D([S_0] - [\tilde{S}_r]) = \frac{V_{max}[\tilde{S}_r]}{K_M' + [\tilde{S}_r]}$$

where $[\tilde{S}_r]$ is the steady state substrate concentration.

Thus:

$$\frac{V_{max}}{D} = \frac{K_M'([S_0] - [\tilde{S}_r])}{[\tilde{S}_r]} + ([S_0] - [\tilde{S}_r])$$

Thus the steady state substrate concentration $[\tilde{S}_r]$ is dependent upon the amount of enzyme present (V_{max}), the dilution rate (D), the K_M' and the input substrate concentration $[S_0]$.

The output per unit volume of the reactor is given by $D([S_0]-[\tilde{S}_r])$

Thus, by using an enzyme of known K_M' we can adjust D, V_{max} and $[S_0]$ to produce a system of the desired output.

Let us try a few calculations.

SAQ 5.4

1) We have a CSTR of total volume of 100 l. We would like to use this to treat a solution containing 1 mol l^{-1} substrate at a rate of 5 lh^{-1} and achieve 50% conversion. If the K_M' of the enzyme is 1 mmol l^{-1}, how much enzyme should we use in the reactor? (ie calculate the required V_{max})

2) If the substrate flow rate was increased to 10 lh^{-1}, how much enzyme would we require?

3) How much enzyme would we require if we use a different enzyme with a K_M' of 100 mmol l^{-1} and a flow rate of 10 litres per hour?

Immobilised enzyme reactors

In SAQ 5.4, you will have demonstrated that the higher the K_M', the greater the amount of enzyme required. This is most pronounced when $[S_r]$ are similar to, or smaller than K_M'.

5.5.2 Consequences of diffusion limitation

If we want to achieve high levels of conversion (that is, we would like $[[\tilde{S}_r]$ to be low), then this may be achieved by:

- using a long residence time (slow flow rate);
- using a high level of activity (high V_{max});
- using enzymes of high affinity (low K_M').

These factors will of course all tend to reduce the effectiveness factor of the system because they are precisely the conditions which tend to lead to diffusion limitation. Since substrate diffusion is given by $k_L([S_b]-[S_s])$ where k_L = mass transfer coefficient, and $([S_b]-[S_s])$ is the differences between the substrate concentration in the bulk liquid and in the enzyme matrix, reaching a low value of $[S_b]$ (= $[\tilde{S}_r]$) means that we are likely to encounter substrate limitation.

However, with CSTRs we are in a good position to calculate the effectiveness factor. Since we are operating the reaction under a steady state concentration of substrate, η should also remain constant.

You should recall that the velocity of the reaction is given by the enzyme in free solution multiplied by η.

Thus in circumstances where η is less than 1, our balance equation for an CSTR becomes:

$$D([S_0] - [\tilde{S}_r]) = \eta \left(\frac{V_{max}[S_r]}{K_M' + [S_r]} \right)$$

Let us do a calculation using this relationship.

∏ A substrate is added to a reactor at a dilution rate of 0.05 l^{-1}. The input substrate concentration is 1 mmol l^{-1} and we wish the steady state substrate concentration to be 0.1 mmol l^{-1}. Given that the effectiveness factor for the immobilised system is 0.5 at this concentration and the K_M' of the enzyme is 0.1 mmol l^{-1}, how much enzyme do we need to include in the reactor?

Using the relationship we have just derived, we can substitute in the appropriate values:

$$0.05(1 - 0.1) = 0.5 \frac{(V_{max} \times 0.1)}{0.1 + 0.1}$$

thus $V_{max} = 0.18$ mmol l^{-1} h^{-1}.

In other words, we would need to add sufficient enzyme to give us a maximum velocity of 0.18 mmol l^{-1} h^{-1} per litre of the reactor.

A diffusion limited CSTR is to a certain extent self adjusting. If the substrate concentration rises, the substrate diffusion will increase (and thus η will increase) and then the velocity of the reaction will increase. This will lead to a reduction in the concentration of the substrate in the reactor. If the substrate level falls, the rate of reaction will fall because the rate of diffusion will be slowed and thus η becomes smaller. This reduction in the rate of substrate consumption will allow the substrate concentration to begin to rise again.

Now try the following SAQs.

SAQ 5.5

1) The substrate concentration entering a CSTR containing an immobilised enzyme is 1 mol l^{-1} and the dilution rate is fixed at 0.1 hr^{-1}. We wish to convert 70% of the substrate to product and thus the required steady state substrate concentration will be 0.3 mol l^{-1}. At this substrate concentration the effectiveness facter (η) of the enzyme immobilised in the form of beads, is estimated to be 0.6. The K_M' of the enzyme is 0.01 mol l^{-1}.

   a) How much enzyme should be used in the reactor?

   b) If the volume of the reactor is 20 litres, how much product is made hr^{-1}?

2) If in the system described in 1) the dimensions of the immobilised system are altered such that the effectiveness factor is decreased to 0.4, how much enzyme should then be added to achieve the same amount of substrate conversion?

5.5.3 Comparison of CSTRs and PBRs

In a CSTR, all of the enzyme is exposed to a relatively low substrate concentration (that is [$\tilde{S}_r$]) and a relatively high product concentration (that is [$\tilde{P}_r$] where [$\tilde{P}_r$] = ([S_0]-[$\tilde{S}_r$]). This may have a significant effect on the effectiveness factor (η) and could mean that all of the enzyme in the reactor is being used ineffectively.

more enzyme may be needed in CSTRs than PBRs

In contrast, in a packed bed system, the initial part of the reactor column will be operating at high substrate concentration (close to [S_0]), thus η will be high. It is only in the final sections of a PBR that the enzyme is exposed to low substrate concentrations and high product concentrations. Thus it is only in the lower parts of the reactor that the effectiveness factor may be decreased. What this means is that we may need more enzyme (ie greater V_{max}) in an CSTR than in an PBR. Thus it would appear that PBR's are more efficient especially in diffusion limited systems. In practice, however, this is not always the case. Since we can achieve turbulent flow in stirred tanks, we are less likely to encounter diffusion limitation. In plug flow systems, we use laminar flow and therefore are more likely to encounter diffusion limitation. In many instances CSTRs give as good a performance as PBRs. They have the added attraction that it is relatively easy to monitor and control physical parameters such as pH and temperature in CSTRs.

mixing is an important factor

The inhibitor effects of high substrate and product concentration observed with some enzymes may also cause deviations from the relationships we have established. In these cases, the effects of substrate and product inhibition are different in the two systems. The reaction mixtures in CSTRs are well mixed. This means that the substrate concentration is the same as the output substrate concentration. The same is true for the product concentration. Thus the enzyme in an CSTR is subjected to minimal substrate inhibition but maximal product inhibition. In contrast, in the PBR there are concentration gradients of substrate and product. Enzyme molecules close to the input are subjected to maximal substrate inhibition and this only reaches a minimum at the output. The gradient of product concentration means that enzyme molecules close to the input are minimally inhibited by product whilst those near the outlet are maximally inhibited by product. Thus if the enzyme is subject to inhibition by substrate but not by product then the CSTR system is likely to be most effective. If, on the other hand, the enzyme is subject to product inhibition but not to substrate inhibition, a PBR system will be more effective. Mathematical relationships have been derived for product and substrate inhibited systems and for reactions which are freely reversible. We will not derive them here but we have summarised them in Table 5.1 We have provided these as a reference source. We would not expect you to remember them in detail.

---

substrate-inhibited PBR

$$\frac{V_{max}}{\phi} = X[S_0] - K_M \ln(1-X) + [S_0]^2 (2X - X^2)/2K_s$$

substrated-inhibited CSTR

$$\frac{V_{max}}{\phi} = X[S_0] + K_M [X/(1-X)] + [S_0]^2 (X - X^2)/K_s$$

product-inhibited PBR

$$\frac{V_{max}}{\phi} = X[S_0](1 - K_M/K_p) - K_M \ln(1-X)(1 + [S_0]/K_p)$$

product-inhibited CSTR

$$\frac{V_{max}}{\phi} = X[S_0] + K_M \{X/(1-X)\}(1 + X[S_0]/K_p)$$

reversible reaction in a PBR

$$\frac{V}{\phi} = X[S^*] + K \ln(1-X)$$

reversible reaction in a CSTR

$$\frac{V}{\phi} = X[S^*] + K \{X/(1-X)\}$$

Table 5.1 Equations describing the performances of CSTRs and PBRs. (Adapted from Chaplin, M.F. and Bucke, C (1990). Enzyme Technology, Cambridge University Press and Cambridge ISBN 0 521 34884 6).

Key:

V_{max} = maximum rate of reaction in the reactor (mol s^{-1})

ϕ = flow rate (l s^{-1}; m^{3}s^{-1} etc)

X = fraction conversion = $\frac{[S_0] - [S]}{[S_0]}$ for an irreversible reaction

X = fraction conversion = $\frac{[S_0] - [S]}{[S_0] - [S_{0b}]}$ for a reversible reaction

continued

$[S_0]$	=	input substrate concentration
$[S]$	=	output concentration of substrate
K_s	=	substrate inhibition constant (mol l^{-1})
K_p	=	product inhibition constant (mol l^{-1})
K_M	=	Michaelis constant
v	=	velocity of a reversible reaction $= \dfrac{(K_{eq}+1)}{K_{eq}} \times \left(\dfrac{K_M^p}{K_M^p - K_M^s}\right) \times v^f$

$$K = \frac{K_M^s K_M^p + (K_M^p + K_M^s K_{eq})[S_{0b}]}{K_M^p - K_M^s}$$

where K_M^s, K_M^p are the K_M values for substrate and product:

K_{eq}	=	equilibrium constant
$[S_{0b}]$	=	substrate concentration at infinite time (ie at equilibrium)
v^f	=	maximum velocity of the forward reaction
$[S^*]$	=	concentration of substrate that can react in a reversible reaction (ie the difference between $[S_0]$ and $[S_{0b}]$)

5.5.4 Reactor kinetics and reversible reactions

Throughout our discussions of the kinetics of immobilised enzymes and reactors containing immobilised enzymes, we have so far assumed that the reaction follows Michaelis-Menten kinetics. Strictly speaking, Michaelis-Menten kinetics only relate the initial reaction velocity (that is v_0 at $t = 0$) to V_{max} and K_M.

Thus if we represent the interaction of enzyme and substrate to form a product in the following way:

$$E + S \underset{k_{-1}}{\overset{k_1}{\rightleftharpoons}} ES \underset{k_{-2}}{\overset{k_2}{\rightleftharpoons}} P + E$$

It is assumed that $k_{-2}[P][E] = 0$.

In other words, the relationship $= \dfrac{V_{max}[S]}{K_M + [S]}$ only applies when $k_{-2}[P][E] = 0$.

Now let us consider two situations. In the first, let us assume that the equilibrium position for the reaction greatly favours the formation of product. Thus $K_{eq} = \dfrac{[P]}{[S]}$ gives a high value. This implies that the value of k_{-2} is very low compared to k_2. In practical terms, this means that during the course of the reaction $k_{-2}[P][E]$ remains very low (almost 0) compared to $k_2[ES]$. Thus the Michaelis-Menten relationship approximates the kinetics of the reaction. It is only as we approach equilibrium that the concentration of $[S]$ and therefore the concentration of $[ES]$ will have fallen to such a low value and $[P]$ risen to such a high value, that $k_{-2}[P][E]$ has a comparable value to $k_2[ES]$. In other words, for reactions which are (in biological terms) virtually irreversible (K_{eq} = very high value), the reaction behaves according to Michaelis-Menten kinetics until equilibrium is almost reached. We are, therefore, justified in using Michaelis-Menten kinetics in describing the kinetics of enzyme-catalysed reactions in situations where the product:substrate ratio does not approach K_{eq}.

Not all enzyme-catalysed reactions have K_{eq} values that favour the formation of product so strongly. In freely reversible reactions (ie reactions in which K_{eq} is close to 1), we must take into account the back reaction.

In these cases, $k_{-2}[P][E]$ is only 0 when $[P] = 0$ (ie at the beginning of the reaction). As product is made, $k_{-2}[P][E]$ will build up and the reaction no longer follows Michaelis-Menten kinetics. The kinetics of such reactions are quite complex and we will not derive them here. (A derivation is given in Chaplin, M.F. and Bucke, C (1990) Enzyme Technology Cambridge University Press). We will, however, give you some of the relationships.

You might anticipate that the overall reaction rates will depend upon the relative rates of the forward and reverse reactions and the effects of [S] and [P] on these reaction rates. From theoretical considerations, the net forward velocity of a reversible reaction catalysed by an enzyme is given by:

$$v = \frac{v\,[S*]}{K + [S*]}$$

where:

$$v = \frac{(K_M^P v^f + K_M^s v^r)}{K_M^P - K_M^s} = \frac{K_M^P v^f + K_M^s v^f/K_{eq}}{K_M^P - K_M^s}$$

$$K = \frac{K_M^s K_M^P + (K_M^P + K_M^s K_{eq})\,[S_{0b}]}{K_M^P - K_M^s}$$

and K_M^s and K_M^P are the Michaelis constants for substrate and product in the forward (K_M^s) and reverse (K_M^P) reaction; v^f = maximum velocity of the forward reaction; v^r = maximum velocity of the reverse reaction and $[S_{0b}]$ = substrate concentration at equilibrium.

You will notice that although the net forward velocity of the reaction (v) has the same general form as the Michaelis-Menten relationship, v and K have quite different meanings from V_{max} and K_M.

We do not anticipate that you will remember this relationship in detail. The reason we have included this discussion is to ensure that you do not assume that all enzyme catalysed reactions which display Michaelis-Menten kinetics at t = 0 will do so throughout incubation. It is only under circumstances where we do not approach equilibrium that Michaelis-Menten kinetics approximate the kinetic behaviour of enzyme catalysed reactions.

SAQ 5.6

Identify which of the following circumstances will be approximated by the Michaelis-Menten relationship. Assume in all cases, that the enzyme displays Michaelis-Menten kinetics when $t = 0$ (ie when $[P] = 0$). Read our response carefully as it emphasises an important point.

1) The reaction $S \rightarrow P$ has an equilibrium constant $K_{eq} = 10^4$. We propose to start the reaction by adding the substrate (concentration 1 mol l^{-1}) to an enzyme and to collect the product when 80% of the substrate has been converted.

2) The reaction $S \rightarrow P$ has an equilibrium constant $K_{eq} = 3$. We propose to start the reaction by adding the substrate (concentration 1 mol l^{-1}) to the enzyme and to collect the product when 60% of the substrate has been converted.

3) The reaction $S \rightarrow P$ has an equilibrium constant $K_{eq} = 10^2$. We propose to start the reaction by adding the substrate (concentration 1 mol l^{-1}) to the enzyme and to collect the product when the substrate concentration has fallen to 15 mmol l^{-1}.

SAQ 5.7

Which of the following enzymes are most likely to catalyse reactions which follow reversible reaction kinetics rather than Michaelis-Menten kinetics.

1) Transaminase.

2) Glucose isomerase.

3) ATPase.

4) Acetyl CoA thiolase.

5.5.5 Productivity of PBRs and CSTRs and enzyme stability

In discussing the kinetics of PBRs and CSTRs, we have assumed that the enzyme has remained active. We know from experience that this is rarely the case. We described inactivation of enzymes in Chapter 3, where we explained that inactivation often occurs according to first order kinetics.

Combining the first order inactivation kinetics and the relationships relating substrate conversions with flow rate for PBRs and CSTRs allows us to compare the productivity of these reactors at different times.

We remind you that for a PBR:

$$V_{max} t = K_M \frac{[S_0]}{[S_t]} + ([S_0] - [S_t])$$

where t = residence time θ.

But V_{max} changes with time according to a first order inactivation:

$$V_{max}^t = V_{max}^0 e^{-k_d t}$$

Immobilised enzyme reactors

Combining these equations followed by integration gives:

$$\ln \frac{[S_0] - [S_t^0] + K_M \frac{[S_0]}{[S_t^0]}}{[S_0] - [S_t^t] + K_M \frac{[S_0]}{[S_t^t]}} = k_d t$$

where $[S_0]$ and $[S_t^0]$ are the input and output substrate concentrations when the PBR is first run (ie $t = 0$); $[S_0]$ and $[S_t^t]$ are the input and output substrate concentrations after the PBR has been run for time t and k_d is the first order inactivation constant for the enzyme.

For a CSTR, we remind you that:

$$\frac{V_{max}}{D} = K_M \frac{([S_0] - [\tilde{S}_r])}{[\tilde{S}_r]} + ([S_0] - [\tilde{S}_r])$$

Again applying $V_{max}^t = V_{max}^0 e^{-k_d t}$ and integrating gives:

$$\ln \frac{\left\{ ([S_0] - [\tilde{S}_r^0]) + K_M \left(\frac{[S_0] - [\tilde{S}_r^0]}{[\tilde{S}_r^0]} \right) \right\}}{\left\{ ([S_0] - [\tilde{S}_r^t]) + K_M \left(\frac{[S_0] - [\tilde{S}_r^t]}{[\tilde{S}_r^t]} \right) \right\}} = k_d t$$

where $[S_0]$ and $[\tilde{S}_r^0]$ are the input and output concentrations of substrate when the reactor is first set up; $[S_0]$ and $[\tilde{S}_r^t]$ are the input and output concentrations of substrate when the reactor has been run for time t and k_d = first order inactivation constant of the enzyme.

SAQ 5.8

A CSTR has been set up such that the input substrate concentration 1 mol l^{-1} and the output substrate concentrations is 0.1 mol l^{-1}. However, the enzyme used is progressively inactivated by first order kinetics at a rate of 0.1 day^{-1}. What will be the output substrate concentration be after the reactor has been run for 10 days? The K_M of the enzyme is 10 mmol l^{-1}.

5.6 Fluidised bed reactors (FBRs)

FBRs have some characteristics of CSTRs and PBRs

In some ways, these reactors behave in a manner that is part way between a PBR and a CSTR. They consist of a bed of immobilised enzyme which is fluidised by the upward flow of reaction media. One way of describing these is to consider them as PBRs in which the substrate flow is so large that the bed expands and is no longer packed. Alternatively they may be considered as CSTRs in which the immobilised enzyme has partially settled to form a loose, fluid bed.

It is perhaps not surprising that the kinetics of these reactors behave somewhere between that of PBRs and CSTRS. Depending on the operating conditions, some behave more like PBRs, others more like CSTRs.

FBRs useful for reactions involving gases

FBRs are particularly useful for conducting reactions involving gases. For example if a reaction involves O_2, we will have to deal with the fact that O_2 is not very soluble and therefore it will be difficult to supply it in a solution as we do for conventional, water-soluble products. By bubbling gaseous oxygen through the bed, it fluidises the bed and comes into close contact with the water film surrounding the immobilised enzyme particles. Usually small particles are used (giving a good surface area) but they must be fairly dense otherwise they are lifted out of the reactor and will be lost. Typically the particles used have diameters in the range 25-50 μm.

A disadvantage of the FBR system is that it is difficult to scale-up, therefore, their use is mainly confined to small-scale, high-priced operations.

5.7 Membrane and hollow fibre reactors

In our discussion of the industrial use of soluble enzymes in Chapter 3, we introduced the concept of membrane reactors. You should recall that the main principle underpinning membrane reactors is that a membrane acts as a selective barrier, preventing the free movement of substrates and products. We can, as an alternative, use an immobilised enzyme. This can be either attached to the membrane or presented in particulate form within the membrane compartment. This has the advantage that we can use membranes with large pore sizes since the size differences between solutes and immobilised enzymes is greater than the size differences between solutes and soluble enzymes. Such membranes would allow easier movement of solutes and be less susceptible to clogging.

A variety of configurations is possible. A common configuration is to use the semipermeable membrane in the form of hollow fibres which are immersed in the reaction mixture (see Figure 5.4). Such reactors may be used in batch mode or in continuous mode.

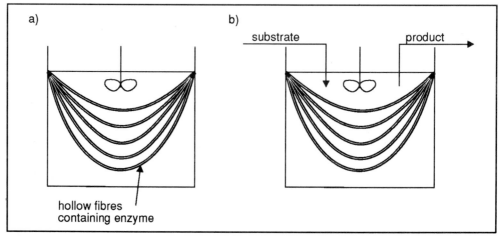

Figure 5.4 Hollow fibre (membrane) reactors: a) operated in batch mode; b) operated in continuous mode.

The kinetics of membrane reactors are similar to batch STRs when they are used in batch mode. If, however, they are used in continuous mode, the kinetics resemble those of CSTRs. A particularly attractive feature of these reactors is that they can function as a

Immobilised enzyme reactors

reaction surface between two different liquid phases and avoid the need to form emulsions when water insoluble solutes are used. The techniques for attaching enzymes to the membranes are similar to those used to attach enzymes to other supports.

A variety of arrangements for delivering substrates and removing products are shown in Figure 5.5.

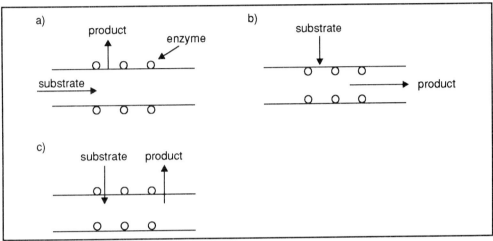

Figure 5.5 Presentation of substrate and removal of product in membrane reactors with enzymes attached to the membrane.

Π Which of the arrangements shown in Figure 5.5 would be most suitable for using immobilised lipase which catalyse the hydrolysis of glycerides.

$$\begin{array}{c} CH_2-O-\overset{O}{\underset{\|}{C}}-R^I \\ | \\ CH-O-\overset{O}{\underset{\|}{C}}-R^{II} \\ | \\ CH_2-O-\overset{O}{\underset{\|}{C}}-R^{III} \end{array} + 3H_2O \longrightarrow \begin{array}{c} CH_2OH \\ | \\ CHOH \\ | \\ CH_2OH \end{array} + \begin{array}{c} HO\overset{O}{\underset{\|}{C}}-R^I \\ HO\overset{O}{\underset{\|}{C}}-R^{II} \\ HO\overset{O}{\underset{\|}{C}}-R^{III} \end{array}$$

We would suggest that either arrangement a) or b) could be suitable. The substrate is not very water-soluble and could be supplied dissolved in an organic (water immiscible) solvent. The substrate would then diffuse across the membrane into the enzyme compartment which contains an aqueous medium. After hydrolysis, the products are more water-soluble and would be removed from the membrane compartment.

Thus:

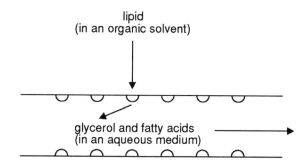

5.8 Satisfying cofactor and cosubstrate requirements

A large number of enzymes require a variety of cofactors and/or cosubstrates in order to catalyse the conversion of substrate to product. For an enzyme to be useful industrially, these requirements have to be met economically.

inorganic cofactors are usually easy to supply

With inorganic cofactors (for example Mg^{2+}, Ca^{2+}, K^+), there is usually little difficulty in supplying these. They are usually only required at quite low concentrations and soluble salts of such cofactors are relatively cheap. The actual amount of such inorganic cofactors used in a reactor will depend on the requirement of the enzyme and for an immobilised system we must take into account any partitioning of the required ion between the enzyme matrix and the bulk liquid.

It is with organic cofactors and cosubstrates that we meet with most difficulties. We will not review all of these here, but, commonly, enzymes require such cofactors/cosubstrates as NAD^+ (NADH), $NADP^+$(NADPH), FAD(FMN), coenzyme A, ATP, pyridoxal phosphate, biotin and thiamine pyrophosphate.

If the cofactor is attached to the enzyme (for example biotin) and this is regenerated during the reaction (that is, it is a true cofactor not a cosubstrate), then again in principle we do not face too many difficulties. In circumstances where the cofactor is not attached to the enzyme, then we have to supply it in a solution. Invariably such cofactors are very expensive (just take a look at the catalogue of any biochemical supplier!) and we have to arrange to use the minimum amount of cofactors possible. Furthermore some so-called cofactors are really cosubstrates and are 'consumed' during reaction. For example, consider the reaction catalysed by alcohol dehydrogenase.

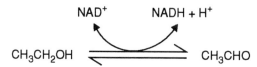

Under normal operating conditions, we would require amounts of NAD^+ equimolar to the amount of alcohol we wish to dehydrogenate. Clearly this is prohibitively expensive. ATP is another good example of a prohibitively expensive cosubstrate.

∏ Can you think of a general way in which the cofactor requirement of an enzyme might be fulfilled without too much additional cost?

Immobilised enzyme reactors

use of whole cells — We anticipate that you might have thought of using whole cells, rather than isolated enzymes, as the biocatalyst. Use is then made of the cofactor content of the cells or their ability to regenerate the cosubstrate. This is not always a good solution for the reasons given earlier (eg multiplicity of metabolic products; key enzyme(s) represents only a small proportion of the biomass). A key question is, therefore, can we find a way of supplying the required cofactor/co-substrate in sufficient quantities so that we can use purified enzymes on an industrial scale?

Usually we try to find a way of regenerating the required cofactor/cosubstrate within the reactor. In this section, we will describe some of the strategies we may use to achieve this. We will predominantly focus on the regeneration of oxidised pyridine nucleotides ($NAD^+/NADP^+$) to explain these strategies.

We can broadly divide these strategies into:

- enzymatic methods;
- electrochemical methods;
- chemical methods.

5.8.1 Enzymatic methods of cofactor generation

These methods can be divided into two subgroups:

- those which use coupled enzymes;
- those which use coupled substrates.

We will consider coupled enzymes first.

Coupled enzymes

Consider two dehydrogenases, one oxidises substrate A to form B, the other oxidises C to form D. If we incubate these two enzymes in the presence of A and D then we could envisage a coupled reaction of the type:

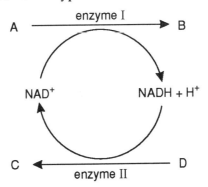

Models using this type of system have been established using, for example, alcohol dehydrogenase and pyruvate dehydrogenase. In practise these do, however, present difficulties because the cofactor tends to diffuse out of the immobilised enzyme matrix and thus the concentration of the cofactor in the vicinity of the enzymes is low. It also demands a good and even distribution of the two enzymes on the support. As might be anticipated such a coupled system works best in a micro-encapsulated system providing the pores in the capsules are too small to allow the cofacter to escape but large enough for the substrates and products to leave!

An alternative to using a single, second enzyme is to use a functional multienzyme complex such as the NADH oxidase system from *Escherichia coli* to regenerate the cofactor.

Thus:

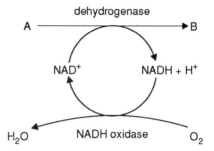

Coupled substrates

Using substrate coupling potentially reduces the loss of cofactor from the support. In this case a single enzyme is used and the reaction can be written as:

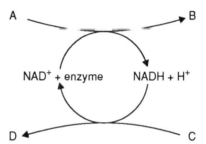

This type of system works provided the enzyme will work with a variety of substrates. An example is alcohol dehydrogenase which will catalyse the dehydrogenation of geraniol and ethanol. Thus:

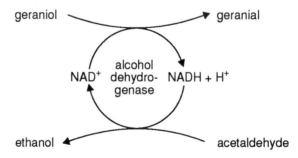

Incubation of the enzyme in the presence of geraniol and acetaldehyde, leads to the production of geranial and ethanol.

5.8.2 Electrochemical methods

In this strategy, the required form of the cofactor is generated by an electrode. We will illustrate this by using alcohol dehydrogenase as an example.

Immobilised enzyme reactors

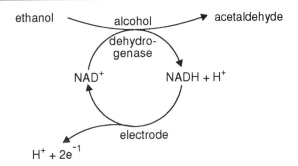

In practise this type of system only works well on a small scale in which the enzyme is attached to the surface of the electrode. Its main application is in biosensors. We will describe these in more detail in Chapter 8.

5.8.3 Chemical methods

In these methods, a chemical is used to convert the cofactor into the desired form. For example we can use the oxidising agent phenazine methosulphate to regenerate NAD^+.

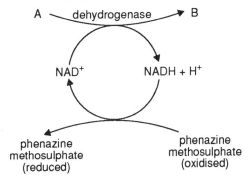

5.8.4 The advantages and limitations of the methods available for regenerating cofactors

Π Before reading on make a list of the advantages and limitations you might envisage for the methods of cofactor regeneration described in Sections 5.8.1, 5.8.2 and 5.8.3.

You probably came up with many limitations. Here we will confine ourselves to some of the main ones.

Enzymatic coupling. This method necessitates immobilising two enzymes and two substrates. This, of course, makes preparation of the immobilised system more complex and increases the difficulty of separating substrate and products. We also have to find a support that is compatible with both enzymes (unless we encapsulate these enzymes). Because of losses of NAD^+ from the enzyme matrix, the number of cycles the system can be used for is limited.

Substrate coupling. The limitation here is that we have to find an enzyme which will act on two substrates, one which we wish to convert and the other which provides a cheap way of regenerating the appropriate cofactor. In practise no industrial process has been developed using this approach.

Electrochemical methods. Only practicable on a small scale. Difficult to visualise on a large scale. Mainly confined to biosensors.

Chemical methods. Economically these appear very attractive. There is, however, a need to separate the desired product from a more complex chemical mixture. Because of the cost of the chemicals involved, this is only applicable to small-scale, high-priced commodities.

SAQ 5.9	

1) Pyridoxamine-5 phosphate is a prosthetic group attached to many enzymes. One such enzyme is transaminase. Does the requirement of pyridoxamine-5 phosphate as a cofactor in the transamination reaction pose a problem in the use of transaminase in an immobilised form? If so how would you overcome this difficulty?

2) An enzyme used in an immobilised enzyme reactor requires ATP to carry out its catalytic process. Which of the following offers the best solution for supplying ATP to the reactor?

   a) add aliquots of a concentrated ATP solution during the reaction;

   b) add the enzyme acetate kinase, acetyl phosphate and a small amount of ADP;

   c) add mitochondrial extracts, a substrate which will pass electrons onto the electron transport chain and small amounts of ADP.

5.9 Reactor development is a team activity

In this chapter, we have briefly examined some operational issues concerning the use of immobilised enzymes on a large scale. We have examined technical issues concerning the effective use of enzymes without going into too many details concerning the issues of process technology. We have not for example discussed the issues of heat generation/loss, or instability of the immobilised enzyme both of which have significant bearing on the operation of the process. These are of course relevant to the final design and economic operation of the process. Nor have we considered the commercial issues (capital costs, running costs, market volume, added value etc). These are all essential factors and call upon specialist expertise. Here our focus has been on considering the process from an enzymological perspective and to think in terms of what sorts of information the process technologist will need in order to design a process.

The process technologist will need, for example, the data necessary to calculate the effectiveness factor of the immobilised catalyst and data relating to the stability of the enzyme under the operational conditions. If the enzyme is unstable then V_{max} will change over time and this will influence conversion efficiencies. From what you have learnt you should be able to attempt SAQ 5.10. Read our response carefully as it emphasises several important points.

SAQ 5.10 Assume you are an enzymologist working with a process technologist involved in the design of a reactor using an entrapped enzyme. Make a list of the information the technologist may need from you in order for him to calculate the effectiveness factor for various configurations of the biocatalyst.

It should be self evident from our discussion that the design of an industrial-scale enzyme based process is not a simple matter. The success of such an enterprise depends not only upon good enzymological data but also upon good process engineering and sound commercial judgement.

Summary and objectives

In this chapter we have described the main reactor configurations which have been developed to use immobilised enzymes on an industrial scale. We discussed the performances of each and made a comparison between packed bed and continuous flow stirred tank reactors. We also discussed the importance of taking into account enzyme stability when analysing the productivity of immobilised enzyme reactors. Towards the end of the chapter, we also explained the problems associated with the regeneration of cofactors/cosubstrates required by many enzymes.

Now that you have completed this chapter, you should be able to:

- make drawings of the main reactor types in which immobilised enzymes are used and describe their principal features;

- explain why laminar flow is more likely to lead to diffusion limitation in stirred tank reactors than is turbulent flow;

- calculate the amount of enzyme required to achieve specified conversions using supplied data for PBRs and CSTRs;

- explain why the effectiveness factor (η) of an immobilised system is likely to change with time in a batch reactor or with distance in a packed bed reactor but is more likely to be constant in a CSTR;

- compare the [S] and [P] regimes in PBRs and CSTRs;

- identify situations where the kinetics of an enzyme-catalysed process will deviate significantly from the Michaelis-Menten relationship;

- use data relating to enzyme inactivation kinetics to calculate changes in the performances of reactors over time;

- identify appropriate strategies for fulfilling the cofactor/cosubstrate requirements of enzymes.

Use of enzymes in large-scale industrial applications

6.1 Introduction	150
6.2 Industries which use enzymes	150
6.3 Enzymes in the food industry	151
6.4 Enzymes used to modify carbohydrates	151
6.5 Use of proteases in the food industry	161
6.6 Lipases	162
6.7 Amino acids	164
6.8 The use of enzymes in the pharmaceutical and healthcare industries	166
6.9 Other industries employing the use of enzymes	174
Summary and objectives	179

Use of enzymes in large-scale industrial applications

6.1 Introduction

In the four previous chapters, we discussed the principles underpinning the preparation of enzymes and their use in large-scale industrial processes. In this chapter, we will extend this discussion to provide you with an overview of these large-scale operations in a variety of business sectors. It is not, however, our intention to give you an in-depth appreciation of all of these processes. More detailed discussions of specific processes are provided in the 'Innovations' texts in the BIOTOL series. For example, the use of chymosin in cheese manufacture, pectinases and cellulases in fruit juice extraction, amylases and isomerases in the production of high fructose corn syrups and the use of enzymes in the manufacture of stereospecific amino acids are described in a series of case studies in the BIOTOL text, 'Biotechnological Innovations in Food Processing'. (A list of BIOTOL texts is given at the front of this text).

We begin this chapter by identifying the major enzyme-using business sectors and then examine each of these business sectors. We will confine this discussion to what we might regard as bulk applications. You should realise, however, that the enzyme business is enormous and is an area of expansion on a massive scale. The examples we have selected will give you an appreciation of the diversity and scale of industrial enzymology.

In Chapters 7 and 8, we will examine the use of enzymes as reagents used for analysis and, in Chapters 9 and 10, we will examine the use of enzymes in the manipulation of genetic material to generate novel genes and gene combinations. The use of enzymes in analysis and in gene manipulation (genetic engineering) are themselves important industrially.

6.2 Industries which use enzymes

In Chapter 3, we provided a table (Table 3.1) listing some important industrial enzymes and their sources. From the information provided in Table 3.1? You should already have some idea of the industrial sectors which employ enzymes.

∏ List three industrial sectors which use enzymes.

We anticipate that you will have listed the food industry, the pharmaceutical/health industry and the detergent industry. We could add to this many other industries including the leather/wood trade, fine chemicals and environmental management.

By far the largest use of enzymes is found in the food industry. Many of the processes involved in food manufacture are dependent upon enzymes.

Enzymes are important in the production of many dairy products (for example, cheese and low lactose milk), for products used in confectionary, in the processing of meat, the preparation of materials used in baking and in brewing and in fruit juice manufacture. Because of this importance, we will begin our examination of enzymes in industry by discussing their use in food processing.

6.3 Enzymes in the food industry

For the purposes of our discussion here, we can sub-divide the industrial use of enzymes in the food industry into:

- those enzymes which modify carbohydrates;

- those enzymes which modify proteins;

- those enzymes which modify lipids.

In our discussion, you will see that enzymes used in the food industry may be employed either in a soluble or in an immobilised form. The choice depends upon the nature of the process being undertaken, the required specification of the product and economic considerations.

6.4 Enzymes used to modify carbohydrates

The main carbohydrates handled by the food industry are starch, cellulose, sucrose, maltose, lactose, glucose and fructose. There are, however, a great range of other carbohydrates which are important ingredients in some food products.

The major sources of carbohydrate are the storage and structural components of plants. The main purpose of using enzymes is to convert these primary products into more useful forms with higher market value. We will use the modification of starch as an example of such modifications.

6.4.1 Starch

Starch is a common storage compound in plants and is usually produced in the form of granules. Although starch is itself a valuable commodity (found for example in flour), its hydrolysis into low molecular weight carbohydrates (saccharification) plays an important role in many food processes. For example, yeasts cannot metabolise starch and in order to generate alcohol (brewing) or CO_2 (baking), the starch first needs to be hydrolysed. Starch can be hydrolysed by acid hydrolysis. This has been used in the past but gives low yields and coloured products. Now, the industry is dependent upon using enzymes to achieve hydrolysis.

importance of
saccharification
of starch

The structure of starch is shown in Figure 6.1. It contains linear chains of glucose linked by α1-4 links (amylose) and branched chains of glucose linked by α1-4 and α1-6 links (amylopectin). Starches from different sources differ from each other in terms of the lengths of the chains and in the degree of branching. It is not surprising, therefore, that the amounts and combinations of enzymes used to hydrolyse starch are different in different operations.

amylose and
amylopectin

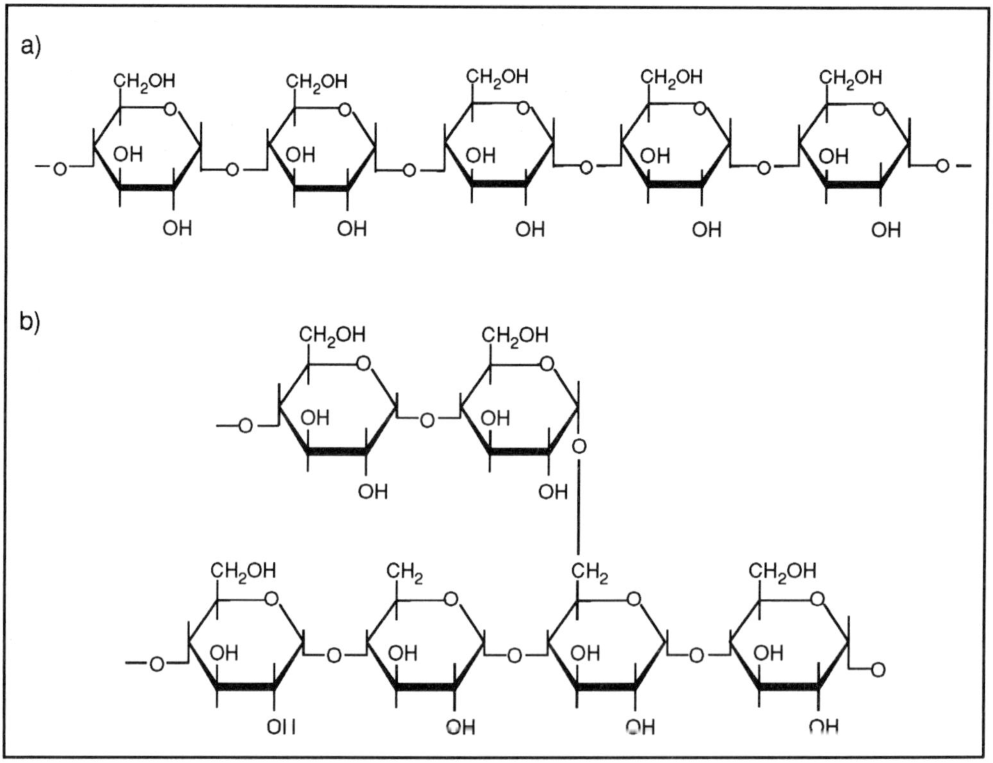

Figure 6.1 The structure of starch: a) shows a short segment of the linear chains of amylose; b) shows a branch point found in amylopectin. Amylose and amylopectin are large molecules composed of many glucose moieties.

Π Examine Figure 6.1 and identify the sorts of enzymes which should be used to hydrolyse starch.

It should be self-evident that we need enzymes which will catalyse the hydrolysis of α1-4 and α1-6 links. Some examples of these are given in Table 6.1

Π Will the enzymes listed in Table 6.1 be used in a soluble or in an insoluble form?

pre-treatment of starch

Since starch is particulate and the enzymes are relatively cheap, then you should have concluded that they are used in a soluble form. However, we must point out that where purified starch granules are used, the starch granules are often pre-treated to bring them into a soluble form. This process involves two stages (gelatinisation and liquefaction) and is illustrated in Figure 6.2. The final hydrolysis of the liquified starch yields a variety of products depending on the enzymes used in the final incubation.

Π Use the information in Table 6.1 to identify what the main product would be if: a) fungal α-amylase was used; b) fungal β-amylase was used.

Your answer for a) should be maltose and short oligosaccharides and for b) glucose.

Enzyme	Source	Action and products
α-amylase	*Aspergillus niger* *Aspergillus oryzae*	hydrolyses α1-4 links to give maltose and short oligosaccharides
	Bacillus amyloliquefaciens	hydrolyses α1-4 links in oligosaccharides to produce maltose and very short oligosaccharides
	B. licheniformis	hydrolyses α1-4 links to produce maltose
	B. subtilis	hydrolyses α1-4 links to produce maltose and glucose
β-amylase	*Aspergillus niger*	hydrolyses α1-4 and α1-6 links from non-reducing end to yield glucose
pullulanase	*Bacillus audopullulyticus*	hydrolyses α1-6 links

Table 6.1 Enzymes used for starch hydrolysis. (After Chaplin, M.F and Bucke, E. 1990 Enzyme Technology Cambridge University Press). Note that these are major bulk enzymes and are relatively cheap.

Using schemes based on that illustrated in Figure 6.2, a variety of products can be made from starches from various sources. These include glucose syrups and maltose groups.

Fungal α-amylases find use in the baking industry to promote the release of glucose from flour starch. This enables yeast to metabolise the glucose released to produce CO_2 which is essential to cause the dough to 'rise' and to generate the desired texture.

release of sugars during malting

In brewing, the hydrolysis of starch in barley was, historically, dependent upon the action of hydrolytic enzymes produced by barley during the malting. In this process, the barley is first allowed to germinate, and is then baked and milled. The milled grain is then steeped in liquor to release the solubilised sugars. It is now common practice to supplement the naturally-occurring enzymes in germinating barley by bacterial and fungal enzymes. This not only speeds up saccharification, it is also needed to hydrolyse other starch-containing materials which are added to increase the amount of fermentable sugars. Barley also contains polymers of glucose (glucans) which are linked by β1-3 and β1-4 bonds. The hydrolysis of these is achieved by using β-glucanases.

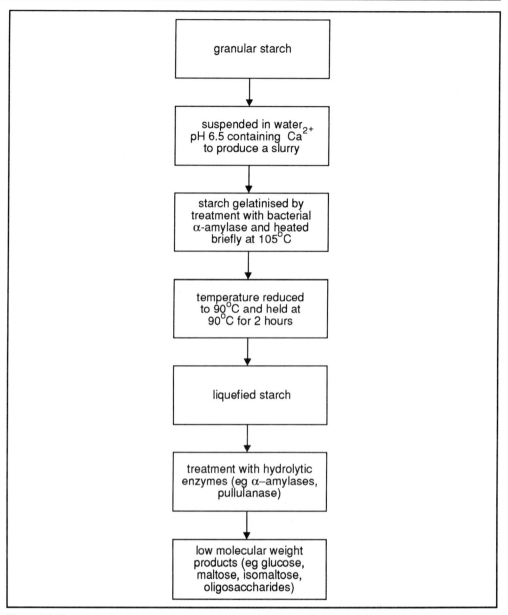

Figure 6.2 The generalised process for converting starch to low molecular weight sugars.

SAQ 6.1

Which of the following would produce glucose from amylose most rapidly?

1) α-amylase (1 dosage).

2) Pullulanase (1 dosage).

3) β-amylase (1 dosage).

4) β-amylase (½ dosage) and pullanase (½ dosage).

6.4.2 Sucrose

Sucrose is used as a sweetener. It may be sold in granular form or as a syrup. In this latter form it has a tendency to crystallise. The hydrolysis of sucrose to form glucose and fructose produces a more stable syrup. The enzyme invertase (the hydrolysis of sucrose is called 'inversion') from yeast has been used to carry out this process.

invertase

$$\text{sucrose} \xrightarrow{\text{invertase}} \text{glucose} + \text{fructose}$$

Although the enzyme is inhibited by high sucrose concentrations, it is still commercially viable because the enzyme is relatively cheap to produce. In the past the enzyme was simply produced from *Saccharomyces cerevisiae* by allowing the yeast to autolyse (achieved by temperature shock) and the autolysate used directly. However, invertase concentrates are now available commercially. Invertase has also been used in an immobilised form.

removal of hazes in sucrose solutions

The presence of high molecular weight carbohydrates as contaminants of sucrose extracted from sugar cane and sugar beet poses problems. They tend to clog filters used to clarify the sucrose extracts and produce hazes in sucrose solutions. These problems may be overcome by using α-amylases and other hydrolytic enzymes to reduce the concentrations of these undesirable components. Usually thermostable enzymes are used that will withstand the conditions (high temperatures and pH) used in the sucrose extraction process.

6.4.3 Fructose

An important enzyme in the food and drink industry is glucose isomerase, which is responsible for the conversion of glucose to fructose. The question that needs to be asked is why is fructose more valuable than glucose when they are both hexose sugars? The answer lies in the sweetness ratings that these two carbohydrates possess.

sweetness ratings

The commonest known form of sugar is sucrose and traditionally this carbohydrate has been used to sweeten many products in the food and drink industry. As a sweetener sucrose is employed as a standard (with a value of 1.0) against which all other sweeteners are measured.

sweetness of fructose

Sucrose is a disaccharide comprised of one molecule of glucose and one molecule of fructose. Glucose and fructose are both monosaccharides with the former possessing a sweetness rating of 0.7 while the value for fructose can vary from 1.2-1.7 depending upon the conditions employed (see Table 6.2). The greater sweetening capacity of fructose compared to both sucrose and glucose means that, as a sweetener, it is industrially important.

Carbohydrate	Relative sweetness*
lactose	0.2
maltose	0.3
galactose	0.7
glucose	0.7
sucrose	1.0
fructose	1.4

Table 6.2 Relative sweetness of carbohydrates. *Values depend on the conditions employed in the evaluation.

⊓ From which natural sources is fructose obtained?

Although fructose has obvious advantages over glucose and sucrose regarding its sweetening capacity it is not as readily available as the other two carbohydrates. Fructose does occur in fruits, but it is not economically viable to grow crops in the quantities required to harvest the sugar.

potential sources of fructose

Traditionally sucrose has been available from both sugar cane and sugar beet, both of which can be farmed in large quantities. Glucose is not commonly available as the monosaccharide but, as we have seen, is available in abundance in the form of starch which is a plant storage carbohydrate. Starch can be obtained from both maize and potatoes, crops which can be produced in large quantities due to modern farming methods.

glucose isomerase

In order to obtain fructose for industrial purposes it is first necessary to convert starch to the glucose monosaccharide and then to convert glucose to fructose using the enzyme glucose isomerase. We have already shown how enzymes play an important part in generating glucose from starch, so we will move directly onto the glucose → fructose conversion step.

chemical conversion of glucose to fructose

Before enzymes were readily available as biocatalysts, glucose was traditionally converted to fructose by chemical techniques using alkaline isomerisation methods. However, this was not an ideal method as the conversion resulted in the presence of a number of unwanted sugars and coloured side products which meant that further purification steps were needed in order to 'clean-up' the main product. Once again, the introduction of biocatalysts has led to a more efficient method for the conversion of glucose to fructose. After the glucose molecules have been obtained from the original starch starting material it is possible to convert them to fructose via the enzyme glucose isomerase as shown below:

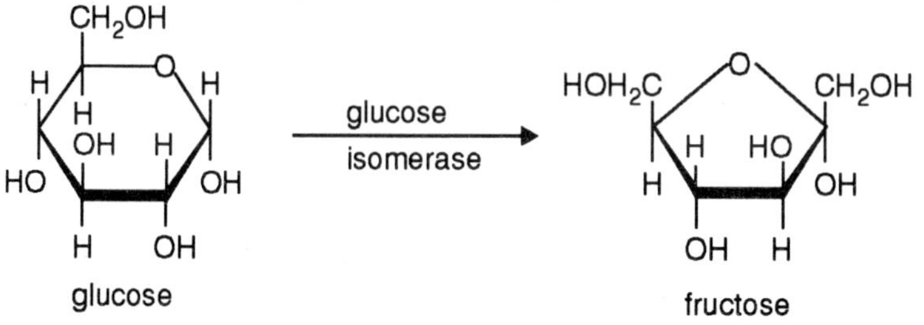

glucose →(glucose isomerase)→ fructose

Use of enzymes in large-scale industrial applications

xylose isomerase

The enzyme glucose isomerase is actually a xylose isomerase. In the late 1950's scientists who were looking for enzymes that would convert glucose to fructose discovered that the xylose isomerase from some bacteria would perform this biotransformation. Since then a number of bacteria have been screened to determine whether they possess glucose isomerase activity. To produce an active glucose isomerase enzyme, it is necessary to provide xylose in the growth media of the bacteria in order to induce expression of the xylose isomerase gene. When using glucose isomerase, it is also necessary to add magnesium ions and to use alkaline conditions to achieve maximal enzyme activity. The xylose isomerase enzymes which are used to provide glucose isomerase activity are intracellular and are often used in the immobilised state either as immobilised whole cells or in the form of immobilised purified enzymes.

The enzymes used commercially are produced by *Actinoplanes missouriensis*, *Bacillus coagulans* and *Streptomyces spp*.

∏ Since these enzymes are intracellular, they are quite expensive to produce. What do you think this means about the way they are used?

They are used in an immobilised form. Originally this was done in batch processes but now isomerisation is done in PBR's.

A comparison between batch and continuous PBR reactors is given in Table 6.3.

Feature	Batch using a soluble enzyme	Batch using an immobilised enzyme	Continuous PBR
reactor volume (m^3)	1100	1100	15
enzyme consumption (tonnes)	180	11	2
activity half life (hr)	30	300	1500
residence time (hr)	20	20	0.5
temperature (°C)	65	65	60
product refinement	ABCD[1]	B,C,D	B
cost per tonne converted ($)	1000	60	10

Table 6.3 Comparison of glucose isomeration procedures. (Adapted from Chaplin M.F. and Bucke C. 1990 Enzyme Technology Cambridge University Press). [1] A = filtration; B = activated carbon; C = cation exchange D = anion exchange.

∏ Examine Table 6.3 carefully as it gives a clear indication of the advantages of using PBRs especially in using an enzyme which catalyses a reversible reaction. Make a list of these advantages.

The principal advantage is that it is much cheaper than the alternatives. This reduction in cost is the result of being able to use a smaller reactor, shorter residence time and simpler downstream processing.

At equilibrium, about 50% of the glucose is converted to fructose. But remember that to achieve equilibrium, a long incubation time is required. A compromise is reached in which flow rates are adjusted to achieve about 42-46% conversion of glucose in the outflow from the reactor.

HFCS

The product made by this process is often called high fructose corn syrup (HFCS) as corn (maize) is the usual source of the starch used to generate the glucose used in the process. Usually HFCS is sold as a 42% fructose syrup (about 2 million tonnes per annum). In some cases, syrups with higher fructose contents (55%) are required. These are produced using large chromatographic columns of exchange resins or zeolites to purify the fructose. The fructose produced in this way is blended with the 42% fructose syrup to produce 55% fructose.

SAQ 6.2

Which of the following favours the production of HFCS rather than invert sugar as a source of fructose?

1) Starch is cheaper than sucrose.

2) Enzymes can be used to manufacture HFCS.

3) Invert sugar contains 50% fructose.

4) Three separate enzymes are required for HFCS production.

SAQ 6.3

Identify the appropriate enzymes from the list provided to complete the labelling of the diagram.

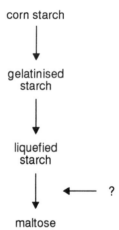

Enzyme list:

glucose isomerase; α-amylase; pullulanase; β-amylase; invertase.

Use of enzymes in large-scale industrial applications

SAQ 6.4

The flow diagram below represents the production of HFCS 55 (55% fructose) from high glucose syrup. From the list below, add labels to the diagram to complete it.

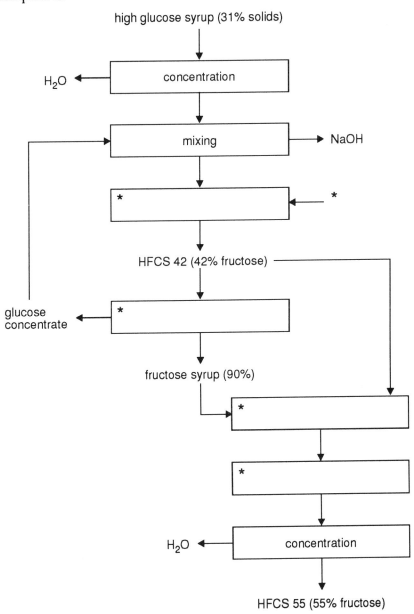

Word list:

blending; glucose isomerase; chromatography; activated charcoal treatment; isomerisation.

6.4.4 Cellulose

Cellulose is the most common organic chemical on earth. It is a major component of all plant cell walls. It is made up of long chains of glucose linked together by β1-4 links. You might anticipate, therefore, that cellulose might be a good source of glucose. In practice this is not so.

∏ List two reasons why cellulose is not a good commercial source of glucose.

The two main reasons are that:

- cellulose is invariably produced associated with a large number of other cell wall constituents (eg lignin, hemi-cellulose and other polymeric carbohydrates);

- the β1-4 links are relatively stable and are difficult to hydrolyse enzymically.

Cellulase is in fact a complex of enzymes which are needed to hydrolyse cellulose. For example, cellulase from *Trichoderma reesei* consists of a number of enzymes which work synergistically. These enzymes are:

- endo-1,4β-glucanase;

- 1,4β glucosidase;

- exo-1,4β-glucosidase;

- 1,4β-cellobiosidase;

- exo-cellobiohydrolase.

The hydrolysis of cellulose to release glucose requires these enzymes to work sequentially on the crystalline cellulose found in nature. First the crystalline cellulose is acted on by the endo-1,4β-gluconase to form a non-crystalline form and to break the cellulose chains into shorter segments and then the other enzymes hydrolyse the broken chains to release cellobiose (glucose-β1-4-glucose) and glucose. Although impure cellulose (plant cell walls) is cheaper than starch, it costs more to produce glucose from this source. Purified cellulose (paper pulp) is more expensive than starch. Thus, so far, a production processes starting from cellulose has not proven itself to be a cost-effective way of producing glucose.

6.4.5 Lactose

lactose intolerance — Lactose is the major carbohydrate of milk where it occurs in concentrations of about 4.7% (w/w). Unfortunately a large proportion of the World's population, especially in the Far East, is unable to metabolise this carbohydrate. Such individuals are said to be 'lactose intolerant'. Over 90% of the population of China and the Thai peninsula are lactose intolerant. The presence of lactose in milk and other dairy products such as cheese, therefore, limits the use of these products both for adults and, especially, young children.

lactase — The hydrolysis of lactose to produce glucose and galactose by the enzyme lactase offers opportunities to overcome this problem. It also has the advantage that the products of hydrolysis are sweeter (x 4) and are more soluble.

The enzyme is prepared from the yeast *Kluyveromyces fragilis* or the fungi *Aspergillus niger* and *A. oryzae*. Although the enzyme is fairly expensive to produce, it has found use in the treatment of milk to be used for infant feeds and in the treatment of whey produced during cheese manufacture. It is also used in the production of ice cream and condensed milk. It is most frequently used at a cool temperature (5°C) and sufficient enzyme is added to achieve about 50% hydrolysis after about 16-24 hours incubation.

6.4.6 Miscellaneous enzymes involved in carbohydrate modification

Glucose oxidase. This is a highly specific enzyme which oxidises glucose to gluconic acid using molecular oxygen. Hydrogen peroxide is a product of the reaction. The enzyme, usually obtained from *Aspergillus spp* or *Penicillium spp*, may be used to remove oxygen or traces of glucose in order to improve storage. The hydrogen peroxide produced is also a potent bactericide. If hydrogen peroxide accumulation is to be avoided, catalase is also added unless the material already contains this enzyme. Glucose oxidase finds particular application in the removal of glucose from egg white used in the baking industry.

glucose oxidase to improve storage and to remove glucose from egg white

Pectinases. Pectin is a major component of plant material. Its presence in fruit extracts causes gelatinisation and, in wine manufacture, causes hazes. The presence of pectins in a fruit extract reduces the amount of juice than can be extracted from these sources. Pectinases (a mixture of pectin lyase, pectin esterase and polygalacturonase) are extensively used in the fruit juice and wine industries. Also important in this context are arabinases and galactanases which hydrolyse arabans (polymers of arabinose) and galactans (polymers of galactose). The use of these enzymes is described in detail in the BIOTOL text, 'Biotechnological Innovations in Food Processing'.

pectinases used in fruit juice extraction and processing

6.5 Use of proteases in the food industry

There has been a long tradition of using proteases in the food industry. The classical example is rennet (from the stomach of calves) used in the manufacture of cheese. Proteases have also found use in the tenderisation of meat and for producing protein hydrolysates.

We can divide the industrial use of proteolytic enzymes into two subgroups:

- those which cause hydrolysis of specific links;

- those which cause general hydrolysis.

Rennet is a good example of the first group. The coagulation of milk requires a specific hydrolysis: (-phe105-met106-) in κ-casein. The traditional source of this enzyme (chymosin) is limited and expensive (fourth stomach of unweaned calves). Recently microbial sources of an enzyme with similar activity have been used. The major sources of these substitute enzymes are *Mucor miehei*, *M. pusillus* and *Endothia parasitica*. Using genetic engineering techniques, the gene coding for the active ingredient (chymosin-like enzyme) has been transferred to organisms more acceptable to the food industry (*Kluyveromyces lactis*, *Aspergillus niger* and *Escherichia coli*). (Note that the genetic engineering steps and the problems and merits associated with these new sources are discussed in the BIOTOL text, 'Biotechnological Innovations in Food Processing,' and will not be enlarged upon here).

microbial enzyme replacements for chymosin

It should be noted that a protease from *Bacillus amyloliquefaciens* and lipases from *Mucor miehei* or *Aspergillus niger* are also included in some cheese formulations. These modify the texture and flavour of the mature cheese.

non-specific proteases

Non-specific proteases are used to recover protein from parts of animals and fish that would be wasted. Usually this material is attached to bones. The bones are crushed and incubated in the presence of proteases which loosen the meat. The slurry is separated from the bone material and used in canned meat products and soups.

The texture of bread depends on the proteins (gluten) present in flour. Thus, when a thin, even texture is required (for example, in biscuit manufacture) proteolytic enzymes may be added to hydrolyse the gluten.

meat tenderisation

Finally in this section we will mention the use of proteolytic enzymes in tenderising meat. The proteolytic enzyme (usually papain derived from the paw paw fruit (*Carica papaya*) is injected in an inactive form just prior to slaughter. After slaughter, reducing conditions are generated in the muscles and these reducing conditions convert the inactive enzyme into its active form. There is, however, an increasing public lobby against this practise. Also within the industry, it is recognised that the process reduces the commercial value of internal organs such as heart, liver and kidneys.

SAQ 6.5

Which of the following is (are) true?

1) α-amylase leads to increased loaf volume during baking by increasing the concentration of amylose.

2) Proteases are only applied to wheat flour with a high gluten content.

3) Proteases and starch hydrolysing enzymes can each improve the leavening of bread.

6.6 Lipases

The major components of oils and fats are triglycerides and their physical properties depend upon the structure of their fatty acid moieties. Natural oils and fats can be used directly in products either individually or as mixtures. In many cases, it is necessary to modify their properties, particularly their melting characteristics, to make them suitable for particular applications.

Lipases catalyse the hydrolysis of oils and fats to give diglycerides, monoglycerides and free fatty acids. The reaction is reversible and consequently microbial lipases also catalyse the formation of glycerides from glycerol and free fatty acids. This process can be used to produce new triglycerides with specific combinations of fatty acids (Figure 6.3).

Use of enzymes in large-scale industrial applications

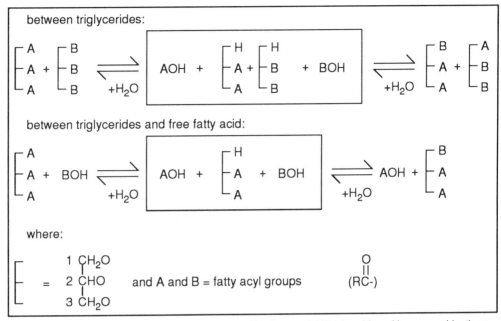

Figure 6.3 Mechanisms of lipase-catalysed interesterifications to yield triglycerides with new combinations of fatty acids.

Different lipases show different specificities. Some will exchange any of the fatty acids. Others only work on the outer (1- and 3-) positions (they display regio-specificity). By carefully selecting an appropriate enzyme, the desired triglyceride may be generated. The enzymes used are extracellular microbial lipases from a variety of sources.

Let us take a specific example. Palm oil is readily available. Its triglyceride composition is given in Table 6.4. For many purposes, cocoa butter has more desirable properties (it imparts a 'melt in the mouth' feel to foods). Its composition is also given in Table 6.4.

Π Before reading on, see if you can draw a flow diagram for an interesterfication process in which palm oil is to be incubated with stearic acid in the presence of a regio-specific lipase to increase the number of stearic acid residues in the 1 and 3- positions of the triglycerides.

In a typical reaction, a feed stream of refined palm oil and stearic acid is dissolved in petroleum ether and is saturated in water. This is pumped through a bed of regio-specific lipase (usually from *Mucor miehei*, supported on diatomaceous earth). The enzyme catalyses the exchange of the fatty acids and the triglycerides leaving the bed have a higher stearic acid content (see Table 6.4). You will notice that we have used a non-aqueous medium in this process. We will discuss the use of non-aqueous systems in the final chapter of this text.

Triglyceride	Palm oil mid fraction (% dry weight)	Enzymatically produced fat (% dry weight)	Cocoa butter (% dry weight)
StStSt	5	3	1
POP	58	16	16
POSt	13	39	41
StOSt	2	28.5	27
StLnSt	9	8	8
StOO	4	4	6
Others	2	1.5	1
StStSt	⎡ stearate ⎢ stearate ⎣ stearate	POP	⎡ palmitate ⎢ oleate ⎣ palmitate
POSt	⎡ palmitate ⎢ oleate ⎣ stearate	StOSt	⎡ stearate ⎢ oleate ⎣ stearate
StLnSt	⎡ stearate ⎢ linoleate ⎣ stearate	StOO	⎡ stearate ⎢ oleate ⎣ oleate

Table 6.4 Composition of cocoa butter equivalent prepared using enzyme technology.

SAQ 6.6

Which of the following favours using enzymatic interesterification of oils and fats rather than using a chemical process?

1) Lipases can have specificity for the 1- and 3- positions of triglycerides.

2) Lipases can be non-specific for the 1-, 2- and 3- positions of triglycerides.

3) Lipases operating at low water concentrations show limited hydrolysis of triglycerides.

4) Lipases work well at relatively low temperatures.

6.7 Amino acids

L-amino acids are important food supplements. Chemical synthesis of amino acids invariably leads to the production of racemic mixture of the D and L forms (the exception is glycine). Because of the stereospecificity of enzymes, these may be used to produce the desired stereoisomer.

aminoacylase

One of the most common processes involves using aminoacylase immobilised by adsorption onto an anion exchange resin. These are used in PBRs and, if the enzyme from *Aspergillus oryzae* is used at about 50°C, these reactors have an operational half-life of over two months. The re-charging of the reactor is quite straightforward by simply adding more enzyme.

The reaction is as follows. A racemic mixture of the N-acyl amino acid is made chemically and passed through the PBR. The acyl group is hydrolysed from the L-amino acid derivative but not from the D-amino acid derivative. The products are quite easy to separate by crystallisation. The N-acyl-D amino acid is racemised (enzymatically or chemically) and re-used. We can represent this process in the following way:

$$\text{N acyl-DL-amino acid} \xrightarrow[+ H_2O]{\text{aminoacylase}} \text{L-amino acid} + \text{N acyl-D-amino acid} + R'-COO^-$$

$$\text{N acyl-D-amino acid} \xrightarrow{\text{racemase or chemical racemisation}} \text{N acyl-DL-amino acid}$$

Since the production of amino acids (L configuration) as additives for food are dealt with in depth elsewhere in this series ('Biotechnological Innovations in Food Processing') we will not prolong the discussion here. However, we will list some of the important applications of enzymes in the generation of stereospecific amino acids.

aspartic acid production

Aspartic acid is used extensively in food and pharmaceuticals and in the production of the synthetic sweetener aspartame. It is produced by using entrapped *Escherichia coli* in PBRs. Ammonium fumarate is perfused through the reactor and converted to aspartic acid via the enzyme aspartate ammonium lyase:

$$\text{fumarate} + NH_4^+ \xrightleftharpoons[]{\text{aspartate ammonium lyase}} \text{L-aspartate}$$

This L-aspartate is combined with phenylalanine (also made enzymatically) to form aspartame. The aspartate acid can also be converted to L-alanine using the enzyme L-aspartate β-decarboxylase.

D-aminoacids and urocanic acid

Stereospecific enzymes are also used to generate D-amino acids, some of which are important components of antibiotics and insecticides. Urocanic acid is produced from L-histidine by histidine ammonia lyase. Urocanic acid is a sun-screening agent.

SAQ 6.7

The enzyme L-aspartate-β-decarboxylase catalyses the decarboxylation to form L-alanine.

$$^{-}OOC-CH_2-CH(N^+H_3)-COO^- \longrightarrow CH_3-CH(N^+H_3)-COO^-$$

L-aspartate → L-alanine

A suitable source of this enzyme is the bacterium *Xanthomonas oryzae*. With this knowledge, describe how L-alanine may be produced from fumarate on an industrial scale.

6.8 The use of enzymes in the pharmaceutical and healthcare industries

In Section 6.2, we mentioned a number of enzymes which have application in the health industries. We can broadly divide these into three groups:

- the use of enzymes to produce therapeutic agents;
- the use of enzymes for use as therapeutic agents;
- the use of enzymes in diagnosis.

Here we will mainly focus on the first group but will also list some therapeutic enzymes. The use of enzymes in diagnosis will be discussed in a later chapter.

6.8.1 Use of enzymes to produce therapeutic agents

steroids and semi-synthetic antibiotics

There are two main areas in the pharmaceutical industry where enzymes play an important role, namely in the production of steroids and semi-synthetic antibiotics. The stereochemistry of steroids makes them difficult to produce by conventional chemical methods. Enzymes are therefore useful tools in steroid manufacture due to their stereospecificity. The ability of microbial enzymes to specifically incorporate hydroxyl groups at many positions on the steroid nucleus is probably the best example of steroid transformation (Table 6.5).

Reaction	Substrate	Product	Micro-organism
1α-hydroxylation	androst-4-ene-3,17-dione	1α-hydroxyandrost-4-ene-3,17-dione	*Penicillium* sp.
11α-hydroxylation	progesterone	11α-hydroxy progesterone	*Rhizopus*
11β-hydroxylation	11-deoxycortisone	hydrocortisone	*Curvularia lunata*
17β-hydroxylation	androstane-3,11-dione	17β-hydroxy-androstane-3,11-dione	*Wojinowicia graminis*
21-hydroxylation	A-nor-pregn-3-ene-2,20-dione	21-hydroxy-A-nor-pregn-3-ene-2,20-dione	*Aspergillus niger*
18β-hydroxylation	11β,21-dihydroxy-pregn-4-ene-3,20-dione	11β,18,20-Trihydroxy-pregn-4-ene-3,20-dione	*Corynespora cassiicola*

Table 6.5 Examples of steroid hydroxylations by microbial enzymes. (Adapted from Biotechnology - The Science and the Business (1991), Moses, V and Cape, R.E. (Eds), Horwood Academic Publications, Chapter 17, by Neidleman, S L., Table 16, Pages 306-307).

Use of enzymes in large-scale industrial applications

∏ What do you think is the major problem that needs to be overcome in using enzymes to hydroxylate steroids?

The major problems is the low solubility of steroids in aqueous systems. It means that if aqueous systems are used, the substrate can only be supplied in low concentrations. This, of course, has important consequences for the economics of the process. We will return to this problem in the final chapter. Here we will focus on the use of enzymes to generate new antibiotics especially in the diversification of penicillins.

6.8.2 Semi-synthetic antibiotic products - penicillins

The structures of the two naturally produced penicillins are shown in Figure 6.4. By 'naturally produced', we mean products that are made in cultures of micro-organisms.

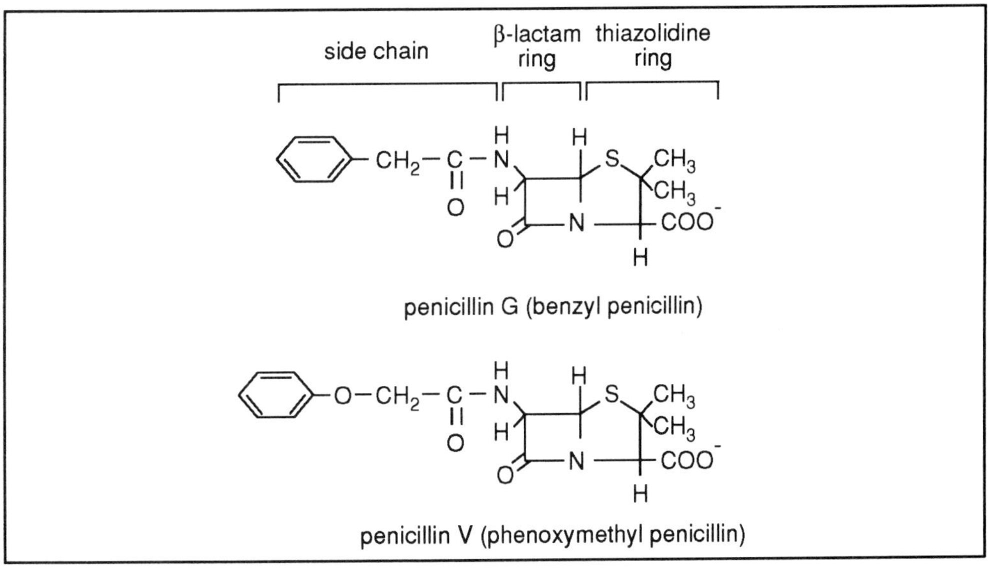

Figure 6.4 The structure of two naturally produced penicillins.

Semi-synthetic penicillins are penicillin derivatives which are produced from naturally made penicillins using *in vitro* techniques. These *in vitro* techniques make use of enzymes to bring about the desired modifications.

∏ What is the main drawback with using natural penicillins to fight bacterial infections and what advantages can semi-synthetic penicillins have over natural penicillins?

resistance to antibiotics

Although these natural penicillins, particularly penicillin G, have been very useful in fighting bacterial infections there is an increasing number of bacteria which are becoming resistant to the penicillins. The bacteria in question produce an enzyme, beta lactamase (β-lactamase), which cleaves the beta-lactam ring of the antibiotic rendering it inactive. To overcome this problem scientists have turned their attention to the production of semi-synthetic forms of penicillin, where the side chain attached to the β-lactam ring is altered. These semi-synthetic penicillins can resist β-lactamase attack and are, therefore, effective against a greater range of bacteria. Alternative side chains may confer other benefits, such as acid stability, thus enabling the antibiotic to be administered orally.

All penicillins have the same basic structure which consists of 6-aminopenicillanic acid (6-APA):

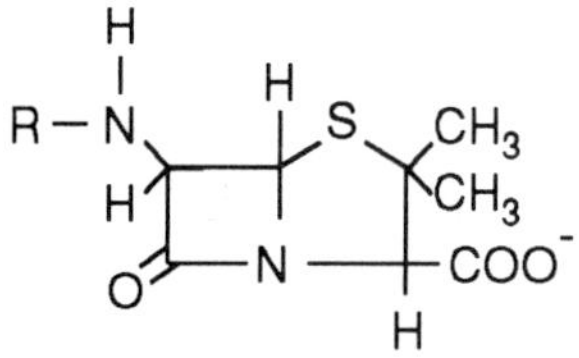

6-aminopenicillanic acid

(R=H)

The penicillins differ in the structure of the R-side chain.

To produce semi-synthetic penicillins, it is necessary to remove the side chain from penicillin G or V and replace it with a new side chain.

Penicillin acylases

Π Write the chemical equation for the deacylation of benzyl penicillin (penicillin G) by penicillin acylase.

Essentially you should have drawn a reaction sequence in which the benzyl group is removed from penicillin G to produce 6-aminopenicillanic acid. The structure of these are given in Figure 6.4 and in the text.

Initially the removal of the side chain from a penicillin molecule (deacylation) was performed chemically using a complex series of reactions which required low temperatures (-40° to -50°C):

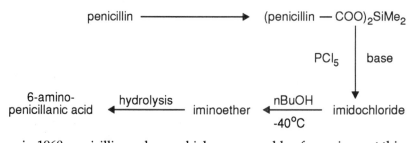

penicillin acylase

However, in 1960, penicillin acylases which were capable of carrying out this reaction in one step were discovered and over the following decade they were developed so as to be able to perform the task in question on an industrial scale:

penicillin $\xrightarrow[\text{35°C, pH 7-8}]{\text{penicillin acylase}}$ 6-aminopenicillanic acid + side chain

Production of penicillin G or V by fermentation is followed by treatment of the antibiotic with the penicillin acylase which catalyses the removal of the side chain resulting in the production of 6-aminopenicillanic acid and the separate side chain. The reaction for penicillin G is shown below:

Use of enzymes in large-scale industrial applications

penicillin G (benzyl penicillin)

↓ penicillin acylase, pH 8.0

6-aminopenicillanic acid + phenylacetic acid

recovery of 6-aminopenicillanic acid

To retrieve the 6-aminopenicillanic acid from the reaction mixture, the pH of the solution is lowered to the isoelectric point (pH 4.3) of the 6-aminopenicillanic acid which then precipitates out of solution. A wide range of micro-organisms possess this ability to deacylate penicillins and it is known that penicillin acylases are produced both intracellularly and extracelluarly. In general, bacterial acylases prefer penicillin G as their substrate while fungal acylases act best upon penicillin V.

There is, therefore, a wide range of sources of penicillin acylases and the preferred enzyme varies from company to company depending upon their individual requirements. The penicillin G acylase gene has been cloned to a high copy number and expressed in *E. coli*.

There are a variety of ways in which penicillin acylases can be used. The enzyme can be used in solution in the free state, inside the non-proliferating microbial cell or, in the preferred mode, as an immobilised enzyme.

Π List four different methods by which penicillin acylase may be immobilised.

The choice of immobilisation technique by a company will differ depending upon which method best suits their application. Immobilisation techniques that are employed include adsorption, covalent binding to CNBr-activated Sepharose, physical entrapment and glutaraldehyde cross-linked cells (see Chapter 4 for enzyme immobilisation techniques). The use of an immobilised penicillin acylase means that, due to lack of enzyme leakage, the 6-aminopenicillanic acid produced is very pure and therefore the possibility of any allergenic properties introduced during production are greatly reduced. By using immobilised penicillin acylase it is possible to produce 6-aminopenicillanic acid with a yield of 90% and a purity of 98%.

acylation

Having obtained the 6-aminopenicillanic acid, the next step is to add on the new side chain. Acid chlorides can be used to achieve this but the penicillin acylase enzyme can also be used in the reverse reaction (acylation) using a lower pH to facilitate the reaction. The following example shows the manufacture of the semi-synthetic penicillin, ampicillin:

6-aminopenicillanic acid + phenyl glycine

$\xrightarrow{\text{penicillin acylase, pH 5.0}}$

ampicillin + H$_2$O

There are a variety of side chains that can be employed to produce semi-synthetic penicillins. Some of them are shown in Table 6.6.

Antibiotic	Side chain structure
ampicillin	C$_6$H$_5$–CH(NH$_2$)–CO–
carbenicillin	C$_6$H$_5$–CH(COONa)–CO–
cloxacillin	(2-Cl-C$_6$H$_4$)–C(=N–O–)–C(=CH$_3$)–CO– (isoxazolyl)
methicillin	(2,6-di-OCH$_3$-C$_6$H$_3$)–CO–

Table 6.6 Some of the side chains employed to manufacture semi-synthetic penicillins. (Adapted from Enzymes in Industry and Medicine (1987), Bickerstaff, GF (Ed), Edward Arnold (Publishers) Ltd, 41, Bedford Square, London, Figure 5.3, Page 73).

The resulting products show changed characteristics. Many resist deactivation by β-lactamase producing organisms and they show different adsorption and physical stabilities. Thus, by producing a diversity of penicillins, they can be used to treat a wider spectrum of infections.

6.8.3 Semi-synthetic antibiotic products - cephalosporins

The cephalosporins are another group of β-lactam ring-containing antibiotics that have benefitted from the knowledge of microbial acylases. These enzymes have again enabled the manufacture of semi-synthetic derivatives of the natural cephalosporin (cephalosporin C).

$$^-OOC-\underset{NH_3^+}{CH}-(CH_2)_3-\underset{O}{\overset{}{C}}-N\cdots \text{[β-lactam ring]}\cdots CH_2-O-\underset{O}{\overset{}{C}}-CH_3 \quad \text{cephalosporin C}$$

As with some of the natural penicillins, cephalosporin C has limited usefulness in pharmaceutical applications. Therefore, the ability to produce semi-synthetic derivatives is of importance in fighting microbial infections.

The cephalosporins are very similar in structure to the penicillins and can be produced by fermentation which gives rise to cephalosporin C, the enzymatically deacylated form of which is 7-amino-cephalosporanic acid (7-ACA). Cephalosporins can also be produced by chemical modification of the five-membered ring of the penicillins to produce the six-membered ring. Enzymatic deacylation of these chemically synthesised cephalosporins can also occur but this time the nucleus formed is 7-aminodesacetoxy cephalosporanic acid (7-ADCA). Enzymatic acylation of either 7-ACA or 7-ADCA can then be used to introduce new side chains onto these molecules, thus creating some of the semi-synthetic cephalosporins.

enzymatic acylation of ACA and ADCA

In Table 6.7, we have cited a number of examples of antibiotics which are produced by using immobilised systems. Examine this table closely as it shows examples of both isolated-enzyme and whole-organism catalysis. Note also that a wide range of supports have been used to immobilise these systems.

∏ Are the methods of immobilisation used in the systems described in Table 6.7 all dependent upon the adsorption of the enzyme system onto the support as a result of electrostatic interaction?

The answer is no. DEAE-cellulose is a typical ion exchangers which will adsorb oppositely charged entities: hydroxylapatite also binds enzymes by adsorption. Polyacrylamide and calcium alginate on the other hand, are typical gel-formers and are used to entrap enzymes.

Antibiotic	Feed stocks	Enzymes or microbial cells	Supports for immobilisation
amoxycillin	6-APA and p-hydroxyphenyl-glycine methyl ester	penicillin amidase	cellulose triacetate
ampicillin	6-APA and D-phenylglycine methyl ester	*Bacillus megaterium* (penicillin amidase)	DEAE-cellulose
		Achromobacter aceris (penicillin amidase)	DEAE-cellulose
		Achromobacter liquidum (penicillin amidase)	DEAE-cellulose
		Kluyvera citrophila (penicillin amidase)	polyacrylamide
		succinylated penicillin amidase	DEAE-sephadex
bacitracin	nutrient medium (starch, eptone, meat extract)	*Bacillus* sp. (multi-enzymes)	polyacrylamide
cephalexin	7-ADCA and D-phenylglycine methyl ester	*Achromobacter* sp. (cephalosporin amidase)	DEAE-cellulose
		Achromobacter sp. (cephalosporin amidase)	hydroxylapatite
		Bacillus megatrium (cephalosporin amidase)	celite
		cephalosporin amidase	cellulose triacetate
cephalosporin	7-ADCA and D-phenylglycine methyl ester	*Xanthomonas citri* (cephalosporin amidase)	polyacrylamide
cephalosporin C	3-(N-Morpholino) propane sulfonic acid	*Streptomyces clavuligeus* (multi-enzymes)	polyacrylamide
cephalothin	7-ACA and 2-thiophene acetic acid methyl ester	cephalosporin amidase	celite
nikkomycin	nutrient medium (malt extract, peptone, starch, mannitol)	*Streptomyces tendae* (multi-enzymes)	calcium alginate
patulin	glucose	*penicillium urticae* (multi-enzymes)	carrageenan
penicillin G	glucose	*penicillium chrysogenum* (multi-enzymes)	polyacrylamide calcium alginate
tylosin	nutrient medium (malt extract, peptone, starch, mannitol)	*Streptomyces* sp. (multi-enzymes)	calcium alginate

Table 6.7 Synthesis of antibiotics by immobilised biocatalysts. (Adapted from Enzymes and Immobilised Cells in Biotechnology (1985), Laskin, A.I. (ED), The Benjamin/Cummings Publishing Company, Inc., Menlo Park, California 94025, USA, Table 3.2, Pages 52.53).

Use of enzymes in large-scale industrial applications

SAQ 6.8

1) Ampicillin has the structure:

[structure of ampicillin showing phenyl-CH$_2$-C(NH$_2$)-C(=O)-NH- linked to the penicillin nucleus with S, CH$_3$, CH$_3$, N-H, COO$^-$]

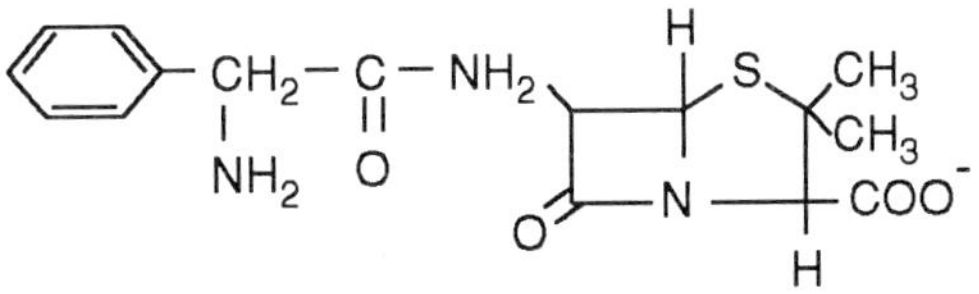

You have available a source of penicillin G and phenyl glycine which have the corresponding structures:

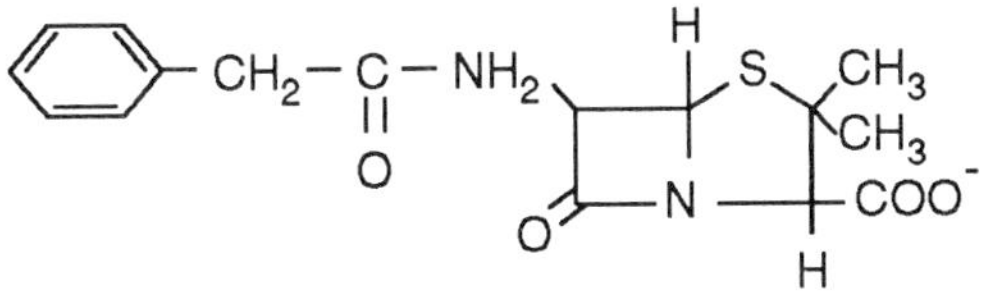

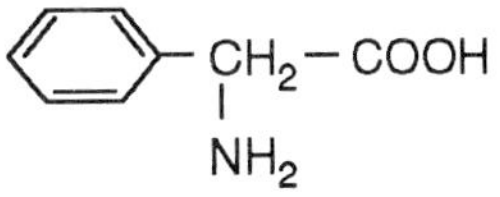

[structure of penicillin G: phenyl-CH$_2$-C(=O)-NH- linked to penicillin nucleus]

[structure of phenyl glycine: phenyl-CH$_2$-COOH with NH$_2$]

Explain how a single enzyme might be used to convert the penicillin G into ampicillin.

2) The side chain of the semi-synthetic penicillin, methicillin, has the structure:

[structure showing a benzene ring with two OCH$_3$ groups (ortho positions) and a -C(=O)- group]

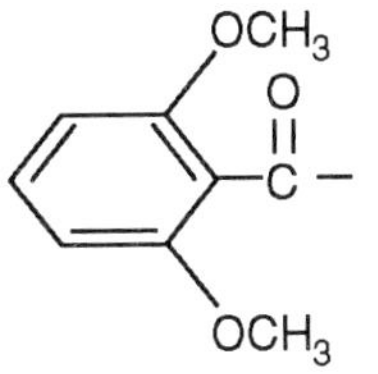

Assume methicillin is to be produced from 6-amino penicillanic acid by a) a chemical method; b) an enzymic method. What reagents will be needed?

6.8.5 Enzymes used as therapeutic agents

The production of enzymes for use as therapeutic agents is an important growth industry. Some examples are listed in Table 6.8.

∏ From the examples given in Table 6.8, what are the most common uses of enzymes as therapeutic agents?

The most common uses are as topical (surface) applications, in the removal of toxins and in the treatment of life-threatening disorders. In principle, enzymes should also be useful in the treatment of metabolic disorders.

Enzyme	Reaction	Medical problem
asparaginase	L-asparagine → L-aspartate	leukaemia
collagenase	hydrolyses collagen	skin ulcers
glutaminase	L-glutamine → L-glutamate	leukaemia
hyaluronidase	hydrolyses hyaluronic acid	heart attack
thiosulphate sulphur transferase (Rhodanase)	$S_2O_3^{2-} + CN^- \rightarrow SO_3^{2-} + SCN^-$	cyanide poisoning
RNase	hydrolysis of RNA	viral infections
streptokinase	plasminogen → plasmin	blood clots
urokinase	plasminogen → plasmin	blood clots

Table 6.8 Some examples of the use of enzymes as therapeutic agents.

Π See if you can list 2 or 3 reasons why, in practice, enzymes have found limited use in the treatment of metabolic disorders.

The main restrictions are:

- metabolism occurs within cells. It is difficult to envisage ways in which exogenously applied enzymes can gain access to these sites of metabolism through the plasma membrane. Also, different cells and different organs carry out different metabolic functions. Thus this is not just a problem of getting enzymes into cells, they also have to be targeted at particular cells;

- enzymes are proteins and are capable of stimulating the immune system (they are immunogenic). Thus introducing an enzyme into the bloodstream will stimulate the production of antibodies. Subsequent doses of the enzyme will be inactivated quickly by the antibodies;

- the circulatory lifetime of introduced proteins is usually short and they are often degraded.

Nevertheless, much research is being conducted into finding ways of targeting enzymes and stabilising them within the body. This work has met with some success (for example in the treatment of lung embolism with urokinase) but there is still much to be done.

6.9 Other industries employing the use of enzymes

So far we have focused on the use of enzymes in the food and health-orientated industries. The use of enzymes is not, however, confined to these two sectors.

6.9.1 Potential uses of enzymes in the petrochemical industries

Π Before reading on see if you can list the potential advantages of using enzymes in the petrochemical industry compared to conventional chemical techniques.

Use of enzymes in large-scale industrial applications

issues of safety

In the petrochemical industry many of the starting materials employed are potentially flammable and/or explosive and so the traditional chemical conversion techniques which often employ high temperatures and/or pressures are energy consuming and dangerous. However the ability of enzymes to carry out biotransformations at low pressures and ambient temperatures means that they can be employed in the petrochemical industry in a much safer way.

multistage reactions

Micro-organisms are a very useful source of enzymes as they can be used as non-proliferating whole cells. As micro-organisms contain many enzymes this means that a sequence of reactions can be performed concurrently. The non-proliferating whole cells catalyse multiple step reactions and these can be performed at ambient temperatures and low pressures. This again illustrates some of the advantages of using micro-organisms as biocatalysts compared to conventional chemical methods. In this section we shall look at some of the potential applications of enzymes as biocatalysts in the petrochemical industry.

Benzene cis-glycol production

One compound which is of interest to industry is benzene cis-glycol (5,6-dihydroxycyclohexa-1,3-dione). This is because derivatives of this compound can form high molecular weight polymers following polymerisation. These precursor polymers are soluble in organic solvents and can be used to produce films, fibres and coatings by being converted into polyphenylene when heated to 140-240°C. A benzene 1,2-dioxygenase enzyme is present in a strain of *Pseudomonas putida* which is capable of oxidising benzene to benzene cis-glycol (BCG).

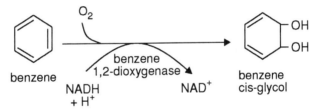

Using mutant strains of this *Pseudomonas* which are unable to metabolise BCG leads to substantial accumulation of this chemical in the media.

Π Why is it necessary to use a co-substrate such as ethanol or acetate in the production of benzene cis-glycol?

Because of the complexity of this enzyme and its requirement for the cofactor NADH, intact non-proliferating cells are used in the presence of a co-substrate (for example ethanol or acetate) which is oxidised to CO_2 but facilitates the regeneration of the NADH.

The enzymatic route to produce BCG and polyphenylene offers a real alternative to conventional chemical synthesis and overcomes many of the problems associated with the fabrication of this thermally stable plastic.

Adipic acid production

Adipic acid is a precursor in the production of nylon. There are obvious dangers that arise from reacting volatile and explosive organic compounds at high temperatures and pressures. The explosion at Flixborough in 1974 is just one example and arose from a process that required the oxidation of cyclohexane at high temperatures and pressures. Biological alternatives potential reduce the risk of such explosions. Cyclohexane can be used to produce adipic acid and here again microbial enzymes have shown that they are capable of the task. A species of the bacterial genus *Xanthobacter* is known to be capable of growth on cyclohexane as the sole source of carbon and energy. This organism is able to bring about the oxidation of cyclohexane to adipic acid via the sequence of steps shown in Figure 6.5. The important first step in the pathway is the initial oxidation of cyclohexane to cyclohexanol which is brought about by the enzyme cyclohexane hydroxylase which, like benzene 1,2-dioxygenase, is a complex enzyme requiring reduced pyridine nucleotide as a cofactor.

Once the six-membered carbon ring has been attacked, further oxidation of the cyclohexanol occurs readily. Again this biotransformation can be brought about using non-proliferating whole cells of *Xanthobacter* sp. in the presence of a suitable co-substrate to facilitate cofactor regeneration.

Use of enzymes in large-scale industrial applications

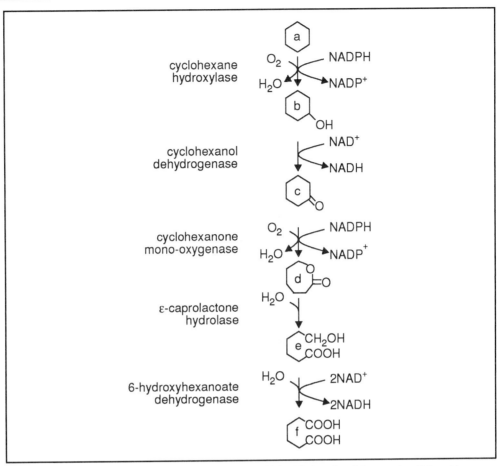

Figure 6.5 Outline of the reaction sequence for the microbial oxidation of cyclohexane to adipic acid. Cofactors shown are those for *Xanthobacter* sp. Compounds: a) cyclohexane; b) cyclohexanol; c) cyclohexanone; d) ε-caprolactone; e) 6-hydroxyhexanoate; f) adipic acid. (Adapted from Biotechnology and Genetic Engineering Reviews (1986), volume 4, Russell, GE (Ed). Published by Intercept Ltd., Newcastle upon Tyne. Chapter 9, by Griffin, M. and Magor, A.M. Figure 17, Page 284).

6.9.2 Use of enzymes in the leather industry

Leather is an important commodity. After removal from the animal, the hide has undesirable hair. This hair has to be removed and this historically involves the use of corrosive and toxic chemicals including sulphides. These not only present dangers to workers in the industry but are, in effluent water, environmentally damaging. The replacement of these traditional techniques by the use of proteases alleviate many of these problems. Figure 6.6 illustrates the sites of actions of the proteases used to remove the hair from hides.

undesirable consequences of chemical dehairing of hides

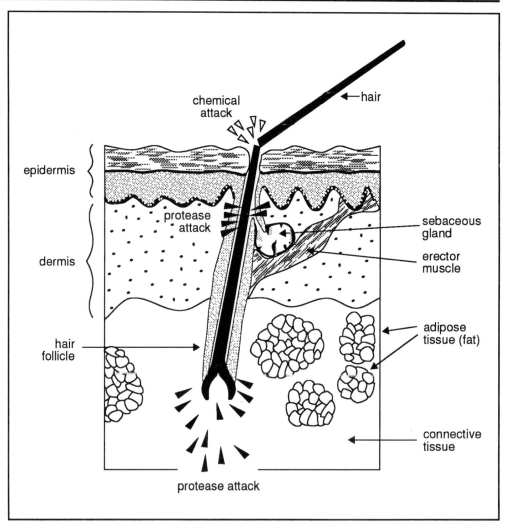

Figure 6.6 A diagram showing the use of chemicals and the use of proteases to remove hair from a skin. (Redrawn from 'Enzymes in Industry and Medicine', (1987), Bickerstaff, G.F. (Ed) Edward Arnold, London, P84).

Summary and objectives

In this chapter, we have reviewed some of the applications of enzymes in bulk industrial processes in a variety of industries. We have not attempted to cover all of the processes that are currently operational but, instead we have selected examples to illustrate the range of processes being undertaken. Some use soluble enzymes, others use immobilised enzymes or whole cells. The main emphasis of the discussion was on the use of enzymes in the food industry although the application of enzymes in the pharmaceutical and fine chemical industries were also described.

Now that you have completed this chapter, you should be able to:

- identify suitable enzymes for the hydrolysis of polymeric carbohydrates;

- explain why, in specific cases, methods employing enzymes are preferred to chemical methods;

- draw out flow diagrams showing the sequence of events in a number of industrial processes which use enzymes;

- explain why the therapeutic use of enzymes is largely limited to topical applications and to the treatment of life-threatening conditions and is not used to correct metabolic disorders.

Use of enzymes in analysis

7.1 Introduction	182
7.2 Why use enzymes in analysis?	182
7.3 Range of compounds assayed using enzymes	184
7.4 Measuring enzyme reactions	186
7.5 General strategies for substrate estimation using enzymes	187
7.6 The use of enzymes to assay activators and inhibitors	197
7.7 Formats in which enzymes may be used in analysis	198
7.8 Diagnostic dipstick tests	198
7.9 Enzyme-linked immunosorbent assay (ELISA)	200
7.10 Enzyme multiplied immunoassay technique EMIT	204
Summary and objectives	207

Use of enzymes in analysis

7.1 Introduction

In the previous chapters, we have examined the application of enzymes in bulk processes. In this and the following chapter we turn our attention to another important application of enzymes, their use in analysis. Enzymes are reagents in a large number of analytical procedures but find greatest use in clinical analysis, in analysis in the food industry and also in procedures involving environmental analysis. In all of these instances, enzymes often need to be able to identify minute amounts of an individual compound, sometimes a contaminating substance, which is present in a complex solution containing many other compounds (eg blood, fermentation broth, etc). Both soluble and immobilised enzymes are used in analysis; immobilised enzymes are more favoured particularly in the food and fermentation industries where process control is a common requirement.

In this chapter, we will examine and discuss the use of enzymes in analysis and describe the different ways in which enzymes can be employed as analytical tools. We will first explain why they are used before examining the practical issues involved in enzyme-based analysis. In the next chapter, we will describe a particular form of enzyme-based analysis, the so-called biosensors.

7.2 Why use enzymes in analysis?

Π What properties do enzymes possess that make them useful in analytical procedures?

specificity

The properties of enzymes that make them useful as analytical tools are essentially two-fold. The principal property is that they often possess a high substrate specificity. This is important since many of the solutions that are analysed contain a variety of solutes; the enzyme must therefore be able to target the correct substance. Examples of multicomponent solutions include industrial reactor broths and physiological fluids. It is therefore important that the enzyme is able to detect and select the right substrate which may only be present in $\mu mol\ l^{-1}$ concentrations. The second important property is that enzymes are often extremely sensitive, which means that they are able to detect low concentrations of substrates.

sensitivity

The precise degree of specificity varies from enzyme to enzyme, and this has considerable bearing on the usefulness of a given enzyme for analysis. Thus bacterial proteases often display low specificity and hydrolyse proteins at a variety of peptide bonds. In contrast, urease is highly specific, it only hydrolyses urea (NH_2-CO-NH_2) and fails to act on structural analogues (eg NH_2-SO-NH_2). Since estimation of urea is important in medical analysis, it is not surprising that urease is used in analysis.

Π Pyruvate can be estimated chemically on the basis of its reaction with phenylhydrazine, as follows:

$$\begin{array}{c} CH_3 \\ | \\ C=O \\ | \\ COO^- \end{array} + H_2N-N\overset{H}{|}-\langle C_6H_5 \rangle \longrightarrow \begin{array}{c} CH_3 \\ | \\ C=N-N \\ | \quad\quad | \\ COO^- \;\; H \end{array}-\langle C_6H_5 \rangle + H_2O$$

The product, a phenylhydrazone, is yellow and absorbs strongly at 410 nm whereas neither pyruvate nor phenylhydrazine absorb at this wavelength. Thus a calibration curve of absorbance against pyruvate present may be easily prepared using appropriate pyruvate standards to enable pyruvate to be estimated.

If you require to estimate pyruvate in serum or in tissue extracts, can you foresee any problem in obtaining accurate, reliable results using this procedure?

The major problem is one of specificity. The reaction is common to other compounds containing oxo-groups (also known as keto- or aldo- groups) such as oxaloacetate and α-ketoglutarate whose structures are:

$$\begin{array}{c} COO^- \\ | \\ CH_2 \\ | \\ C=O \\ | \\ COO^- \end{array} \quad\quad\quad \begin{array}{c} COO^- \\ | \\ CH_2 \\ | \\ CH_2 \\ | \\ C=O \\ | \\ COO^- \end{array}$$

oxaloacetate α-ketoglutarate

Since these are metabolites involved in the Krebs cycle, they are likely to be present in samples which also contain pyruvate. Thus estimates of pyruvate by the phenylhydrazine reaction will give high values when other reacting molecules are present.

We shall see shortly how pyruvate could be estimated enzymatically utilising the specificity of an enzyme.

The sensitivity of enzyme reactions can be described using a specific example. A widely used method for assaying glucose employs the enzyme glucose oxidase. In a typical procedure, concentrations in the range 0 to 0.25 mmol l^{-1} are detected colorimetrically using samples of up to 1 ml. Thus amounts as low as 0.05 μmol may be easily detected.

SAQ 7.1

1) Two enzymes which act on glucose and which may, in principle, be used to assay glucose have been isolated from microbial sources. These enzymes are glucose oxidase and hexokinase. Solely on the basis of the specificity data shown below and assuming visualisation procedures were available, which enzyme would you recommend for the assay of glucose?

	Glucose oxidase*	Hexokinase*
β-D-glucose	100	100
galactose	0.2	15
mannose	0.8	55
maltose	6.0	5
sucrose	0	0
fructose	0.5	8

*activities expressed relative to that of β-D-glucose.

2) Could glucose-6-phosphate dehydrogenase in a cell homogenate prepared from a fungus be used to assay glucose-6-phosphate?

7.3 Range of compounds assayed using enzymes

∏ Bearing in mind the normal functions of enzymes, what do you think are the most common compounds that are assayed using enzymes?

assay of metabolites and metabolic products

The most common compounds that are assayed by enzymes are metabolites or products of metabolism. It is logical that compounds which are produced as a result of the action of enzymes must themselves be substrates of enzymes. Thus, you should have anticipated that we are most likely to find an enzyme most suitable to use in analysis if we are trying to assay naturally occurring compounds. In contrast, we are less likely to find enzymes capable of using man-made (synthetic) compounds especially if they are not related to biologically-produced molecules.

Some examples of the compounds assayed using enzymes are given in Table 7.1. Although this list is quite long, it in no way attempts to give a complete list. You may not, at this stage, be familiar with all of the enzymes listed. Note that we have included one example of a man-made compound that may be assayed using enzymes. In this latter case, the assay is based on the effects of the analyte on the enzyme (analyte = compound being analysed).

You will also notice from the information given in Table 7.1 that in some cases a combination of enzymes is used. We will return to the reasons behind this in more detail later. We suggest you try to remember the information given in Table 7.1 as we will refer to it several times in this chapter.

Compound (analyte)	Enzyme	Comment
acetate	acetate kinase	
amines	monoamine oxidase	freshness of meat
ATP	luciferase	detection of contaminating micro-organisms
cholesterol	cholesterol oxidase	serum cholesterol
citrate	citrate lyase malate dehydrogenase lactate hydrogenase	
creatine	creatine kinase pyruvate kinase lactate dehydrogenase	tissue levels of creatine
ethanol	alcohol dehydrogenase	blood levels of ethanol, food stuffs
fructose	hexokinase phosphoglucose isomerase glucose-6-phosphate dehydrogenase	food stuffs
glucose	glucose oxidase or hexokinase and glucose-6-phosphate dehydrogenase	blood, urine, food stuffs
gluconate	gluconate kinase 6-phosphogluconate dehydrogenase	
L-glutamate	glutamate dehydrogenase	food stuffs
glycerol	glycerol dehydrogenase or glycerol kinase pyruvate kinase lactate dehydrogenase	monitoring media
L- or D-lactate	L- or D-lactate dehydrogenase	food manufacture serum analysis
L-malate	malate dehydrogenase	
maltose	maltase hexokinase glucose-6-phosphate dehydrogenase	food manufacture
pyruvate	lactate dehydrogenase	serum analysis
raffinose	α-galactosidase galactose dehydrogenase	food manufacture
saccharose	invertase hexokinase glucose-6-phosphate dehydrogenase	food manufacture
sorbitol	sorbitol dehydrogenase	food manufacture
urea	urease	urea in serum
uric acid	urate oxidase	uric acid in serum
organophosphate insecticides/nerve poisons	acetyl cholinesterase	detection of toxic agents

Table 7.1 Enzymes used for the assay of metabolites and other compounds. Note that compounds being measured are often referred to as analytes.

∏ Apart from the specificity of an enzyme and its sensitivity, what other criterion must be applied to the successful application of enzymes for analysis?

We hoped that you would realise that it is essential that we can readily measure the reaction catalysed by the enzyme. If it is difficult to measure this reaction, then it becomes very difficult to use enzymes for analysis. We will briefly examine the options for measuring enzyme-catalysed reactions in the next section.

7.4 Measuring enzyme reactions

The principle underpinning the use of enzymes for analysis is that an enzyme converts a substrate to a product and that this process can be monitored.

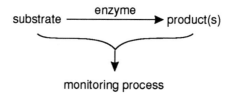

The process by which a substrate is converted to the appropriate product can be monitored in different ways. It is possible to measure the decrease in concentration of the substrate for which the enzyme is specific or alternatively to record the increased presence of reaction product(s).

∏ From your knowledge of the strategies which are used for the assay of enzymes (which are discussed in other BIOTOL texts, including 'The Molecular Fabric of Cells' and the companion text to this volume 'Fundamentals of Enzymology for Technological Applications'), list the different ways in which an enzymatic reaction may be followed.

Your list might include the following:

- spectrophotometry;
- colourimetry;
- turbidity;
- fluorimetry;
- calorimetry;
- amperometry;
- potentiometry;
- radiometry.

Use of enzymes in analysis

The first two are easily the most widely used, reflecting their ease, convenience and accuracy. Others are used as the need dictates. Note that the use of fluorimetry and of radioactively-labelled substrates are particularly advantageous when extreme sensitivity is required.

We shall assume that you are familiar with the more commonly used procedures for enzyme assay. If you are not, we recommend that you read (or re-read) the relevant sections of the texts mentioned above.

monitoring NADH (NADPH) — In some cases it may not be the substrate or product directly that is monitored but rather a cofactor such as one of the nicotinamide adenine dinucleotides (NADH, NADPH) which can be monitored spectrophotometrically at 340 nm. In all cases, the substance that is being analysed is referred to as the analyte.

With all these different methods of analysis, it is the specificity of the enzyme which enables it to recognise the correct substrate that is the key to the analytical process.

7.5 General strategies for substrate estimation using enzymes

kinetic and end-point methods — When soluble enzymes are employed in analytical procedures there are two different methods that can be used to measure the amount of substrate present, namely the kinetic method and the end-point or equilibrium method. The kinetic method relies on undertaking a time-dependent assay whilst the equilibrium method does not.

7.5.1 Kinetic method

The kinetic method relies on an enzyme-catalysed reaction obeying the Michaelis-Menten relationship whereby the rate of an enzyme-catalysed reaction is dependent on the concentration of its substrate [S] until all the catalytic sites of the enzyme are occupied at which point the enzyme is working at its maximum velocity (V_{max}).

Figure 7.1 shows such a relationship. It is a figure which, by now, you should be familiar with.

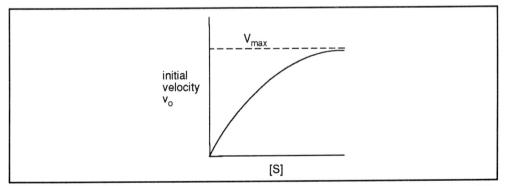

Figure 7.1 Effect of substrate concentration [S] on the initial velocity of an enzyme-catalysed reaction which displays Michaelis-Menten kinetics.

Providing assays are conducted under controlled conditions, and only the substrate concentration varies from sample to sample, then the data represented in Figure 7.1 can be used as a calibration curve. The observed initial velocity in the presence of a test sample can then be used to determine the substrate concentration in that sample.

Thus in setting up an analysis of this type we can identify a number of key steps. These are:

- standardise (optimise) assay conditions (eg temperature, pH, salt concentration);
- check linearity of initial reaction (v_o) with time over a range of substrate concentrations, in order to establish an appropriate incubation time;
- construct a calibration (reference) curve of v_o against substrate concentration;
- measure the initial reaction velocity using the test sample;
- use the calibration curve to determine the amount of substrate in the test sample.

range of [S] that can be used

Note, however, that the sensitivity of an enzyme to increasing substrate concentration decreases markedly when $[S] > K_M$ so that $[S] = K_M$ is about the upper limit for the use of the kinetic method. This method also suffers from other drawbacks since the rate of reaction is affected by a number of parameters including enzyme concentration, temperature, pH and contaminating material in the sample. The method is also complex in that it requires several determinations of the parameter which is being measured (for example, decrease in substrate concentration or increase in product concentration) at different times. A plot of change against time must then be constructed from which the initial linear rate of reaction is calculated as shown below.

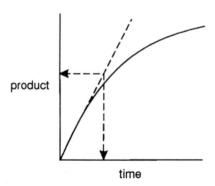

Clearly a continuous assay procedure is desirable for a kinetic assay of a substrate. The introduction of kinetic analyzers to clinical laboratories reflects the fact that many of these tasks can be automated.

Use of enzymes in analysis

SAQ 7.2

The following diagram shows a graph of initial velocity (rate) against substrate concentration for an enzyme-catalysed reaction (S → P). On the graph, three zones have been identified (A, B, C).

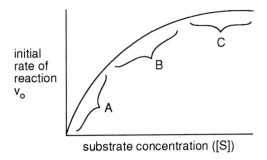

1) Identify which of the regions A to C is:

   a) predominantly first order;

   b) predominantly zeroth order;

   c) a mixture of zeroth and first order.

2) Which region will be most useful for measuring [S]?

3) Which region will be of no value for measuring [S]?

4) Does the K_M of the enzyme impose any limits on the range of substrate concentration which may be determined?

5) Identify factors which must be controlled if unknown samples are to be analysed.

6) Outline in what respect the relationship between velocity and substrate concentration would be different if this was not an enzyme-catalysed reaction.

Π We earlier assessed the suitability of a chemical method (based on phenylhydrazine) for the estimation of pyruvate and found it liable to interference when applied to biological samples. From your knowledge of biochemistry and the commonly used methods of enzyme assay, propose an enzyme-based procedure for the estimation of pyruvate which should avoid the specificity problem identified earlier.

Whilst several enzymes utilise pyruvate, the most attractive for analytical purposes is lactate dehydrogenase. You may recall that this enzyme catalyses the reversible reduction of pyruvate to lactate, employing NADH as reducing agent.

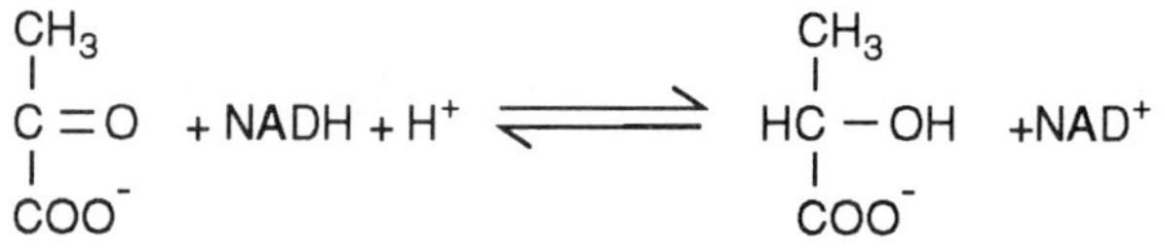

$$\begin{array}{c} CH_3 \\ | \\ C=O \\ | \\ COO^- \end{array} + NADH + H^+ \rightleftharpoons \begin{array}{c} CH_3 \\ | \\ HC-OH \\ | \\ COO^- \end{array} + NAD^+$$

pyruvate　　　　　　　　　　　　　lactate

You should also recall that NADH absorbs strongly at 340 nm (the extinction coefficient is 6220 lmol^{-1} cm^{-1}), whereas NAD$^+$ has no absorbance at this wavelength. Thus the activity of this enzyme may be measured by the decrease in [NADH] as pyruvate is reduced to lactate. This routinely involves a continuous assay at 340 nm using a spectrophotometer.

(Note, if the sample contains lactate, this will interfere with the assay).

SAQ 7.3

In the kinetic assay of pyruvate using lactate dehydrogenase and NADH, the following initial velocities were observed using a range of pyruvate concentrations. Use these data to determine the concentration of pyruvate in sample X.

[pyruvate] (mmol l^{-1})	initial velocity (μmol min^{-1})
0	0
0.25	7.7
0.5	14.3
1.0	25.0
1.5	33.4
2.0	40.0
X	18.4

7.5.2 End-point or equilibrium method

The end-point or equilibrium method is based on measuring the concentration of the product present in the reaction after the reaction has been allowed to go to completion (equilibrium). A knowledge of the time required to reach this end-point is therefore important before undertaking such assays.

This can be calculated from the integrated form of the Michaelis-Menten equation such that:

$$t = \frac{2.3 \, K_M}{V_{max}} \log_{10} \frac{[S_o]}{[S_t]} + \frac{[S_o] - [S_t]}{V_{max}} \qquad \text{Equation 7.1}$$

where t = time of reaction

[S$_o$] = the initial substrate concentration

[S$_t$] = substrate concentration after time t.

If one accepts that 99% conversion of substrate to product is adequate then $\log_{10}[S_o]/[S_t]$ is approximately 2 and $[S_t]$ can be ignored relative to $[S_o]$.

Hence:

$$t = 2.3 \times 2 \frac{K_M}{V_{max}} + \frac{[S_o]}{V_{max}}$$

$$= 4.6 \frac{K_M}{V_{max}} + \frac{[S_o]}{V_{max}} \qquad \text{Equation 7.2}$$

Hence for any given substrate concentration and percentage conversion the time necessary for an enzyme-catalysed reaction to reach completion can be calculated.

Π Assume that an end-point assay is to be used with an enzyme having a K_M for the analyte of 2×10^{-3} mol l^{-1}. Assuming also that the reaction is irreversible and that the maximal enzyme activity in the reaction mixture is 2 IU ml^{-1}, how long will it take for the reaction to go to 99% completion when 1) $[S_o] = 2 \times 10^{-4}$ mol l^{-1}, 2) $[S_o] = 4 \times 10^{-3}$ mol l^{-1}? (Note that 1 IU ml^{-1} = 1 μmol ml^{-1} min^{-1}).

(Attempt these on a sheet of paper before reading on).

We can use Equation 7.2, but you must adopt uniform units. V_{max} is replaced by the activity present in the reaction mixture.

Thus:

1) $$t = 4.6 \frac{K_M}{V_{max}} + \frac{[S_o]}{V_{max}}$$

$$= \frac{4.6 \times 2 \times 10^{-3} \text{ mol l}^{-1}}{2 \times 10^{-6} \text{ mol ml}^{-1} \text{ min}^{-1}} + \frac{2 \times 10^{-4} \text{ mol l}^{-1}}{2 \times 10^{-6} \text{ mol ml}^{-1} \text{ min}^{-1}}$$

$$= \frac{4.6 \times 2 \times 10^{-6} \text{ mol ml}^{-1}}{2 \times 10^{-6} \text{ mol ml}^{-1} \text{ min}^{-1}} + \frac{2 \times 10^{-7} \text{ mol ml}^{-1}}{2 \times 10^{-6} \text{ mol ml}^{-1} \text{ min}^{-1}}$$

$$= (4.6 + 0.1) \text{ min} = 4.7 \text{ min}$$

2) t is calculated in the same manner as in 1).

thus:

$$t = \frac{4.6 \, K_M}{V_{max}} + \frac{4 \times 10^{-6} \text{ mol ml}^{-1}}{2 \times 10^{-6} \text{ mol ml}^{-1} \text{ min}^{-1}}$$

$$t = (4.6 + 2) \text{ min} = 6.7 \text{ min.}$$

For reasons of accuracy, the preferred operating range for substrate concentration is low relative to the K_M.

Π **Looking at the result of the previous activity, what is notable about Equation 7.2 and the required reaction time when [S] << K_M?**

In case 1) above, when [S] was only one tenth of K_M, the contribution of the second term in Equation 7.2 was negligible. Thus if we arrange that analyses are conducted at low substrate concentrations relative to the K_M, (the reaction is first order), we may ignore the second term in Equation 7.1, which then simplifies to:

$$t = 2.3 \frac{K_M}{V_{max}} \log_{10} \frac{[S_o]}{[S_t]}$$

Equation 7.3

Providing that the analyte concentration is known to be low relative to the K_M (ie the reaction will be first order), this simplified form is perfectly adequate.

end-point methods better for coupled systems
End-point determination methods are also more amenable to coupled systems where the primary reaction cannot be monitored and therefore has to be coupled to a second or detecting enzyme (see Section 7.4.3).

$$S \xrightarrow[\text{primary reaction}]{\text{primary enzyme}} P_1 \xrightarrow[\text{detecting reaction}]{\text{detecting enzyme}} P_2$$

The main disadvantages of the end-point determination method are the time involved in reaching reaction completion and the need to use extremely pure enzymes and cofactors in order to avoid non-specific side reactions.

A list of the advantages and disadvantages of the kinetic and equilibrium methods is given in Table 7.2. This is quite an extensive list; read through it carefully. You will notice that some of the comments refer to the accuracy and sensitivity of the two approaches. Others refer to factors such as the cost of equipment, reagents and time required. In practice the choice is greatly influenced by the nature and concentration of the substrate that is to be assayed, the availability of suitable monitoring (measuring) opportunities and the K_M of the available enzyme(s).

	Equilibrium method		Kinetic method	
	Rating	Comments	Rating	Comments
Change followed, (eg A)	+++	Relatively large and accurate measurement. More sensitive	++	Very small but within ability of sophisticated kinetic analysers. Electronic noise can be a problem
Change in pH	+	May alter K_{eq} but unlikely to cause error unless K_{eq} is small	-	Any change from optimun pH may cause low values - especially if more than one enzyme is used in the assay
Change in T	+++	Will alter K_{eq} but no serious error unless K_{eq} is small and T change is several degrees	+	Good temperature control ($\pm 0.1\%$) is essential
Presence of inhibitors	+	Lengthened assay time, check reaction terminated	-	Low values
Presence of activators	+++	Equilibrium reached more quickly	-	High values
Influence of K_M	+	Low K_M reduces times required for assay	+	The higher the K_M the higher S_o that can be assayed
Maximum [S_o]	+	$S_o < K_M$	+	$S_o < K_M$
Linearity with [S_o]	+	Linear but may get deviation at low K_{eq}		Nonlinear, therefore calibration curve needed
K_{eq}	+	Problems with low K_{eq} of solubility, inhibition or background absorbance	+	With low K_{eq} assay may be too insensitive
Time required	+	Several minutes	+++	Generally < 1 minute
Cost of enzyme	-	Relatively high to complete assay in reasonable time	+	Relatively low. Require sufficient amount to give adequate signal to noise ratio
Cost of cofactors	-	Can be higher than cost of enzyme especially if K_{eq} is low	+	Relatively low. Concentration must not be rate-limiting
Product formation	+	May cause inhibition or reversal of reaction	+++	Too little to cause effect
Unstable product	+	Cannot measure product	+	No problems if assay is rapid
Turbid or coloured sample	+	Blank required	+++	Should not affect reaction rate
Equipment	+++	Sophisticated equipment not required but dedicated equipment being produced	+	Requires sophisticated equipment but this may be used for other analyses too

Table 7.2 Comparison of equilibrium and kinetic methods for the assay of substrates. Table adapted from Handbook of Enzyme Technology (1985) Wiseman, A. (Ed). Second Edition, Ellis Horwood Limited, Chichester, England. Table 5.2 Page 215. Key: A = absorbance, T = temperature, K_M = the Michaelis constant for the substrate, S_0 = initial substrate concentration, K_{eq} = the equilibrium constant for the reaction.

SAQ 7.4

An enzyme whose K_M is 0.4×10^{-4} mol l^{-1} is to be used to estimate the concentration of its substrate. The reaction may be readily assayed by absorbance changes and a continuous assay is available. Comment on the suitability of kinetic and end-point methods to estimate the substrate concentrations when the concentration in the reaction mixture is expected to be:

1) 0.1×10^{-4} mol l^{-1}.

2) 8×10^{-4} mol l^{-1}.

Think particularly about the likely accuracy of the determination.

SAQ 7.5

Assume that you are setting up an end-point assay using an enzyme whose K_M for the analyte is 4×10^{-3} mol l^{-1}. The reaction may be assumed to proceed under first order kinetics and to be irreversible. It is important that the reaction is completed within 5 minutes.

1) If 99% completion is accepted, will the addition of sufficient enzyme to give a maximum activity of 6 IU ml^{-1} in the reaction mixture be sufficient, if the substrate concentration is 10^{-6} mol ml^{-1}, to complete the reaction within 5 minutes?

2) What is the minimum activity of enzyme in the reaction mixture which would give 99% completion within 5 minutes?

7.5.3 Assays involving multiple enzymes

Many enzyme-catalysed reactions are not easy to monitor because their substrates or products have no distinctive, and easy to measure properties. For example, the enzyme maltase (an α-glucosidase) hydrolyses maltose to form glucose. Both substrate and product are reducing sugars and have similar reactivities with chromogenic reagents.

maltose →(maltase, + H$_2$O) 2 glucose

The use of maltase to assay maltose is, by itself, therefore rather worthless. However the product of the reaction can be assayed enzymatically, for example by using glucose oxidase or hexokinase and glucose-6-phosphate dehydrogenase.

Let us draw a reaction scheme for this latter situation.

Use of enzymes in analysis

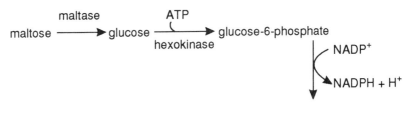

☐ What is the easiest way of monitoring this sequence of reactions?

The most obvious and simplest way of monitoring this sequence of reactions is to measure the increase in absorption at 340 nm (remember NADP$^+$ does not absorb at this wavelength whilst the reduced form, NADPH, absorbs strongly at this wavelength).

☐ Using our model of maltose and maltase as an example, design coupled-enzyme assay schemes for measuring: a) alanine; b) glycerol; c) cellobiose using a selection of enzymes from those available below. In each case you should design your system such that the reaction can be monitored by measuring its absorption at 340 nm.

Enzymes

lactate dehydrogenase
  pyruvate + NADH + H$^+$ ⇌ L-lactate + NAD$^+$

glycerol kinase
  glycerol + ATP ⇌ glycerol-3-phosphate + ADP

transaminase
  alanine + α-ketoglutarate ⇌ pyruvate + glutamate

hexokinase
  glucose + ATP ⇌ glucose-6-phosphate + ADP

β-glucosidase
  cellobiose + H$_2$O ⇌ 2 glucose

pyruvate kinase
  ADP + phosphoenol pyruvate ⇌ ATP + pyruvate

glucose-6-phosphate dehydrogenase
  glucose-6-Ⓟ + NADP$^+$ ⇌ 6-phosphoglucono-δ-lactone + NADPH + H$^+$

This exercise has been designed to enable you to see how combinations of enzymes might be used to produce systems which can be applied to assay a variety of metabolites. In many cases, more than one option is possible and the final choice will depend upon factors such as the cost and stability of the enzymes and their requirement of cofactors.

For alanine 1) we would suggest a combination of transaminase and lactate dehydrogenase might provide a suitable assay system. Thus:

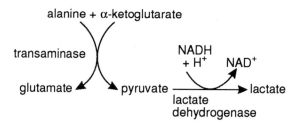

Thus in this instance, we would incubate the alanine in the presence of α-ketoglutarate, NADH and the two enzymes and monitor the decrease in absorption at 340 nm.

For glycerol 2), the situation is a little more complex. Given the enzymes in our list, the combination we would suggest is, glycerol kinase, pyruvate kinase and lactate dehydrogenase. Thus the reaction sequence would be:

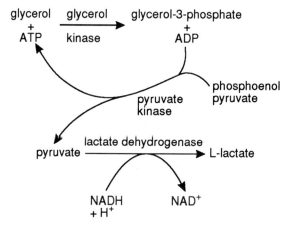

In this case we would need to add the three enzymes, ATP, phosphoenolpyruvate and NADH. The amount of NADH oxidised depends upon the amount of ADP generated from the phosphorylation of glycerol and thus, indirectly, NADH oxidation reflects the amount of glycerol present.

For cellobiose, we would suggest a combination of β-glucosidase, hexokinase and glucose-6-phosphate dehydrogenase. Thus:

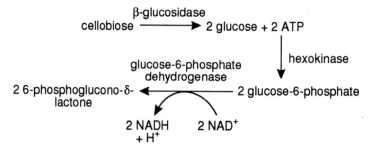

This exercise gives some idea of the flexibility of using enzymes for analyses. Nevertheless there are some important limitations as to which enzymes will work together. For example, the enzymes need to have similar pH optima (or at least there must be a pH value at which both enzymes display a certain degree of activity).

Use of enzymes in analysis

We must also bear in mind that the more enzymes we use, the greater is the probability of encountering difficulties in interpreting data. For example, in the assay of cellobiose described above, the presence of other β-glucosides, glucose and glucose-6-phosphate in samples would interfere with the assay. There are also three enzymes present in the assay system and each of them might be susceptible to inhibition by factors within the samples.

> ∏ Are assay systems using multiple enzymes most suitable for kinetic analysis or a end-point?

In principle, they can be used in either way although, as we will describe below, some difficulties may be encountered in interpreting kinetic data.

Consider the reaction scheme.

$$A \xrightarrow[k_1]{\text{enzyme 1}} B \xrightarrow[k_2]{\text{enzyme 2}} C$$

complex kinetics of coupled reactions

If we use substrate A at low concentration (ie well below the K_M of enzyme 1), then this reaction will proceed according to first order kinetics. [B], initially at 0, will also be converted to C by first order kinetics unless [B] rises to values approaching the K_M for enzyme 2. There is, therefore, a rather complex relationship since the reaction rate for step 2 will at first rise as [B] rises. But the rate of formation of B declines with time since [A] also declines with time (according to $[A_t] = [A_o]e^{-k_1 t}$). Thus if enzymes 1 and 2 were present in similar amounts and they had similar K_M values, we would anticipate that the rate of formation of C would rise steadily for a while and then decline. The precise nature of these changes depends on the exact ratio of the two enzymes and on their respective K_M values. In practice, this problem may be reduced by ensuring that either the second enzyme is present in vast excess compared to the primary enzyme or that its K_M is much lower than that of the primary enzyme.

By doing this, we have produced a system which 'scavenges' B as it is formed. Thus, although there will be a brief delay at the beginning of the reaction, the subsequent rate of the reaction B → C is the same as that for A → B.

We will meet several examples of coupled reaction systems which measure the kinetics of the reactions in later sections. However, analyses using coupled enzymes are also commonly employed for end-point (equilibrium) analysis. The same criteria apply to these systems as to analysis based on the use of single enzymes. Sufficient time must be allowed for the reaction(s) to go to completion.

7.6 The use of enzymes to assay activators and inhibitors

assay of factors which affect enzymatic reaction kinetics

So far we have discussed the use of enzymes to assay the amounts of substrates. We can, however, also use enzymes to measure the concentrations of compounds which modify the activities of enzymes. For example, we could determine the concentration of magnesium in a sample using enzymes which are dependent upon magnesium ions. A good example of such an enzyme is isocitrate dehydrogenase. This enzyme can be readily assayed using the increase in absorption at 340 nm resulting from the reduction of the pyridine nucleotide co-substrate. Similarly, many insecticides can be assayed by their inhibitory effects on acetyl cholinesterase or carbonic anhydrase.

For assays depending on activation or inhibition of an enzyme, we first set up a calibration curve in which the rate of reaction is assayed in the presence of known concentrations of the test agent. Then we carry out the reaction, under identical conditions, in the presence of the sample to be assayed. You should note that in many instances, a non-linear relationship exists between inhibitor (activator) concentration and the extent of inhibition (activation) of the reaction. For example, inhibition of acetyl cholinesterase by the insecticide parathion is a direct function of log [parathion] rather than of [parathion]. In this case a plot of % inhibition against log [parathion] gives a linear plot and is much more useable as a calibration curve.

SAQ 7.6

Below are data concerning the inhibition of an enzyme in the presence of an inhibitor. From this data, determine the concentration of the inhibitor in sample X.

Inhibitor concentration (μmol ml^{-1})	Activity of the enzyme (μmol min^{-1} ml^{-1})
0	1.7
1	1.46
2	1.22
4	0.98
8	0.74
16	0.50
X	1.054

7.7 Formats in which enzymes may be used in analysis

There are numerous formats in which enzymes may be employed as analytical tools. In the previous sections, we have discussed their use in solution to assay metabolites, enzyme activators and inhibitors. This is not the only format in which we can use enzymes for analysis. Other ways include their immobilisation onto supports in the form of analytical 'dipstick' tests, their use in conjunction with electrodes (biosensors) and as 'labels' to enable quantification of other processes. A good example of the latter group is in the use of enzymes linked to antibodies for use in enzyme-linked immunosorbent assays (ELISA) and enzyme multiplied immunoassays (EMIT). In this and the following sections we will discuss dipstick tests, ELISA and EMIT. The design and performance of biosensors will be dealt with in the next chapter.

7.8 Diagnostic dipstick tests

∏ What are the advantageous features of diagnostic dipsticks?

clinical use of dipsticks

The development of diagnostic dipsticks and dry reagent chemistry has had a dramatic effect on the world of clinical analysis. These easy to use tests have enabled patients to monitor their own blood or urine for the presence of a variety of substances. Probably

use of dipsticks is by patients suffering from diabetes. These people
...onitor their own blood or urine for the presence of glucose and to
...ereby controlling their condition. The ease with which these tests can
...t they can perform these tests at home and therefore avoid frequent
...centres. Dipsticks have also found a useful role in the doctors'
...ity to test for increased or decreased concentrations of a particular
...mple glucose, cholesterol) in a patient's fluids can give an early
...diagnosis of a patient's condition. Previously doctors had to send
...ir patients to clinical laboratories. Now they are able to carry out
...ir surgeries using dipsticks. This has many obvious advantages to
...ice and to patients.

...ped for the testing of glucose concentrations are probably one of
...used methods of dry reagent technology and therefore will be
...detail here. A typical dipstick contains a small cellulose fibre pad
...nto it two enzymes, namely glucose oxidase and peroxidase. In
...which is chromogenic, such as 3,3'5,5'-tetramethyl benzidine
...pad is covered by a film of ethyl cellulose which prohibits large
..., which might interfere with the reaction from reaching the pad
...'ecules such as glucose to permeate through to the pad

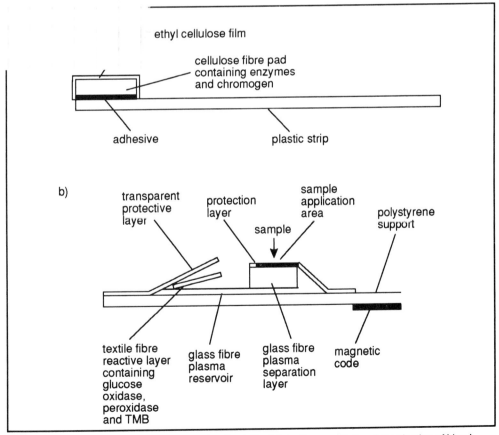

Figure 7.2 a) Simple diagnostic dipstick. b) A refined dipstick for the quantitative determination of blood glucose (see text). Redrawn from Kricka, L.J. and Thorpe, G.H.G. (1986), Trends in Biotechnology, Volume 4, No. 10 [33], Pages 253-258. Figure 3. Page 254.

A drop of blood or urine is placed on the dipstick and the following reaction occurs when glucose is present in the sample.

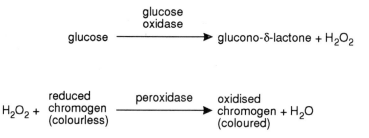

The test is semi-quantitative and the amount of colour produced by the chromogenic reagent after a fixed time can be matched to a colour reference chart supplied with the dipsticks. Thus a visual estimation of the amount of glucose present in the sample is obtained. The basic principle of the glucose dipstick test is also used in other enzyme-based dipstick tests.

More recent developments in dipstick dry reagent technology now allows quantitative determinations of blood glucose concentrations. This has been brought about by combining the use of reflectance photometry with the traditional enzyme-containing dipstick (Figure 7.2b).

The blood sample is placed on the sample application area. The plasma passes through a glass fibre separation layer and enters a reservoir in contact with a reactive layer made of textile fibre. The plasma, containing glucose, rises into this fibre by capillary action. The resulting colour developed in the fibre is measured by reflectance photometry through the transparent protective layer.

7.9 Enzyme-linked immunosorbent assay (ELISA)

limitations of traditional methods for detecting antibody reactions

Antibodies have long been used to detect and measure specific molecules, especially macromolecules (for example specific proteins, complex carbohydrates). A difficulty often encountered in these methods centres on the quantification of the interaction between the antibody and its target compound. Various techniques to achieve this do exist (for example immune precipitation and radioactive immune assays) but many of these have disadvantages (for example lack of sensitivity, lack of quantitative precision, use of hazardous and expensive radioisotopes). More recently, especially since the development of monoclonal antibodies, enzyme-linked immunosorbent assays have taken over centre stage and have become the method of choice.

The ELISA techniques employ the sensitivity of a spectrophotometric assay of an enzyme and the specificity of antibodies. For this purpose, a stable enzyme which can be easily measured and which has a high turnover number is covalently linked to the antibody (or to the antigen). Thus, when an antibody binds with its target molecule, it carries with it, its enzyme 'label'. By measuring the amount of enzyme associated with the target molecule, we obtain a measure of the amount of the target molecule present. The three methods that are used most frequently for the performance of an ELISA are:

- the competitive method;

Use of enzymes in analysis

- the double antibody method;
- the indirect method.

The assays are routinely conducted in plates containing 96 wells thereby permitting extensive replication of both standards and test samples. The overall procedure, including spectrophotometric measurements, is easily automated, thereby enabling large numbers of samples to be rapidly analysed.

The techniques for raising antibodies and attaching enzymes to them are described in the BIOTOL text 'Technological Applications of Immunochemicals'. Here we will examine the techniques using these reagents.

7.9.1 The competitive ELISA technique

use of enzyme 'labelled' antigen

The competitive ELISA technique is schematically described in Figure 7.3. The antigen with a covalently-linked enzyme is allowed to bind to antibodies in the presence of known amounts of the antigen which is not linked to the enzyme. A reference curve relating the amount of enzyme which becomes associated with the antibody to the concentration of unlabelled antigen is prepared. This can then be used to determine the amount of antigen in test samples.

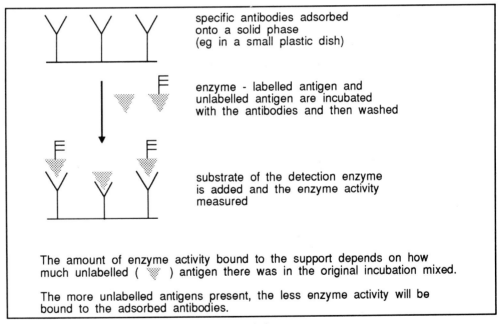

Figure 7.3 The principles of the competitive ELISA technique.

A major disadvantage of this technique is that it requires us to prepare antigen coupled to enzyme for each antigen. This may be quite difficult to achieve especially as we need to link the detection enzyme to each antigen in a (more or less) specific manner. This difficulty may be overcome by using the double antibody and indirect ELISA techniques described below.

7.9.2 The double antibody ELISA technique

This technique is shown, in schematic form, in Figure 7.4. Antigen is added to immobilised antibody. After washing, enzyme-labelled antibody against the same antigen is added. We can then measure the amount of enzyme adhering to the complex. A reference curve is prepared by adding known quantities of antigen in the second step of the procedure.

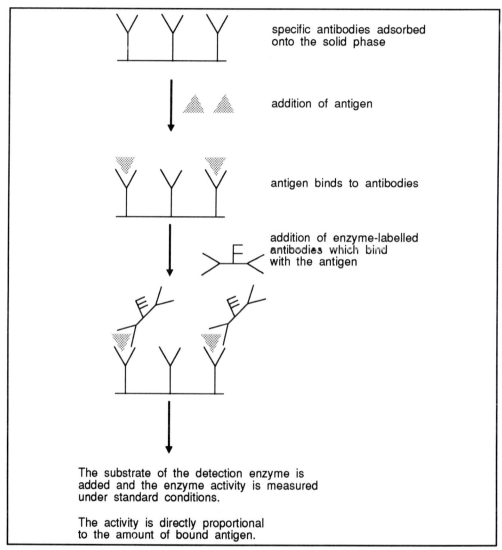

Figure 7.4 The double antibody ELISA technique.

7.9.3 The indirect ELISA technique

We have illustrated this technique in Figure 7.5. Use this figure to follow the description given below.

Use of enzymes in analysis

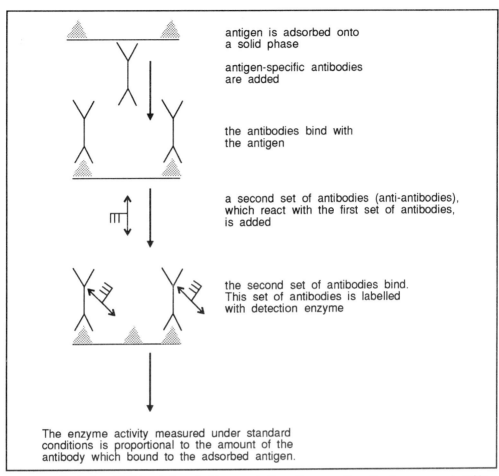

Figure 7.5 The indirect ELISA technique. In this example the antigen-specific antibodies were produced in rabbits. The anti-antibodies were produced in goats by immunising goats with rabbit antibodies (see text for a description).

use of anti-antibodies

In this procedure, the antigen is adsorbed onto the surface of a support. This is usually done at high pH (approx 10-11). The pH is then neutralised and the antibody added. After incubation, the non-bound proteins are removed by washing. Then antibodies which will bind with the first antibodies (so-called anti-antibodies) to which the detection enzyme is coupled are added. After washing, the enzyme activity attached to the complex is measured. Reference curves are prepared by using known quantities of the antigen in the first step. Rabbit antibodies are often employed as the first antibodies and antibodies generated in goats against rabbit antibodies are used as the second antibodies. It is to these that the detection enzyme is coupled.

∏ What is the main advantage of the indirect ELISA technique?

If your knowledge of immunology is limited, you may have had difficulty answering this. The important point is that the enzyme-linked goat antibodies will bind with rabbit antibodies irrespective of the specificity of the rabbit antibodies. Thus we only need to go through the laborious process of linking the enzyme to the antibody once and then we can use this in the assay of many different antigens.

7.9.4 Practical aspects of the ELISA technique

We will focus onto two issues, the nature of the enzyme used and the range of solid supports used.

The enzymes used in ELISA

Π What sorts of enzymes are used in ELISA? List some examples.

There is a wide choice but the key factors are that the enzyme must be stable, easy to assay, with a high turnover number and a substrate whose resulting product has a high extinction coefficient. Examples of such enzymes include dehydrogenases which can be measured by the increase in absorption at 340 nm which accompanies $NAD(P)^+$ reduction (eg glucose-6-phosphate dehydrogenase; alcohol dehydrogenase). Also enzymes which convert chromogenic substrates are employed. For example alkaline phosphatase will hydrolyse p-nitrophenyl phosphate to release the yellow product p-nitrophenol which absorbs strongly at 405 nm. Similarly peroxidase, which oxidises o-phenylene-diamine to nitrobenzene (absorbs strongly at 495 nm), may be used.

Margin notes: dehydrogenase, phosphatase, peroxidase

Solid phase

Many different solid supports have been used including cross-linked dextrans, polyacrylamide beads, filter paper, nitrocellulose filters, polypropylene tubes and polystyrene microtitre plates. The use of microtitre plates has become particularly common. Usually physical adsorption is sufficient but covalent coupling has also been employed. After adsorption, a mild non-ionogenic detergent is added to avoid further non-specific adsorption. The non-specifically bound protein is washed away after each step.

7.9.5 Application of ELISA

The ELISA technique can be applied to virtually all immunological assays and has largely replaced radioimmune assays (RIA) which employs the use of radioactively labelled antibodies or antigens.

ELISAs find particular use in clinical biochemistry for the determination of blood groups, hormones (eg insulin, oestrogen) and the detection of pathogens. They are also employed in the detection of plant and animal viruses. Increasingly they are finding use as a means of quality control. For example, they may be used to detect foreign/contaminating proteins and toxins in food. An example of such use is in the detection of the adulteration of beef products by soya protein or horse meat.

7.10 Enzyme multiplied immunoassay technique EMIT

We will conclude this chapter on the use of enzymes for analytical purposes by discussing the principles of the enzyme multiplied immunoassay technique (EMIT).

Use of enzymes in analysis

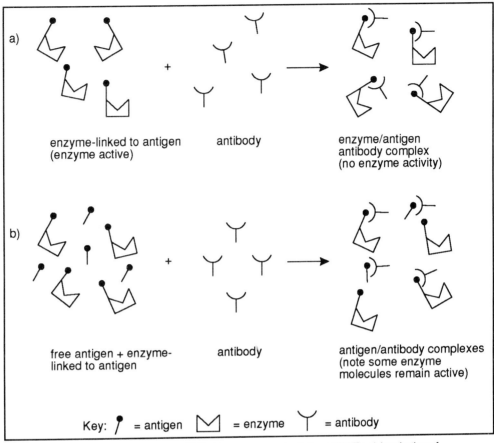

Figure 7.6 Illustration of the enzyme multiplied immunoassay technique (EMIT). a) Incubation of enzyme-linked antigen with antibody leads to the inactivation of the enzyme. b) Incubation in the presence of free antigen leaves active enzyme in solution.

The principle of EMIT is based upon enzyme inhibition and is illustrated in Figure 7.6. An enzyme/antigen complex is prepared in such a way that when the antibody, which was raised against the antigen, is added it binds to the enzyme/antigen complex and in doing so inhibits enzyme activity. Free antigen needs to be present in order to retain the enzyme activity. A series of determinations using standard amounts of enzyme/antigen complex and antibody are set up. Samples containing known quantities of the antigen, are added and the free antigen compete for the antibody, preventing it binding to the enzyme/antigen complex, and so preventing enzyme inhibition occurring. The resulting enzyme activity can then be measured. A standard curve can be constructed from these assays. Then the sample with the unknown concentration of antigen can be tested and the resulting enzyme activity can be related to the unknown antigen concentration using the standard curve.

inhibition of the enzyme prevented by free antigen

This technique finds particular use in detecting small amounts of low molecular weight antigens. For example, in the detection of drug breakdown products in serum samples.

The key features required for the success of EMIT are, like those for ELISA, the specificity of the antibody-antigen interaction, the stability of the enzyme used and its ease of detection.

SAQ 7.7 What features of ELISA and EMIT makes them particularly sensitive and thereby enable us to measure very small amounts of antigens?

SAQ 7.8

1) In an EMIT assay of compound Y, the following data were obtained.

Concentration of unlinked Y used in the assay (nmol ml^{-1})	Enzyme activity remaining after incubation with antibodies (μmol min^{-1} ml^{-1})
0	2.0
20	20.6
10	15.4
5	10.3
2.5	5.2
1.25	3.8

When a sample containing an unknown amount of compound Y was used, the activity remaining after incubation with antibody was 13.8 μmol min^{-1} ml^{-1}.

What is the concentration of Y in the sample?

2) Explain what would happen to the calibration curve prepared in 1) if:

a) twice as much antigen/enzyme complex was used in each reaction mixture;

b) twice as much antibody was used in each reaction mixture.

Summary and objectives

In this chapter, we have outlined the principles involved in the use of enzymes in analysis. We began by explaining why they are useful and listed several examples. We then discussed the strategies applied in using enzymes to assay substrates with particular emphasis on kinetic and 'end-point' methods. We also examined the use of coupled enzyme systems and the use of enzymes to assay compounds which modify the activities of enzyme. We also explained the diagnostic use of enzymes using 'dipsticks'. In the final part of the chapter, we explained how the high specificity of antibodies may be combined with the activities of enzymes to produce potent and sensitive assays for a very wide range of compounds. These techniques (ELISA, EMIT) are particularly valuable for the assay of molecules for which other specific and sensitive assays are not available. These are of particular value in clinical laboratories and in quality control.

Now that you have completed this chapter you should be able to:

- explain why enzymes are useful in analytical procedures;

- explain the differences between kinetic and end-point (or equilibrium) methods used to measure metabolites;

- select suitable enzymes or combinations of enzymes to quantify particular compounds;

- determine the concentrations of compounds using data derived from enzyme-based analysis;

- describe the basic design of diagnostic dipsticks using a specific example;

- explain how enzyme-linked immunosorbent assay (ELISA) and enzyme multiplied immunoassay (EMIT) work and interpret data from such assays.

Biosensors

8.1 Introduction	210
8.2 Generation of an electronic signal	211
8.3 Amperometric enzyme biosensors	212
8.4 Potentiometric enzyme biosensors	220
8.5 Conductimetric enzyme biosensors	223
8.6 Optical enzyme biosensors	223
8.7 Thermal enzyme biosensors	228
8.8 Enzymes in affinity biosensors	230
8.9 Microbial biosensors	240
8.10 Tissue section biosensors	241
8.11 Manufacturing technologies	242
Summary and objectives	246

Biosensors

8.1 Introduction

In the preceding chapters, we have examined the uses of enzymes in large scale industrial processes and their uses as reagents for detecting and measuring a variety of molecular species. In this chapter, we extend this discussion by describing the use of enzymes in a variety of devices collectively referred to as biosensors. For completeness, we will also discuss the design and use of other bio-molecules in biosensors. Note that the term 'enzyme electrode' is a widely used term to describe an enzyme-containing biosensor but would not cover a biosensor involving, for example, nucleic acid detection.

enzyme electrode as a type of biosensor

The field of biosensor technology holds many exciting biotechnological applications for enzymes. It is also a very diverse field, making use of expertise in biochemistry, immunology, optical physics, electrochemistry, materials science, mathematics, computer science and a myriad of other science and engineering disciplines.

We are concerned here with the application of enzymes. Therefore the chapter concentrates on enzyme-based biosensors. However, it is important to address the 'non-biological' aspects of this technology in order to adequately describe what a biosensor actually is. Therefore, the first section in this chapter summarises the types of techniques used to generate an electronic signal. We do realise, however, that if you have little background knowledge of electronics you may find some of the concepts described in these sections rather difficult. Following this, a range of different biosensor types are discussed, arranged broadly in terms of the type of electronic transduction employed. These transduction methodologies are described in more detail in the relevant enzyme biosensor sections. The final section reviews some of the methods used to manufacture biosensors.

biological signals are transduced into electronic signals

So first of all, what actually is a biosensor? Biosensors are a comparatively recent generation of analytical devices in which a biological component is coupled with a transducer to convert a biological signal into an electronic one as shown in the generalised scheme in Figure 8.1. Biological molecules have been used for nearly thirty years to improve the specificity of sensors, and microprocessors can be used to rapidly and accurately process the electronic signal to provide the analytical result. The biological component may be either an enzyme, an antibody or antigen, a nucleic acid probe, a biological receptor molecule, a whole cell or a tissue slice. However, enzymes are by far the most commonly used and most successful biological component applied in biosensors to date. Before moving on to the biological components we will address the question, how can electronic signals be generated?

Biosensors

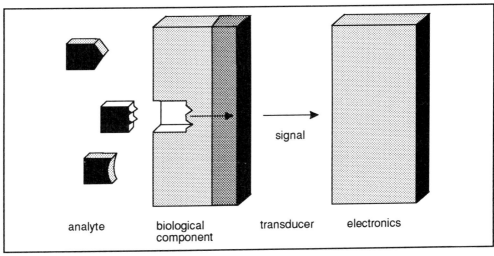

Figure 8.1 A generalised scheme for a biosensor.

8.2 Generation of an electronic signal

A transducer is used in a biosensor to convert a specific biological event (catalytic or binding) into an electronic response that can be displayed directly, or can be further processed by a microprocessor. In order to be successful, this transducer must also be amenable to the immobilisation of the biological component with which it is in intimate contact. Many transducers have been used in the construction of biosensors and some of the most common ones are summarised in Table 8.1. Details of most of these transduction techniques are given, as appropriate, in the following sections.

Transducer	Examples
Electrochemical	
a) Amperometric	Clark oxygen electrode, mediated electrode systems.
b) Potentiometric	Redox electrodes, ion-selective electrodes, field effect transistors, light addressable potentiometric sensors.
c) Conductimetric	Platinum or gold electrodes for the measurement of change in conductivity of the solution due to the generation of ions.
Optical	Photodiodes, waveguide systems, integrated optical sensors.
Acoustic	Piezoelectric crystals, surface acoustic wave devices.
Calorimetric	Thermistor or thermopile.

Table 8.1 Transducers commonly used in biosensors.

Having mentioned some of the ways in which an electronic signal can be generated, let us now look at the biological constituents of biosensors. The range of potential biological molecules, whole cells or even tissues which can be integrated with a transducer to form a biosensor is very large. Enzymes are by far the most commonly used biological components applied in biosensors. The transduction technique favoured in most enzyme-based biosensors is electrochemical transduction, and amperometry the most common configuration. This biosensor type will be considered first.

8.3 Amperometric enzyme biosensors

8.3.1 Amperometric transduction

Amperometry is the measurement of current at a fixed potential. This is classically achieved using a three-electrode system. Here a working (or sensing) electrode is held at a constant potential with respect to a reference electrode, utilising a third electrode, the counter (or auxiliary) electrode, to maintain the potential between the working and reference electrodes by varying its own potential. With this system, very little (effectively zero) current is drawn through the reference electrode.

In many practical devices, the counter and reference electrodes are combined using a chloridised silver electrode. The reference potential tends to drift in this format due to the large current passing through the reference electrode. However, when used only for short periods, such as in single-use, disposable electrodes, these convenient, two-electrode systems have proved to be very successful.

oxidation or reduction at the analyte of the working electrode

The applied potential is chosen so that the species of interest is either oxidised or reduced at the working electrode. This causes a transfer of electrons which results in a current directly proportional to the concentration of analyte at the electrode surface over a wide dynamic range. Since the electroactive species is consumed at the electrode, the signal is dependent on the rate of mass transfer to the electrode surface.

Π What factors influence the rate of transfer of the analyte to the electrode surface?

We can represent the situation in the following way:

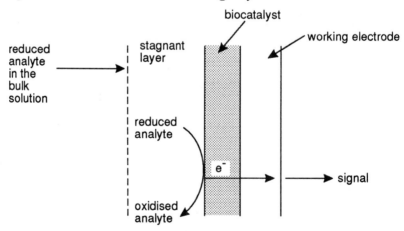

The factors which influence the rate of transfer of the analyte to the surface are:

- the concentration of analyte in the bulk liquid;

- the thickness of the stagnant layer (depends on the mixing of the bulk liquid);

- the properties of the liquid (for example its viscosity);

- the properties of the analyte (for example its molecular mass) which influence its rate of diffusion;

- the temperature of the liquid (which influences diffusion rates).

For any single type of analyte, providing other factors such as mixing rate, viscosity of the medium and temperatures are kept constant, the rate of mass transfer of the analyte should be related simply to the concentration of the analyte in the bulk solution.

Π What factors other than mass transfer rates may influence the size of the signal produced in such a system?

Obviously the activity of the biocatalyst is important. If it was completely denatured, the signal would fall to zero. Except under conditions in which the biocatalyst is present in vast excess, the signal is dependent upon the enzyme reaction kinetics. Also important are the kinetics of the electrode (eg efficiency of transfer of electrons to the working electrode).

optimisation of operating potential is important

Careful optimisation of the operating potential of the working electrode can offer two major advantages. Firstly, it can minimise the effect of small fluctuations in the reference potential on the response and, secondly, it can reduce interference from other components in the sample.

electrodes are usually made of inert metals

Electrodes are commonly made of inert metals, such as platinum, gold or carbon. Carbon electrodes can be either in the form of graphite, glassy carbon or pyrolytic graphite and may be used as a solid or as a paste. These materials are amenable to many advanced manufacturing technologies, allowing inexpensive mass production of amperometric sensors.

A variation on the technique of amperometry is coulometric transduction. Both techniques operate at a constant potential, but coulometry involves the measurement of total charge transfer rather than current. It is therefore an absolute technique which requires no calibration, since the total charge passed is independent of the electrode and of the enzyme reaction kinetics, which may change with time. In order to consume all of an electroactive species generated by an analytical reaction in a reasonable time, it is necessary to carry out the assay in a small volume.

Faraday's Law

The response follows the simple relation defined by Faraday's Law:

$N = Q/nF$

where:

N = number of moles of reactant
Q = total charge passed (coulombs)
n = number of electrons involved in reaction
F = Faraday constant (96 487 coulombs mol^{-1})

The instrumentation required for amperometric analyses has been in routine use for many years. It is inexpensive, simple and sensitive over a very wide range.

SAQ 8.1

1) Consider the following reaction.

$$X_{reduced} \rightarrow X_{oxidised} + 2e^- + 2H^+$$

What will be the total charge passed (in coulombs), if 1 μmole of X is oxidised?

2) In a similar system to that described in 1), the oxidation of $X_{reduced}$ resulted in 77.2 coulombs being passed. How much of $X_{reduced}$ was oxidised?

3) If the $X_{reduced}$ that was oxidised represented all of $X_{reduced}$ in 2 ml of test solution, what is the concentration of $X_{reduced}$ in the test solution (in mol l^{-1}).

Amperometric enzyme electrodes

The first true biosensor was described in 1962 by Clark. An oxygen electrode was used to monitor glucose as it was converted to gluconic acid by glucose oxidase (GOD):

$$\text{glucose} + O_2 \xrightarrow{\text{glucose oxidase}} \text{gluconic acid} + H_2O_2$$

Glucose concentration can be monitored by either measuring the fall in O_2 tension (eg the 'Clark oxygen electrode') or the production of H_2O_2.

double membrane structure to exclude interfering materials

The glucose monitors of the Yellow Springs Instrument company (YSI) utilise this latter approach. Hydrogen peroxide is measured amperometrically by oxidation at 700mV on a platinum electrode. An outer polycarbonate membrane is used to immobilise the GOD close to the electrode surface. This membrane also has a pore size which excludes cells and large molecules from the electrode, whilst allowing the passage of smaller molecules such as glucose. An inner acetate membrane allows the passage of hydrogen peroxide, oxygen, salts and water, but excludes glucose, ascorbic acid and other potentially interfering compounds. The physical organisation of the Yellow Springs Instrument is therefore:

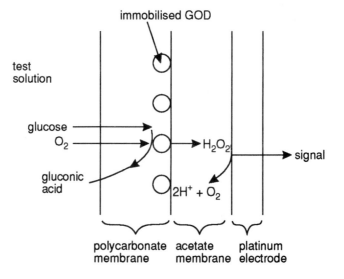

Effectively, this system uses oxygen to transfer electrons from the substrate (glucose) to the electrode. The construction is simple, but is less suited to mass production than other types of biosensor.

glucose sensors used for glycaemic control in diabetes, and for assessing blood glucose-lowering therapeutics

More recently, monitors have been introduced by YSI for the simultaneous measurement of glucose and other analytes utilising similar technology. These units are easily adapted for the measurement of sucrose and starch. The Glucosensor from the company Unitec is the first commercially available portable sensor for the continuous monitoring of blood glucose. Over a twenty-four hour period, 15 to 24ml of blood are continuously withdrawn for measurement. A data storage capacity of thirty-two kilobytes is built into the unit which enables between 3180 and 15 900 glucose values to be monitored over a period of up to 256 hours. In this way, long term glucograms can be obtained under near normal conditions. Advantages of this system include improved glycaemic control of severe insulin-dependent diabetic patients and the realistic assessment of the effect of blood glucose-lowering therapeutics. The amperometric enzyme electrode is connected to a wick which is implanted to equilibrate with the subcutaneous fluid, which can be related to the blood glucose level. The unit weighs 850g and measures 15 x 19 x 7cm.

Despite the success of the above types of amperometric enzyme electrodes, they do suffer from a number of drawbacks. The electrode response is highly dependent upon the initial oxygen concentration and is also subject to the influence of pH which affects the monitoring of hydrogen peroxide. Furthermore, there is a high possibility of interference from other electroactive species in solution, due to the relatively high potentials required to monitor either hydrogen peroxide oxidation or oxygen reduction (about + or - 650mV versus the saturated calomel electrode respectively).

8.3.2 Mediated amperometric enzyme electrodes

One way to avoid the oxygen concentration-dependence is to devise methods to circumvent the necessity to transfer electrons to oxygen.

One possibility which has been investigated is to transfer electrons directly from the prosthetic group of the enzyme to the sensing electrode. Unfortunately, satisfactory direct electron transfer has proved extremely difficult to achieve for all but a few proteins. The major stumbling block to this approach has been the fact that redox enzymes do not undergo oxidation and reduction reactions at sufficiently rapid rates to make practical devices. This is probably due to the fact that direct tunnelling from the active site of the enzyme to the electrode surface is highly unfavourable due to the large distances involved. The probability of electron transfer occurring between two redox centres decreases exponentially with distance.

A solution, which has proved extremely successful, is to use electron transfer mediator molecules to shuttle electrons from the redox centre of the enzyme to the electrode:

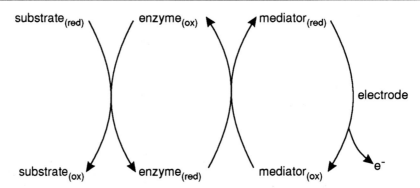

This type of system avoids the involvement of oxygen entirely. A range of these so-called 'mediated amperometric enzyme electrodes' have been developed and a number of commercial products are based on this technology. Some of the commonly used mediators are summarised in Table 8.2.

Π Examine this table carefully, you will notice that the operating potentials used are lower than with oxygen. See if you can explain the advantage of this.

Mediator	Operating potential (mV) (versus standard calomel electrode)
Benzoquinone	350
1,1' Dimethyl ferrocene	100
Ferrocene	181
Ferrocene 1,1'dicarboxylic acid	403
Ferrocene monocarboxylic acid	289
Potassium ferricyanide	450
Tetracyanoquinodimethane	-45
Tetrathiafulvalene	160

Table 8.2 Some commonly used electrochemical mediators.

The working potential of such amperometric enzyme electrodes is determined by the potential of the mediator couple used. A low potential reduces the chance of interference by the oxidation or reduction of analytes other than those of interest. More significantly, provided the mediator is unreactive with oxygen, the measurement is largely independent of the oxygen partial pressure, since oxygen is no longer required as an electron acceptor for the oxidase.

Π See if you can list the properties that a mediator should process to make it useful in biosensors. Then check with our list given below.

In order to be successful, a mediator should ideally possess the following properties:

- rapid electron transfer with redox enzymes;
- reversible electrochemistry;

Biosensors

- reduced form not reactive with oxygen;
- chemically stable, with low volatility;
- low redox potential;
- low toxicity;
- inexpensive to produce.

ExacTech glucose meter (MediSense)

The ExacTech glucose meter from the company MediSense measures glucose in capillary whole blood by a mediated amperometric system. These devices consist of a pen-shaped barrel housing a custom built single chip microprocessor, a sealed power source, a liquid crystal display, an operating button and two connectors, into which the electrode assembly is inserted. The electrode assembly is fabricated in strip form consisting of screen printed electrodes on a plastic substrate. The electrodes are laid down in a multi-print process. Silver impregnated carbon ink is used for electrical connections, a silver/silver chloride layer is used as a reference electrode and the working electrode is comprised of a carbon layer impregnated with glucose oxidase and a ferrocene derivative, which acts as the mediator. The construction is completed by a glucose-spreading layer (gauze) covering the two electrodes, which also helps to minimise evaporation. The gauze also reduces imprecision due to chilling by the latent heat of evaporation of the sample, which is dependent on the ambient relative humidity and temperature.

A two point electrical calibration of the instrument is carried out using two trimmed resistors on a calibration strip. A measurement is made by inserting the disposable electrode (like a nib in the end of a pen) into the instrument, and applying a small drop of blood (about 40µl) from a finger prick to the test surface. No other sample preparation is required. Depression of a button starts the measurement cycle. This applies a voltage of +400mV to the working electrode relative to the reference half cell.

The reaction scheme is as follows:

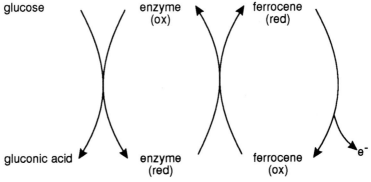

other versions of the MediSense glucose system

Two other versions of the MediSense glucose system have appeared utilising the same basic sensor technology. The first is a modified home monitor in the shape of a credit card, and more recently, the Satellite G system has been launched, which is a benchtop or portable testing station with memory and print-out facilities built-in. The latest addition to the MediSense range of biosensors is a cholesterol sensor, which is again based on mediated screen-printed electrodes. This device incorporates the enzymes cholesterol esterase, to release free cholesterol from cholesterol esters in blood and

cholesterol sensor

cholesterol oxidase, which oxidises cholesterol via a mediated system analagous to that described above for glucose oxidase.

The mediated electrodes described above are simple, one-shot devices. For continuous use, sensors need to be stable over a long period. Short sensor lifetimes due to mediator leaching have led to the development of sensors based on ferrocene polymers, and mediators have been covalently attached to a polymer film.

covalent attachment of redox mediators

An elegant solution to the leaching problem is to modify the enzyme by covalently attaching redox mediator groups to its surface. These are able to shuffle electrons directly from the enzyme to the electrode surface (Figure 8.2).

This technique has been used with the enzymes, glucose oxidase and D-amino acid oxidase, using bound ferrocene as the mediator. Studies show that the electrode response is improved by increasing the concentration of mediator molecules on the enzyme surface.

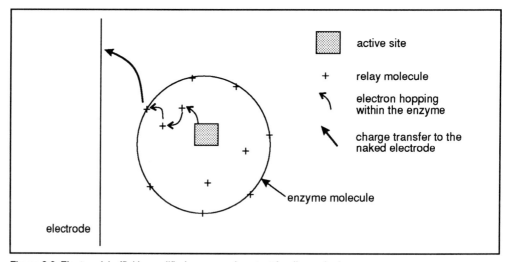

Figure 8.2 Electron 'shuffle' in modified enzymes (see text for discussion).

Biosensors

SAQ 8.2

The enzyme galactose oxidase has been isolated from a bacterium. It has been shown to catalyse the reaction:

galactose + O_2 → H_2O_2 + hexadialdose

It has been shown that this enzyme can use a variety of agents instead of using oxygen as electron acceptor. Amongst these are potassium ferricyanide and ferrocene.

Thus, in principle, two types of biosensors for measuring galactose could be designed using galactose oxidase and these two electron acceptors. These are represented in stylised form below.

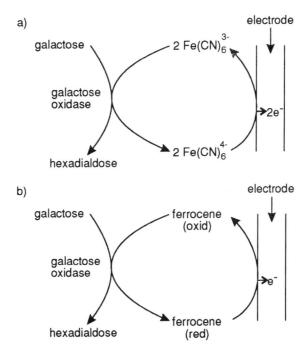

What are the likely advantages and disadvantages of using ferricyanide and ferrocene in these systems? (Hint: ferrocene can be readily polymerised. Also think about the necessary operating potential in each case; Table 8.2 may be helpful).

8.3.3 Organic phase enzyme electrodes (OPEEs)

It has been realised for a number of years that enzymes can retain activity in non-aqueous media (we will discuss this more fully in the final chapter of this text). The use of enzymes, retained in an organic phase in biosensors offers a number of possible benefits (Table 8.3).

Feature	Benefit
Biocatalysis in non-aqueous media	Fats, oils etc can be analysed
Altered substrate specificity	Novel substrates can be detected
Improved thermal stability of enzymes	Increased operational temperature and stability of sensor
Resistance to oxidation and reduction	Wider potential range
Dehydrated enzymes remain active	Fabrication may be simpler
Low heat capacities	Increased sensitivity of enzyme thermistors (Section 8.7)

Table 8.3 Features and benefits of OPEEs.

phenol and cholesterol sensors

The first reported use of a biosensor in which the biocatalyst was in direct contact with an organic phase appeared in 1988 with an amperometric enzyme electrode for the determination of phenols in chloroform. The enzyme was retained in a sufficiently hydrated environment by adsorption onto a nylon membrane which was held in close contact with a carbon electrode. A cholesterol sensor in a solvent mixture of chloroform and hexane has also been described. The enzyme, cholesterol oxidase, was adsorbed onto alumina particles that were located above the sensing element of an oxygen electrode.

retention of water soluble mediators

Furthermore, it has recently been demonstrated that it is possible to use mediators in an organic phase by inverting conventional concepts; utilising a water soluble mediator in a predominantly organic phase. There was little tendency for the mediator (or enzyme) to detach from the electrode in low water conditions.

8.4 Potentiometric enzyme biosensors

8.4.1 Potentiometric transduction

non-destructive transduction technique

Potentiometric transducers measure the difference in potential between the sensing element and a reference electrode under conditions of zero current flow. Under these conditions there is effectively no consumption of analyte and it is therefore a non-destructive transduction technique. In potentiometric measurement, it is essential that the reference potential is accurate and stable, since the working potential is measured relative to this.

Nernst equation

These devices obey a logarithmic relationship between the electromotive force (emf) generated by the electrode and the activity of the ion of interest. This relationship is characterised by the Nernst equation:

$$E_{ise} = E_0 + (RT/zF) \ln a_i = E_0 + (2.303\, RT/zF) \log a_i$$

where: E_{ise} = the potential difference between the working and reference electrodes (V); E_0 = standard electrode potential (ie the potential under standardised conditions) of the ion i (V); R = the gas constant (8.314 kJ kmol^{-1} K^{-1}); T = absolute temperature (K); z = the charge number of the ion i; F = the Faraday constant (96487 coulombs mole^{-1}); a_i = ion activity.

Biosensors

Hence, E_{ise} varies linearly with the logarithm of the activity of the ion i with a slope of $2.303\,RT/zF$, and an intercept of E_0. This corresponds to a voltage change of 59.16 mV for a ten-fold change in the activity of a monovalent ion at 25°C.

∏ What will the voltage change be for a ten-fold change in the activity of a bivalent ion at 25°C?

You should have calculated it to be 29.58 mV (use the Nernst equation, remember that z = 2 as the ion is bivalent).

There are many different potentiometric electrodes with different specificities. For example, potentiometric pH and fluoride electrodes are highly specific, whereas chloride electrode devices tend to be susceptible to interference. The instrumentation required is widely available and is of low cost (a simple pH meter is adequate). Care needs to be taken to avoid electrical interference, but this can be achieved quite simply in most cases by the use of shielded cables.

Before we examine potentiometric enzyme electrodes, we will examine some more familiar potentiometric electrodes.

SAQ 8.3

Below is a graph of E_{ise} against log a_i for an ion of unspecified valency. Determinations were carried out at 27°C. Determine the valency of this ion.

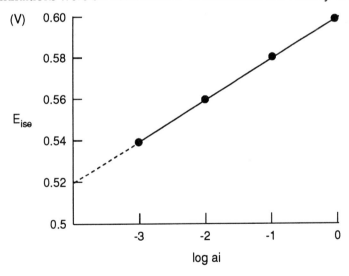

Ion-selective electrodes

The most common type of potentiometric device is the ion-selective electrode, the first example of which was the glass pH electrode. More recently, sodium ion-selective electrodes have also been made from modified glass membranes. A number of other ions (including fluoride, silver, bromide, iodide and sulphide) can be detected by the use of various metal salts as the sensitive elements. Another approach is to use liquid or polymer membranes incorporating ion-exchangers or neutral carriers, for the detection of ions such as potassium, calcium and nitrate. Common support materials are plasticised PVC, polyethylene and silicone rubber.

Coated wire electrodes

An interesting approach that has been gaining in popularity is the use of coated wire electrodes (CWEs), in which a single wire is coated with a polymer membrane material. There are, however, problems with these devices at present, such as poorly reversible and unstable interfaces between the wires and the membranes. Until these problems are solved, their development into commercial products will be limited.

Chemically sensitive field effect transistors (Chem FETs)

It was discovered in the early seventies that a field effect transistor could be used directly as an ion-selective electrode. Essentially in many field effect transistors, the behaviour of the transistor is modified by depositing electrochemically active materials onto the silicon nitride gate layer of the transistor. If the activity of this layer is influenced by the presence of ions surrounding the transistor, then, in principle, the presence of these ions can be detected through the changes in the behaviour of the transistor. This type of device, termed an ion-selective field effect transistor (ISFET), was the first reported type of chemFET. The deposition of a range of electrochemically active materials on to the silicon nitride gate layer of the transistor enables the production of sensors for a wide range of ions.

FET-based devices are very small and this has been considered attractive in sensor design, especially in biomedical applications. Furthermore, the ready availability of micro-electronics technology means that relatively straightforward and inexpensive mass production should be possible. Despite the fact that devices of this type have appeared attractive, they have so far been unable to compete in terms of cost. The simple and inexpensive production of FETs does not necessarily mean that FET-based biosensors can be produced at a similar low cost. A possible area where these devices may have advantages, however, is in multi-sensor production, where many analytes could be measured with a single device containing many sensor elements.

∏ See if you can devise a scheme using ion-selective or coated wire electrodes or Chem FETs which could be used to analyse a specific biochemical. When you have done this, you can check your response with our discussion in the next section.

8.4.2 Potentiometric enzyme electrodes

enzyme is held near to the selective electrode

Potentiometric enzyme sensors can be constructed by holding a biocatalytic substance near the tip of an ion- or gas-selective electrode. The electrode responds when the enzyme substrate diffuses into the catalytic region and is converted into products, one of which can be sensed by the underlying electrode. In most situations, the analysis is complete when the electrode potential reaches steady-state, but enzyme electrodes have been operated in non-steady-state conditions. In most cases, the enzyme is physically trapped or held at the sensing electrode surface, but enzymes can be chemically bound to the membrane surface. A novel example of this approach is the use of perfluoroalkyl groups to modify enzymes and enhance their ability to adhere to fluorocarbon membranes such as those used as the gas permeable barrier in ammonia gas selective electrodes.

As with amperometric enzyme electrodes, the number of permutations of transducers and enzymes is very large. A particularly useful system utilises the fluoride ion-selective electrode and the enzyme peroxidase, which catalyses the breakage of the covalent carbon-fluorine bond in certain organofluorine compounds (X-F) such as 4-fluorophenol, with the elimination of fluoride ions:

$$H_2O_2 + X\text{-}F \xrightarrow{\text{peroxidase}} X\text{-}F_{(ox)} + F^- + H_2O$$

fluoride ion-selective electrodes are highly specific

This system offers many benefits. The fluoride ion-selective electrode is highly specific. The only significant interfering ion is the hydroxyl ion, the effect of which can be removed by operation at low pH. In its simplest form, this system can be used to detect bacterial peroxidase directly. When linked with a disposable fluoride electrode, it can form a convenient biosensor for peroxidase-positive bacteria. A further advantage is that the system can be coupled with many other enzyme-catalysed reactions. Glucose can be monitored by co-immobilising glucose oxidase with peroxidase. The oxidation of glucose produces hydrogen peroxide, which can be linked to the organofluoride oxidation scheme shown above. The release of fluoride ions from the organofluoride compound can thus be measured with a fluoride ion-selective electrode or fluoride-FET, and related to the glucose concentration.

8.5 Conductimetric enzyme biosensors

Many enzyme-catalysed reactions are accompanied by conductivity changes, so that in principle, both enzymes and substrates can be assayed conductimetrically.

use in the measurement of urea

The earliest conductimetric measurements utilised the conversion of urea into ammonium and bicarbonate ions by the action of urease free in solution. Immobilisation of the enzyme close to a platinum electrode enabled the development of a true biosensor. Micro-fabrication techniques have been used to produce interdigitated micro-electrodes in order to increase sensitivity. Such systems have enabled urea solutions (in 150 mmol l^{-1} sodium chloride) at concentrations as low as 1 mmol l^{-1} to give measurable conductance changes in less than two minutes. It is important to carefully select the buffer used in conductimetric assays, since both the basic and acidic forms of the buffer directly participate in the overall conductance change during a reaction.

Another interesting conductimetric biosensor format is to incorporate a dual-sensor configuration, one sensor with enzyme and one without as a control reference. This increases both sensitivity and selectivity. Interdigitated micro-electrodes of platinum on a silicon substrate prepared using micro-electronic fabrication techniques have again proved extremely successful in this type of biosensor.

8.6 Optical enzyme biosensors

8.6.1 Optical transduction

The most simple form of optical transducer relies on a change in the light absorbance characteristics of a reagent layer on interaction with an analyte. This absorbance is characterised by Beer's Law:

Beer's Law

$$A = \log(I_t/I_o) = \varepsilon C l$$

where:

A = Absorbance
(I_t/I_o) = Ratio of transmitted to incident light intensity
ε = Molar absorptivity (l mol^{-1} cm^{-1})
C = Concentration (mol l^{-1})
l = Pathlength (cm)

Illumination may be achieved using a light-emitting diode (LED), with a photodiode as a detector.

We can represent this in the following way:

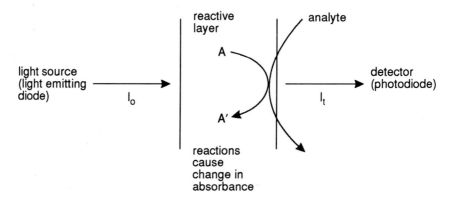

More commonly, an optical fibre can be used as a waveguide to carry light to and from a reaction area.

modal dispersion of light in optical fibres

Optical fibres transmit light by total internal reflection. Light propagates through the fibre in discrete modes, each of which corresponds to a unique incidence angle and has a distribution of electromagnetic and magnetic field intensities such that a given ray does not interfere with itself. These modes propagate through the fibre at rates dependent upon the particular angle between the entering ray and the fibre axis. The larger this angle, the slower the propagation of the mode since it has to travel a greater distance to get to the end of the fibre. This effect is called modal dispersion. It has the undesirable property of causing input pulses to spread as they propagate through an optical fibre. Multimode fibre (ie a fibre which exhibits a great deal of modal dispersion) is the most practical for sensors based on changes in light intensity since it is less expensive and transmits more light. The more expensive, single mode fibre, however, is required for interferometric sensors. Note that the term optrode is often used to describe these fibres.

optrodes

materials used in fibre-optics

If it is necessary to incorporate a reference beam, it is possible to pass the reference beam along the same fibre, thus a separate reference optrode is not required. Materials used in fibres must be both flexible and transparent. Suitable materials include plastics and glasses. Low-melting silicate glasses are inexpensive, but are limited to short-distance transmission in the visible region. High transmission fibres are prepared from highly pure silica doped with borate, germanium oxide or phosphorus pentoxide to modify the refractive index.

There are a number of different formats of optical transduction. Some of the most important are summarised below.

The evanescent wave

When light is reflected at an optical interface where there is a change of refractive index, there is a decay of energy away from the point of reflection into the surrounding medium. This energy field is known as an evanescent wave, which extends into the medium around a waveguide for a distance similar to the wavelength of the light.

We can represent this diagrammatically in the following way.

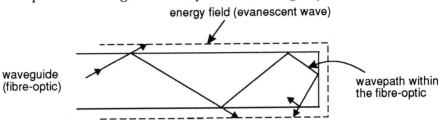

use of evanescent waves enables measurement of analytes close to the surface of optical fibres

Due to the constricted energy field, only optically active compounds in the immediate vicinity of the surface of the fibre will be detected. Unbound compounds in distant regions of the solution will not be illuminated. Therefore, homogeneous assays are possible. Also, since material can be bound directly to the lateral surfaces of a quartz fibre (which provides a large surface area), the waveguide itself becomes a mass concentrating device capable of simultaneously binding, amplifying and detecting any probe molecules bound to the surface. Sensitivity of detection is gained or lost very rapidly as the angle of input or output light approaches or recedes from the critical angle. We remind you that the critical angle is the smallest angle of incidence at which a light ray passing from one medium to another less refractive medium can be totally reflected from the boundary between the two media.

Attenuated total reflection (ATR)

Molecules on the surface of a waveguide can absorb light energy of specific wavelengths. The amount of absorption is proportional to the concentration of molecules at the surface.

Total internal reflection fluorescence (TIRF)

The evanescent wave can be used to excite fluorescent molecules bound to the surface of the waveguide. The molecules in bulk solution are not excited. This can obviate the requirement for the washing step, needed in conventional immunoassays to separate bound from unbound antibody.

We can represent this type of device in the following way:

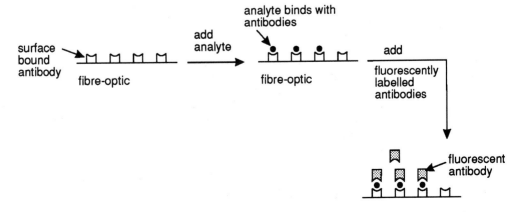

Only the surface-bound fluorescent antibodies are excited by the evanescent wave. Excess fluorescent antibodies in the bulk solution are not activated.

Surface plasmon resonance (SPR)

A surface plasmon is an electromagnetic wave which propagates along the surface of a metal. Optical excitation of a surface plasmon can be achieved if a light beam undergoes total internal reflection at the surface of a glass substrate onto which a thin metal film has been deposited. With the correct choice of metal, at the appropriate thickness, surface plasmon resonance is obtained at a certain angle of incidence of the light beam, which is observed as a sharp minimum of reflectance intensity. This minimum reflects the angle at which an evanescent wave can couple with the electron plasma of the metal.

use of SPR to measure large biomolecules binding to the surface

This angle is very sensitive to variations in the refractive index of the medium within a few hundred nanometres outside the metallised surface. Absorption of large molecules to biological receptors immobilised on the metal surface will change the local refractive index and therefore the resonant angle. The usual mode of operation is to measure the increase of reflected light as the system goes out of resonance.

8.6.2 Enzyme optrodes

The earliest enzyme optoelectronic devices utilised the colour change induced in pH sensitive dyes by the action of enzymes that generate or consume protons. Light can be transmitted down an optical fibre to a sample where it can be reflected and carried back to the detector by the same, or a different fibre.

Thus:

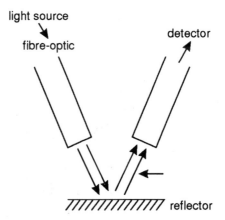

Luminescent systems

luminescence quenched by dissolved oxygen

More recent enzyme optrodes have been most commonly based on luminescent systems. Fibre-optic oxygen systems, termed oxygen optrodes, are based upon the principle that the luminescence intensity of an externally excited luminophore is quenched in direct relation to the concentration of dissolved oxygen. One such luminophore which has proved successful is tris (1,10-phenanthroline) ruthenium (II) cation. When coupled to oxygen-consuming enzymes, such as an oxidase, these oxygen optrodes can be used in a biosensor for a range of analytes.

biosensor for creatinine

A number of such biosensor devices have been fabricated by immobilising oxygen optrodes and enzymes on optical fibres coupled to excitation sources, optical devices and data stations. Oxidase enzymes which have been used include glucose oxidase and sarcosine oxidase. The latter enzyme can be used in combination with two other enzymes; creatinine amidohydrolase and creatine amidinohydrolase to analyse creatinine in blood. This nitrogenous heterocyclic compound is a vital indicator of the renal, muscular and thyroid functions in humans. The reactions involved are:

$$\text{creatinine} + H_2O \xrightarrow{\text{creatinine amidohydrolase}} \text{creatine}$$

$$\text{creatine} + H_2O \xrightarrow{\text{creatine amidinohydrolase}} \text{sarcosine} + \text{urea}$$

$$\text{sarcosine} + O_2 \xrightarrow{\text{sarcosine oxidase}} \text{formaldehyde} + \text{glycine} + H_2O_2$$

Bioluminescent systems

An alternative luminescence approach involves the use of bioluminescence systems, ie enzymes which generate light during catalysis. Two luciferase systems are particularly useful for biosensor applications; the ATP-specific system from the firefly *Photinus pyralis* and the NAD(P)H-specific system from a marine bacteria *Vibrio* species. As both ATP and NAD(P)H are co-enzymes widely involved in many biological reactions, their monitoring appears as an attractive option for the development of novel and versatile biosensors.

Bioluminescent ATP monitoring

The reaction scheme is as follows:

$$\text{ATP} + \text{luciferin} + O_2 \xrightarrow[Mg^{2+}]{\text{firefly luciferase}} \text{AMP} + \text{PPi} + \text{oxyluciferin} + CO_2 + h\gamma$$

where $h\gamma$ represents the emitted light.

This reaction requires the presence of the co-reactants luciferin, Mg^{2+} and molecular oxygen. Under appropriate conditions, the intensity of emitted light ($\lambda max = 560nm$) is proportional to the ATP concentration over a wide range.

Bioluminescent NAD(P)H monitoring

$$\text{NAD(P)H} + \text{FMN} \xrightarrow{\text{NAD(P)H:FMN oxidoreductase}} \text{NAD(P)}^+ + FMNH_2$$

$$FMNH_2 + \text{R-CHO} + O_2 \xrightarrow{\text{bacterial luciferase}} \text{FMN} + \text{R-COOH} + H_2O + h\gamma$$

In the presence of NAD(P)H as target analyte and FMN as co-reactant, the oxidase produces $FMNH_2$ which is a substrate for luciferase. This enzyme requires as cosubstrates, molecular oxygen and a long chain aldehyde, generally decanal. NAD(P)H is stoichiometrically involved in the overall reaction, it can be easily monitored by following the light emission intensity and its concentration related to the intensity of the light emission at $\lambda max = 490nm$.

bioluminescent sensors for NAD(P)H linked enzymes

One successful application for this type of system uses the bacterial enzyme bioluminescence system for the determination of L-lactate in biological fluids by monitoring NADH produced by immobilised lactate dehydrogenase. Co-factors and enzymes are immobilised onto membranes which are stacked onto the tip of a glass fibre bundle. The light generated in response to the lactate substrate is then detected by a photomultipier tube of a luminometer after travelling through the glass fibre.

SAQ 8.4

The enzyme pyruvate kinase catalyses the reaction.

$$\begin{array}{c} \text{COOH} \\ | \\ \text{CO}\,\text{(P)} \\ || \\ \text{CH}_2 \end{array} + \begin{array}{c} \text{ADP} \\ + \text{Pi} \end{array} \xrightarrow[K^+]{Mg^{2+}} \begin{array}{c} \text{COOH} \\ | \\ \text{C}=\text{O} \\ | \\ \text{CH}_3 \end{array} + \text{ATP}$$

phosphoenol pyruvate   pyruvate

Design a biosensor for measuring phosphoenol pyruvate using a bioluminescence detection system.

8.7 Thermal enzyme biosensors

use of thermometers, thermistors, thermopiles

The heat from exothermic reactions can be detected as a temperature change by one of the many types of conventional thermometers, thermistors or thermopiles (arrays of thermocouples). This approach provides a near-universal transducer since the majority of biological reactions are exothermic. An added attraction is that this parameter is also independent of the properties of the sample being treated. One drawback is that the sensor needs to be shielded from external temperature fluctuations.

The conventional approach has involved adiabatic (ie no heat entering or leaving the system) flow but this results in bulky, expensive systems. Recently, the use of some advanced manufacturing technologies such as micro-fabricated thin film thermopiles and thermal optical microfabricated resonating silicon bridges has improved the fabrication of thermal sensors, such that smaller, less expensive devices can be made. Let us examine these in a little more detail.

multiple thermocouples used to increase

Micro-fabricated thin film thermopiles multiply the small voltage change at each thermocouple due to the change in temperature. Since a thin film production technique is used, the thermal mass is also very low. A device of this type has been used to monitor glucose. Multiple thermal junctions connected in series were employed in order to increase the signal. The sensing elements were coated with enzyme, with alternating reference elements left blank, as described in Section 8.5 for conductimetric biosensors. We have represented such a system diagrammatically in Figure 8.3.

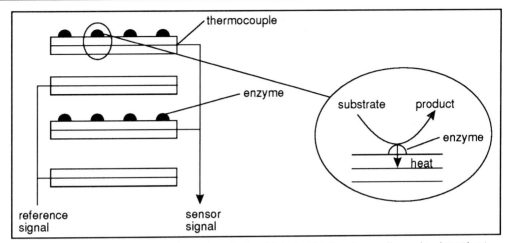

Figure 8.3 Stylised representation of a section of micro-fabricated thin film thermopile used to detect heat released from enzymatically catalysed exothermic reaction. The sensor signal is dependent upon the amount of heat generated at the surface of the thermocouples.

resonance frequency depends on temperature

Thermal effects can also be detected using light. Thermal optical micro-fabricated resonating silicon bridges utilise a microfabricated silicon beam, which resonates when illuminated. Light can be directed at the beam by an optical fibre. Reflected light is modulated by the reasonating beam and can be detected by the same, or a different fibre. The resonant frequency is extremely sensitive to temperature, allowing changes of as low as 0.001°C to be monitored. The device also has a low thermal mass and efficient heat transfer characteristics, thereby maximising sensitivity. Figure 8.4 shows a diagram of such a device, featuring a micro-fabricated resonating silicon bridge, with an etched cell containing an immobilised enzyme layer on the reverse. This type of biosensor can be used to analyse a wide variety of substrates.

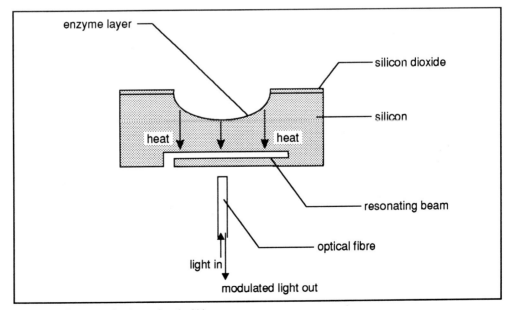

Figure 8.4 Diagram of a thermal-optical biosensor.

Another successful example of a thermal enzyme biosensor is used in fermentation monitoring. Devices based upon glucose oxidase immobilised in calcium alginate on a thermistor have been used for process control during the fermentation of a number of micro-organisms.

lipase based systems

A further promising approach is the use of thermal biosensors based upon OPEEs (organic phase enzyme electrodes). The specific heat capacity of organic solvents is up to three times lower than that of water, making organic phase thermal biosensors inherently very sensitive. An enzyme thermistor for the measurement of lipase in cyclohexane has shown the potential of this approach. Overall, a forty-five fold increase in sensitivity was obtained with both the lower heat capacity and the increased enzyme activity contributing to this increase.

SAQ 8.5

You have an interest in developing a device for measuring the compound shown below using an enzyme biosensor.

$$CH_3-CH(CH_3)-CH_2-[CH_2-CH(CH_3)-CH_2]_5-CH_2-CH(CH_3)-CH_2-CH_2-O-\overset{O}{\underset{\|}{C}}-CH_2-CH_2-CH_3$$

You have an esterase which will catalyse the hydrolysis of this ester. Suggest how you may use this esterase in the design of a biosensor for detecting and measuring the compound of interest.

8.8 Enzymes in affinity biosensors

binding event is detected

Biosensors may be based on affinity reactions of biological molecules. In affinity biosensors, rather than relying on an initial binding of a molecule followed by a biocatalytically induced chemical change for detection, the binding event itself is detected. Although a range of biological molecules with specific binding properties could be used in affinity biosensors, the most versatile and sensitive molecules are antibodies. Antibody-based biosensors are commonly termed immunosensors.

8.8.1 Immunosensors

We dealt with the principles of assays which depended upon the specific binding of antigens by antibodies in Chapter 7 (see Section 7.9), so we will not repeat these in detail here. We will, however, remind you of some of the salient features. The basic principles of all immunosensors are based on classical immunoassay methods which have been routinely used in clinical laboratories and diagnostic tests for a number of years. One of the most common immunoassay methods is the enzyme-linked immunosorbant assay, or ELISA method. This involves the use of a solid phase for the separation of free and bound immunobiological complexes. There are two main types of ELISA; 'competitive' and 'sandwich', both of which most commonly use colorimetric detection.

competitive ELISA

Competitive ELISA is used mainly for small molecules. Here, the antigen of interest is bound to an inert surface, such as polystyrene, in a microtitre plate. The sample (which may, or may not contain antigen) is then added in the presence of antibody-enzyme conjugate. If no antigen is present in the sample, then the antibody-enzyme conjugate forms a complex with the bound antigen and after washing, added substrate will lead to colour formation. If the sample contains antigen, then this competes with the bound antigen for the antibody-enzyme conjugate and reduces the amount of bound complex formation (and hence colour) in the microtitre wells. Therefore, the final measured product concentration is inversely proportional to the concentration of the analyte. A similar, but alternative competitive ELISA technique involves antibody bound to the solid phase and in this situation free antigen in the sample competes with antigen-enzyme conjugate.

sandwich ELISA

The sandwich ELISA approach is used for larger antigen molecules such as proteins or whole cells. The antigen is simply 'sandwiched' between bound and free antibodies. For this, the sample is added along with antibody-enzyme conjugate, to wells containing bound antibody and then washed. Bound complex is subsequently determined by adding a chromogenic substrate for the enzyme. The final product concentration is therefore directly proportional to the antigen present in the sample.

A wide range of monoclonal antibodies are now commercially available for immunoassay, allowing immunological detection of a huge number of molecules. Analytes with molecular masses ranging from 100 up to over 10^6 Daltons, can be detected by immunoassay techniques.

The two transduction methodologies most commonly used in immunosensors are electrochemical and optical. Each will be considered in turn.

8.8.2 Electrochemical immunosensors

Antibodies can be detected electrochemically either directly, or indirectly by the use of enzyme labels. Direct electrochemical detection may permit a sensitivity of as low as 10^{-12} moles due to the extremely high equilibrium association constants attainable with antibodies (10^{10} or greater). However, even lower detection limits are achievable with enzyme-labelled antibodies, as low as 10^{-21} moles in some cases.

Direct labelling of antibodies with electroactive molecules

Labelling of antibodies with electroactive molecules and measurement of the potential change due to binding of an antibody and antigen, allows both amperometric and potentiometric transduction systems to be used. So far, these techniques have generally proved to be too insensitive as practical devices. Work is still progressing in this area using FETs, but commercial devices of this type have yet to emerge.

Enzyme labelling of antibodies

Antibodies can be labelled with enzymes which catalyse a reaction releasing an electroactive product. The enzyme label can also be used as a 'pre-amplifier' which can activate a secondary enzyme system or as a 'power amplifier'. This technique can greatly increase the apparent activity of the enzyme label. This has been demonstrated with a redox cycle which reduces a tetrazolium salt to a strongly coloured formazan dye, which can be measured spectrophotometrically.

An electrochemical immunoassay based on this technique has been demonstrated for prostatic acid phosphatase as shown in Figure 8.5.

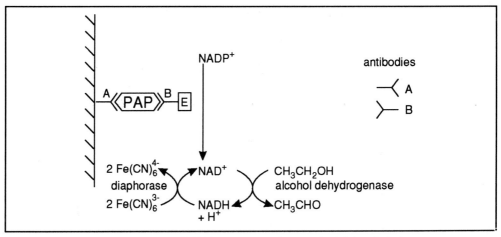

Figure 8.5 Prostatic acid phosphatase (PAP) immunoassay (see text for full description).

'A' represents an immobilised monoclonal antibody directed against prostatic acid phosphatase (PAP). 'B' is another monoclonal antibody directed at a second site on the PAP molecule, which completes the sandwich. The total amount of bound conjugate is directly proportional to the amount of PAP in the sample. 'E' is the alkaline phosphatase label of antibody 'B'. The activity of bound alkaline phosphatase is measured in an amplification assay involving the initial dephosphorylation of $NADP^+$ to NAD^+ and the catalytic cycling of NAD^+ to NADH in a redox cycle generating ferrocyanide. The oxidation of NADH by ferricyanide is catalysed by diaphorase. The reduction of NAD^+ by ethanol is catalysed by alcohol dehydrogenase. The rate of ferrocyanide production can be monitored coulometrically and can be directly related to the NAD^+ produced.

8.8.3 Optical immunosensors

Recently, a great deal of attention has been focused on the use of optical transduction methods for immunosensors. Optical methods can allow the biological recognition (binding) event to be converted directly into a quantitative result. This result can also often be obtained in real-time, unlike some of the electrochemical methods. Of the optical immunosensor technologies, the two currently receiving the most attention are fluorescence and surface plasmon resonance (SPR) devices. Both of these optical methods were mentioned earlier in the chapter, and each will be considered in turn.

Fluorescence-based immunosensors

The most versatile methodology available for immunoassay is one in which an antibody is immobilised onto a solid support. Solid phase immunoassay can allow measurement of both high and low molecular mass analytes in either competitive or sandwich formats.

fluorescein and phycolbili- proteins used as labels

Rather than using either an enzyme label as in ELISA, or radiolabelled derivative of the antibody or antigen as in radioimmunoassay (RIA), fluorescent immunoassays use a fluorescent label. Competitive and sandwich formats of a fluorescent-based immunosensor are shown in Figure 8.6. Fluorescein dyes have been commonly employed in optical immunoassays. These allow a dye detection sensitivity of around 10^{-9} mol l^{-1}. Phycobiliproteins, however, have been used in a number of immunosensor devices. These molecules permit a detection limit of 10^{-12} mol l^{-1}, an approximate thousand-fold increase in total sensitivity over the fluorescein dyes.

Biosensors

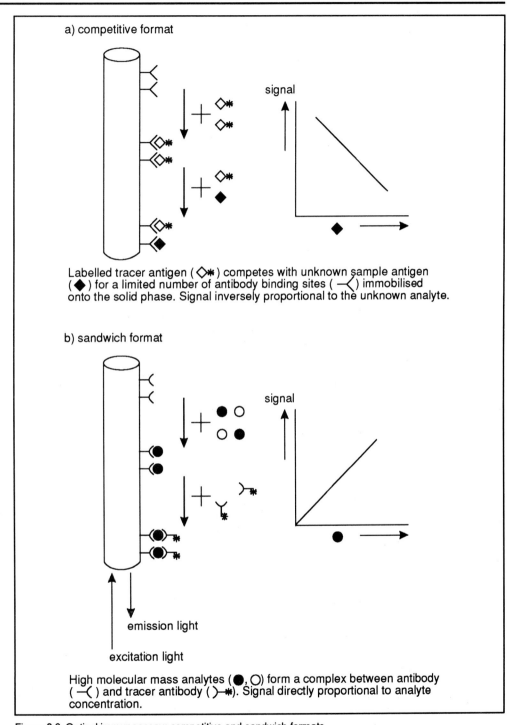

Figure 8.6 Optical immunoassay: competitive and sandwich formats.

sensors for ferritin, digoxin and lidocaine

Several sandwich and competitive immunosensors have been developed for real-time determination of analytes in human serum. In one such device, ferritin, an iron-binding protein of molecular mass 450 000 Daltons, used in the diagnosis of certain anaemias, can be determined within one minute, at concentrations of the order of 10^{-11} mol l^{-1}. The cardiac drugs, digoxin and lidocaine can be rapidly analysed with evanescent wave sensors. The determination time of two minutes compares very favourably with the use of solid phase radioimmunoassay, which takes up to forty-five minutes incubation and process time.

fluorescence capillary fill device (FCFD)

An evanescent wave device which is undergoing commercial development is the fluorescence capillary fill device (FCFD). This consists of two parallel plates, with a 0.1mm separation. All required reagents are contained within the device. The assay capture system is covalently coupled to the baseplate, which also acts as an optical waveguide. Other reagents are dosed onto the top plate in such a way that they dissolve upon addition of sample to the device. On completion of the immunoassay, bound fluorophore can be discriminated from free fluorophore by the evanescent optics of the baseplate waveguide. Two examples of assays which have been developed using the FCFD are analysis of *Rubella* antibody in serum and human chorionic gonadotrophin (HCG) in urine.

FCFDs for Rubella antibody and HCG

Surface plasmon resonance (SPR)-based immunosensors

advantages and disadvantages of SPR-based sensors

SPR offers the advantage that intrinsic properties of the immunocomplex, such as the thickness of the surface layer and the refractive index, are monitored. In theory, this property allows the sensor to be applied to *in vivo* systems where the bulk composition of the measuring fluid (such as whole blood) is variable. Another attractive feature of SPR is that it enables binding reactions to be followed in real-time. The major drawbacks to SPR-based devices are their lack of sensitivity, the difficulty in producing the metal coated slides required and the cost of current instrumentation.

Pharmacia BIAcore instrument

The Pharmacia BIAcore instrument is a research system which uses SPR to measure molecular events which occur during the binding and dissociation of two or more interacting molecules. It provides information concerning the affinity, specificity and kinetics of an interaction. Thus, it offers an alternative to many of the individual assays presently used to characterise biomolecular interactions. Real-time biospecific interaction analysis (BIA) measures interactions directly and by sequential injection of sample. This enables the researcher to monitor the formation of molecular complexes and reveal steric and allosteric effects.

components of the system

The core components of the system are the processing unit, exchangeable sensor chips, immobilisation kit, and a system control and data evaluation software package. The processing unit integrates the optical detection system, autosampler and microfluidic system. The sensing interface of BIAcore is the exchangeable sensor chip. It consists of a glass support coated with a gold film. A biocompatible layer of carboxylated dextran is covalently bound to the surface of the gold film. This increases the sensitivity and capacity of the technique by increasing the effective surface area. It also reduces non-specific binding.

The ligand is immobilised within the dextran layer to form the sensor surface. By using a biospecific ligand, interactions can be studied directly in complex samples without prior purification. The ligand-sensor surface can be regenerated after each injection or after a sequence of injections. Sample loops and valves are situated close to the sensor surface to reduce dead volume. Temperature control is built-in to ensure reproducible analyses.

An immobilisation procedure takes approximately thirty minutes and an average analysis takes from five to ten minutes. The automated procedure is as follows:

- prepare continuous flow buffer and amine coupling kit;
- place a sensor chip in the processing unit;
- place coupling and regeneration chemicals, ligand samples and controls in the autosampler rack;
- start cooling system if required;
- start programmed method and leave;
- review and evaluate results with built-in software.

8.8.4 Nucleic acid biosensors

measuring nucleic acids by hybridisation

In many areas of study, we often need to be able to quantify nucleic acids. There is also often a need to identify and quantify the presence of specific nucleotide sequences. This usually involves the separation of nucleic acids by gel electrophoresis and challenging the separated fragments with radioactively labelled nucleic acid probes. These probes are capable of hydrogen bonding (hybridising) with the target sequence(s) since they contain complementary sequences.

This process is shown in Figure 8.7.

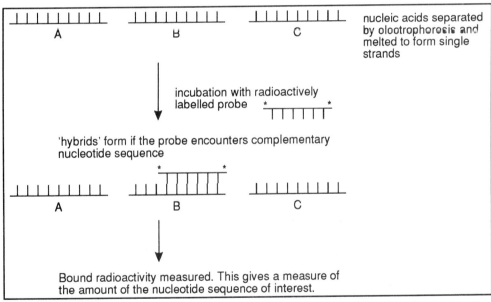

Figure 8.7 An outline scheme for showing the use of radioactively labelled nucleic acid probes for detecting particular nucleotide sequence.

Such methods (eg Southern and Northern blotting) are of course time consuming. The question arises, can biosensors be produced which can determine nucleic acids? In this section, we will describe some of the progress that has been made in this direction.

Electrochemical detection of nucleic acids

There are two distinct possible routes in which electrochemical detection might be applied. These are direct detection, utilising the electrochemical properties of nucleic acids, and indirect methods where hybridisation is detected via the presence of a label or an event which can be detected electrochemically.

Phenyl- and alpha-napthol phosphates have been used as labels in a study of electrochemical nucleic acid detection.

The action of the enzyme alkaline phosphatase results in the cleavage of the phosphate groups from these molecules, liberating an electrochemically active compound from an inactive parent. Conventional immobilised target DNAs are used and hybridisation can be detected by differential pulse voltammetry. A reaction time of twenty minutes give a sensitivity of 1-5ng of DNA, about three orders of magnitude higher than would ordinarily be required. Novo Biolabs have developed an electrochemical system for the detection of nucleic acids. Here, the nucleic acid is labelled with alkaline phosphatase and two enzymes, diaphorase and alcohol dehydrogenase, are used to cycle NAD-NADH. The accumulation of charge is measured over a set time via the mediator, ferricyanide, in a system analogous to that described for the detection of PAP (see Figure 8.5). This system allowed detection of 500pg of DNA. In general, these indirect electrochemical approaches are insufficiently sensitive to be of direct use at present. The advent of PCR (polymerase chain reaction = process for making multiple copies of specific nucleotide sequences), or use of enzyme amplification, may, however, obviate the need for the highly sensitive labels that have previously been thought necessary.

There is a long history of the study of the electrochemistry of DNA. It has been suggested that native DNA adsorbed at an electrode undergoes a conformational change involving strand separation as a prerequisite for reduction. Although the actual mode of application is difficult to foresee, such a phenomenon could be of interest in DNA hybridisation studies.

SAQ 8.6 Draw a schematic diagram to show the sequence of events that occurs in the Novo Biolabs' nucleic acid biosensor. You may find Figure 8.5 helpful.

Light addressable potentiometric sensors (LAPS)

The quantification of single stranded DNA has been achieved using a conjugated system to which urease has been linked and a LAP sensor. We will describe the details of the urease part later.

The LAPS sensor (Figure 8.8) is a silicon-based device which is light activated. A measurement is made by illuminating the back of a silicon chip with one of an array of light emitting diodes (LEDs) which causes a photocurrent to flow. Modulation of the light results in an alternating current, the magnitude of which depends on the potential across the sensor. A chemical reaction at the sensor surface, at one of the discrete chemistry locations, shifts the surface potential and hence the current. By adjusting the external potential to maintain constant current, the rate of the reaction can be monitored by the change in potential.

Biosensors

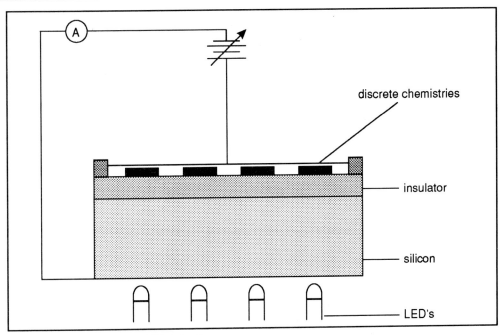

Figure 8.8 Schematic of the LAPS technology.

Threshold system

A commercial LAPS device for use as a total DNA assay system, incorporating an antibody-binding step, is the 'Threshold' system marketed by Molecular Devices Corporation, USA. It is a semi-automated unit in which assays are performed on a disposable stick. Standards, controls or sample replicates are concentrated by active capture onto one of eight measurement sites on a membrane on the stick. The work station can process one to four sticks at a time. The change in potential of the sensor is rapid, and the sensitivity comparable to that of radiolabelling techniques.

Use Figure 8.9 to help you follow the description. In the first step of the assay, the sample is incubated with a single reagent which contains two binding protein conjugates. The single-stranded DNA binding protein is conjugated to a hapten which will be used to make a specific link to the capture membrane. An anti-DNA monoclonal antibody is conjugated to an enzyme label (for example, urease). The labelled DNA complex is transferred to a filtration unit where it is concentrated onto one of the eight measurement sites on a stick. The unbound enzyme conjugate is then washed from the membrane, and the stick is placed in the reader, where it is brought into contact with the sensor surface. The fluid in the reader cavity contains the sensor electrolyte and the enzyme substrate (urea).

The enzyme reaction changes the local pH at each measurement site, which changes the surface potential on the sensor. The rate of change in the surface potential is proportional to the number of enzyme molecules and therefore the amount of DNA at each site.

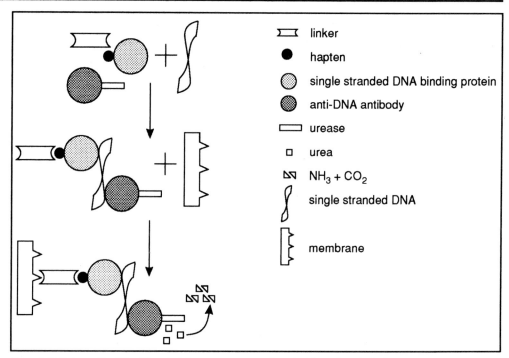

Figure 8.9 Schematic of the 'Threshold' assay sequence (see text for a description).

Optical detection of nucleic acids

detection of nucleic acids using surface plasmon resonance

Of the optical methods, surface plasmon resonance (SPR) has been most successful in the detection of nucleic acids to date. The Pharmacia BIAcore mentioned earlier can also be used for this purpose. In one study, silver was chosen as the metal to obtain the sharpest resonance medium and hence the best response for DNA detection. The metal is deposited as a film, for example, on glass, and must be resilient to the conditions associated with nucleic acid hybridisation, ie high concentrations of salt. In addition to salt tolerance, several other requirements must be met by SPR if it is to be used in nucleic acid probe systems.

∏ See if you can list these.

The main requirements are:

- the nucleic acid target must bind stably to the silver film;

- hybridisation characteristics must remain unchanged;

- non-specific binding to the silver should preferably not occur or should be reduced to an acceptable level.

Piezoelectric detection of nucleic acids

Piezoelectric materials are able to generate and transmit acoustic waves in a frequency-dependent manner. The resonance frequency for optimal wave transmission is highly dependent on the surface of the crystal. Piezoelectric transducers are, therefore, conventionally used to measure small masses of materials in applications such as the vacuum deposition of metals.

the Sauerby equation

The deposition of foreign material on the surface changes the resonance frequency according to the Sauerby equation:

$$\Delta F/F = \Delta m / Art$$

where:

ΔF = frequency change (Hz)
F = resonance frequency (Hz)
Δm = surface mass change (g)
A = crystal area covered by the adsorbed materials (cm^2)
r = density of the crystal (g cm^{-3})
t = thickness of the uncoated crystal (cm)

Bulk wave devices operate by transmitting a wave from one face of a crystal to the opposite face. Surface acoustic wave (SAW) devices transmit waves along a single crystal face from one location to another.

The basic concepts of piezoelectric transduction are relatively simple, but the use of piezoelectric devices in liquids is a relatively new concept, and the precise mechanism of operation is still not completely characterised. However, the simple mass dependency which applies in gaseous phase operation is thought not to apply, but visco-elastic effects are thought to prevail.

use of QCMs to detect Candida albicans

The quartz crystal microbalance (QCM), a piezoelectric-based device, has been used for many years as a mass sensor in vacuum and gas phase experiments. However, it is only recently that the QCM has been used in liquids. The QCM is not a very selective detector. It responds to any interfacial mass change. By monitoring antibody-antigen complex formation, a considerable increase in selectivity can be obtained. For example, silver and palladium coated piezoelectric crystals treated with anti-*Candida* antibody have been used to detect *Candida albicans*.

The sensor was shown to be able to detect the frequency shift resulting from the binding of the yeast in the range 1×10^6 to 5×10^8 cells ml^{-1}. Another device based on a photolithographically deposited interdigital transducer of aluminium on LiNbO$_3$ has been demonstrated for the detection of human immunoglobulins. Possible applications of this device include the detection of HIV, herpes simplex and hepatitis.

piezoelectric-based detection of nucleic acids

sensitivity may need to be improved

A piezoelectric-based system for the detection of nucleic acid hybridisation has also been described. Researchers did not quantify the sensitivity in terms of DNA mass. However, it was claimed that the system was sufficiently sensitive to enable detection of pathogenic bacteria in clinical samples. It is believed that two fundamental hurdles must be overcome in order for this to become a viable detection technique. Firstly greater sensitivity must be achieved, and secondly, it has so far only been demonstrated to work satisfactorily with relatively large pieces of target and probe DNA fragments of the order of several hundred bases. In order to be of widespread use, it will be necessary to demonstrate that the mass changes which occur when small nucleic acid molecules, such as oligonucleotides bind to the quartz crystal can be readily detected.

8.8.5 Biroreceptor-based biosensors

Receptor proteins embedded in lipid membranes are used by organisms for both internal and external chemical recognition. These proteins have potential for use as the selective biological element in a biosensor. The purification of receptor molecules can be a limiting step in the development of these devices. A novel approach is to use a receptor with its associated nerves in the natural organ.

possible use of receptors embedded in bilipid membranes

It is possible to construct a bilipid membrane containing receptors and to carry out electrochemical experiments to detect the opening of ion channels in response to specific molecules. Work is in progress to understand the mechanisms of receptor action with the objective of using them in biosensors. A device has been reported based on the acetylcholine receptor. Much more work is required on the stabilisation of membranes before practical devices can be produced.

In a recent study, the neuroreceptor for nicotinic acetylcholine was immobilised onto fibres following solubilisation in a non-ionic detergent. This receptor has a broad-based specificity for agonists and antagonists, including drugs and toxins that interfere in normal neuromuscular response. Using fluorosceinated alpha bungarotoxin as a label in a competitive assay format, all classes of compounds known to bind to this receptor could be detected.

8.9 Microbial biosensors

Whole cells can be held near to a transducer to form microbial biosensors, in which cell components, such as enzymes, can be used as the biological recognition element of the biosensor. This approach can be advantageous in situations where it may be impractical to use a purified enzyme as the biological component. This may occur if the enzyme is unstable, requires a soluble cofactor, or is simply difficult to purify.

advantages and disadvantages of the lack of selectivity

The variability and multireceptor behaviour of whole cells has advantages as well as disadvantages. The low specificity may be undesirable in some applications, but in others, such as environmental monitoring, it can be highly desirable. The increased stability of whole cell sensors makes them more suitable for continuous operation. However, response times are generally slow, and frequent re-calibration is necessary.

Substrate specificity can often be improved by pre-incubation of the microbial sensor with the desired substrate to be determined.

Π Why may pre-incubation with the desired substrate lead to greater substrate specificity?

The answer is that the production of enzymes by microbial cells is, at least in part, controlled by exogenously available substrates. For example, in the presence of glucose, the bacterium *Escherichia coli* does not express the enzyme β-galactosidase needed to metabolise lactose. On the other hand, in the absence of glucose and in the presence of lactose, β-galactosidase is produced. Thus if we used lactose grown cells, we could use these cells to detect lactose and glucose since the cells would metabolise either sugar. If on the other hand we used glucose grown cells, they could not be used to detect lactose. The point we are making is that by pre-conditioning cells, we influence their metabolic capabilities.

Biosensors

mutation and metabolic inhibitors may improve specificity

We could also consider the use of mutants which have changed metabolic capabilities. For example, we may improve selectivity by using mutants in which undesirable metabolic pathways are blocked. Alternatively we may consider the use of metabolic inhibitors to improve selectivity.

Electrochemical transduction is the favoured technique, with both amperometric and potentiometric devices having been featured. To date, about fifty electrochemical biosensors based on micro-organisms have been developed for the detection of alcohols, ammonia, antibiotics, biochemical oxygen demand, enzyme activities, mutagenicity, nitrate, organic acids, peptides, phosphate, sugars and vitamins.

SAQ 8.7

You have a strain of bacteria that carry a plasmid which enables the bacteria to catabolise phenol. Plasmid-carrying bacteria are capable of using phenol as their sole source of carbon and energy.

Biochemical analysis shows, however, that the cells, even when cultivated in phenol-containing media, produce the enzymes of glycolysis.

You would like to use this bacterium to produce a phenol biosensor, using thermal detection system. Unfortunately, the samples you need to test contain both phenol and glucose. Glucose is known to switch off phenol-metabolising genes. Explain what you should do to make a successful phenol biosensor.

8.10 Tissue section biosensors

Whole sections of mammalian and plant tissues have been studied as an alternative to isolated enzymes and bacterial cells in biosensors. A typical tissue section biosensor is illustrated in Figure 8.10.

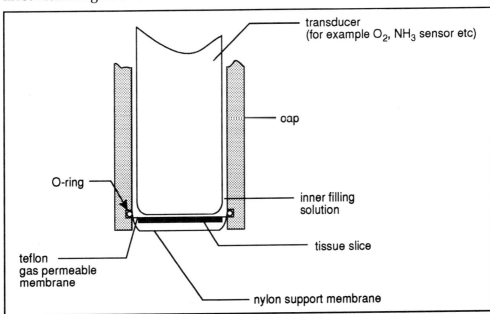

Figure 8.10 Schematic of a tissue section biosensor.

Greater selectivity can usually be obtained from tissue section biosensors than from those based on bacterial cells. This is because tissues cells are only part of an overall living entity and therefore possess a smaller variety of enzymes. A wide range of substrates can be measured using these devices. The sensing elements are very often gas sensors. Table 8.4 illustrates some of these examples.

Substrate	Biocatalytic material	Gas detected
Adenosine	Mouse small intestine mucosal cells	Ammonia
Cysteine	Cucumber leaf	Ammonia
Dopamine	Banana pulp	Ammonia
Glutamate	Yellow squash	Carbon dioxide
Hydrogen peroxide	Bovine liver	Oxygen
Pyruvate	Corn kernel	Carbon dioxide
Tyrosine	Sugar beet	Oxygen
Urea	Jack bean meal	Ammonia

Table 8.4 Examples of tissue section-based biosensors.

8.11 Manufacturing technologies

A variety of techniques, many of which were adapted from the electronics industry, have been used for the production of biosensors. The following sections explain briefly the principles involved in some of the key advanced fabrication processes, and outline how they have been adapted for use in sensor production.

8.11.1 Thin film techniques

vacuum deposition of conduction

A thin film is generally accepted as being a film of less than one micron (μm) in thickness, although film thicknesses can be as high as five to ten microns. Conducting tracks for a wide variety of electronics applications can be formed by the vacuum deposition of thin metal or semiconductor films. The technique can be adapted to lay down conducting tracks for amperometric sensors, or ion-sensitive layers for ion-selective electrodes or FETs.

evaporation aided by heating or sputtering

Evaporation is carried out at a relatively high vacuum, typically below 10^{-5} mbar. At such low pressures, evaporation is possible at lower temperatures and some protection against oxidation is offered. Heating can be achieved either by passing an electrical current through a suitable crucible or boat, or by focusing an electron beam on the material to be evaporated. Sputtering is a related principle in which bombardment of a film of material by (typically argon) ions, under vacuum, volatilises the material of interest. Screens are used to mask areas that are to remain uncoated.

Fine control of the pattern and thickness is possible by the use of chemically etched screens and quartz microbalance thickness monitors. Complicated structures can be built by multiple layer evaporation of a wide variety of materials.

8.11.2 Screen printing

The technique of screen printing consists of squeezing and compressing a paste through a gauze onto an underlying surface. This deposits the paste onto the surface in a film with a controlled pattern and thickness (typically ten to twenty microns). It therefore enables a wide range of electrode geometries to be produced. The different production techniques available are relatively simple, with a high degree of flexibility for fabricating different devices, and well suited to small or larger production runs giving good sensor performances.

the range of electro-conductive pastes

Composite silver pastes, which are widely used in integrated circuit manufacture, are available commercially. Pastes are also available based on graphite, gold, a variety of other metals, dielectrics and resistive materials. Preparation of suitable pastes of solid electrolytes, and metal oxide electroactive materials are the key step for the development of thick film biosensors.

use of alumina

The most common substrate is alumina and, due to its high temperature resistance, it is possible to fire the pastes at very high temperatures, typically of the order of 1000°C. This leaves relatively pure metal on the surface. Obviously, non-ceramic substrates such as plastics cannot be fired at these temperatures, but a variety of pastes are available for low temperature curing. These consist of particles of graphite or metal dispersed in a polymer matrix containing suitable additives in order to obtain the correct dispersion and viscosity for printing.

Multilayer structures can be built up by superimposing any of a wide variety of other layers. For example, the ExacTech glucose electrode is formed by several print runs, with an outer layer containing an enzyme and a mediator.

The Unilever capillary-fill device (CFD) is not yet available commercially, but is described here because its elegant design, based on thick film technology, is suitable for large scale production. The device is easily adaptable for the measurement of a wide range of analytes with the same basic design.

screen printing also includes chemical reagents

The CFD comprises of a small gap (0.2mm) which is formed by sandwiching together a top glass plate, and a bottom ceramic plate which contains screen printed working and auxiliary electrodes. This technology has been used to provide self-contained, disposable measurement cells which can be produced cheaply and reproducibly. The test reagents are screen printed to form a thin layer on the surface of the glass top plate. For a glucose assay, the reagents include buffer salts, glucose oxidase and an excess of potassium ferricyanide (mediator).

Test liquid is then introduced into the cell (which fills by capillary action), whereupon the dried reagents dissolve and disperse throughout the entire cell. Glucose is oxidised to gluconic acid by the glucose oxidase with the concomitant reduction of ferricyanide to ferrocyanide.

The coulometric determination of the ferrocyanide formed is achieved by reoxidising the reduced mediator at the surface of the gold working electrode. The geometry of the cell, and low volume allow rapid coulometric measurement. An advantage in utilising coulometry is that the amount of charge passed during the determination can be related directly to the concentration of glucose without calibration since Faraday's law of electrolysis can be applied. A further advantage is that, since all of the components, including the mediator, are in solubilised form, it is simple to include more than one enzyme in the dry layer in order to carry out analyses which require the sequential activity of two or more enzymes.

SAQ 8.8 Explain how the Unilever capillary-filled device may be adapted to produce a biosensor that could be used to determine a) NADH, b) lactate.

8.11.3 Microlithography

Microlithographic techniques are widely used in the electronics industry for the production of thin layers of patterned metals, dielectrics and semi-conductors. The technique can be used to form cheap, reproducible, and if necessary, sophisticated electrochemical structures.

Radiation sensitive 'resist' layers are used to define the desired structures. Positive resists become more soluble upon irradiation, negative resists become less soluble (Figure 8.11). The coated substrate is irradiated through the mask using visible, ultra-violet or X-ray photons or by an electron or ion beam.

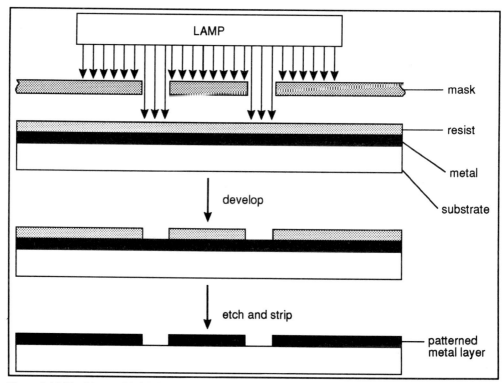

Figure 8.11 The lithographic fabrication process.

The great advantage of this technique for electrochemical applications lies in the reproducibility of the process and the ability to replicate easily the structures many times on the same substrate, so that many devices can be fabricated at one time. Optical masks enable devices to be produced with two micron features. Higher resolution is possible using X-ray and electron beam lithography.

8.11.4 Ink-jet printing

The deposition of biological reagents such as enzymes, can be carried out accurately and reproducibly using a technique originally developed for the printing industry. Ink-jet nozzles were originally developed for the deposition of printing inks, but they can be adapted easily for the deposition of a wide variety of solutions.

The ink-jet nozzle activation principle is based on piezoelectric element contraction induced by an electrical pulse. The nozzle consists of a small chamber containing a hole (typically 50-100 microns in diameter). When chamber contraction is induced, by increasing pressure from the pressure chamber, the reagent solution filling the chamber is pushed into the air and creates a reagent solution drop. After a drop leaves the chamber, the chamber pressure returns to its initial value. The reagent is supplied to the chamber by an ink inlet connected to a reservoir. A solution flow resistance element maintains a constant solution flow rate and appropriate drop size.

Non-contact technology allows fluid to be placed on almost any surface, irrespective of texture, shape or delicacy, giving great flexibility in the design of the finished device.

The Biodot Microdoser is a commercially available machine specifically designed for printing very small droplets of biological fluids onto almost any surface at high speeds. The accuracy of each droplet is adjustable to within fractions of a millimetre, with good reproducibility.

This technique will have great merit in immobilisation of expensive enzymes and in the fabrication of multi-biosensors in which various immobilised enzyme membranes may be deposited on a sensor chip.

Summary and objectives

In this chapter we have provided a discussion of biosensors particularly focusing onto biosensors which employ the use of enzymes. We began by examining the non-biological aspects of biosensor technology. We explained how biological signals are transduced into electronic signals and that key to the success of producing biosensors is the application of appropriate transduction methodologies. We covered, for example, amperometric, potentiometric, coulometric, optical, conductimetric and thermal biosensors. We also included discussion of a range of affinity biosensors including immunosensors, nucleic acid biosensors and bioreceptor biosensors. Biosensors dependent upon the activities of micro-organisms and tissue slices were also described. In the final part of the chapter, we briefly described manufacturing techniques used to produce biosensors.

Now that you have completed this chapter you should be able to:

- describe a wide range of methods used to convert 'biological' signals into electronic signals;

- suggest suitable strategies for producing biosensors using supplied data and information;

- evaluate the advantages and disadvantages of using mediators in amperometric enzyme electrodes;

- explain the potential advantage of using organic phase enzyme electrodes;

- give examples of biosensors for a wide range of biosensor transduction system;

- explain how affinity biosensors for a wide range of biosensor transduction systems work;

- explain in outline, how the techniques of screen printing, microlithography and ink-jet printing may be applied to the production of biosensors.

Use of enzymes in molecular biology and biotechnology: restriction and associated enzymes

9.1 Introduction	248
9.2 General principles of bacterial restriction-modification systems	248
9.3 Restriction-modification system nomenclature	251
9.4 Class I restriction-modification systems	252
9.5 Class II restriction-modification systems	253
9.6 Class III restriction-modification systems	255
9.7 Joining DNA molecules together: DNA ligase	257
9.8 Use of restriction enzymes and DNA ligase in cloning	258
9.9 Use of restriction enzymes in diagnosis of genetic disorders	264
9.10 Genetic fingerprinting	267
9.11 Detection of transcriptionally active genes	270
9.12 Identification and purification of restriction enzymes	273
Summary and objectives	274

Use of enzymes in molecular biology and biotechnology: restriction and associated enzymes

9.1 Introduction

In the previous chapters you have learnt much about enzymes and have been introduced to the concept of enzymes as tools for industrial use. In this and the following chapter we shall review the properties, isolation and modification of enzymes of particular importance in the study and use of DNA for industrial and medical purposes. This chapter will be concerned with the prokaryotic restriction endonucleases and with DNA ligase with respect to their properties and use in cloning and in medical research. We shall see that these enzymes have been of vital importance in the development of genetic engineering.

restriction enzymes

natural role as part of a defence mechanism

protection of host DNA mediated by methylases

In the last 15-20 years, investigation and exploitation of DNA has advanced rapidly, in large part due to the identification and use of a specific group of intracellular prokaryotic enzymes - the restriction endonucleases (or restriction enzymes). These form part of a restriction - modification defence system which bacteria have evolved to protect themselves against bacteriophage attack and other introductions of genetic material (eg by transformation, or conjugation). In general this system has two components, the first of which is the restriction endonuclease activity. These enzymes recognise specific base sequences in DNA and cut (or 'restricts') the DNA within, or at various distances from, these recognition sites. The second component is a corresponding methylase enzyme activity which modifies the endonuclease recognition sequence where it occurs in the bacterial genome, hence protecting it. Only unmodified foreign DNA will thus be restricted if it enters the bacterium.

The restriction-modification systems identified to date may be split into three groups on the basis of their requirements for activity, restriction sites and subunit structure. These groups and their usefulness or otherwise for recombinant DNA work will be discussed in this chapter.

∏ As well as being able to cut up pieces of DNA, what must we be able to do with these fragments to produce new pieces?

DNA ligase

Recombinant DNA technology requires an ability to join together different pieces of DNA ie to produce 'recombinant' molecules. The enzyme used for this purpose is DNA ligase. Two main types of this enzyme are used - that from *E. coli* and that from bacteriophage T4. The properties and use of these enzymes are described in Section 9.7.

9.2 General principles of bacterial restriction-modification systems

The phenomenon of restriction-modification systems was first identified in the 1950s when it was observed that bacteriophages infected different bacterial strains with

Use of enzymes in molecular biology and biotechnology 249

varying efficiencies. Interestingly, phages that had initially infected with low efficiency, subsequently infected the same strain of bacteria with high efficiency (see Figure 9.1).

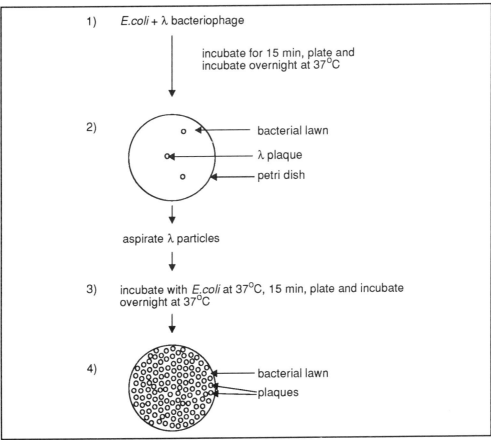

Figure 9.1 Prokaryotic host-mediated restriction-modification. 1) An *E. coli* culture is incubated with λ particles to allow infection. The bacteria are then spread over agar and incubated. 2) After incubation, only very few clear patches are observed on the bacterial lawn, corresponding to incidences where a single λ particle successfully infected an *E. coli* cell in stage 1. 3) Particles from one of the plaques are used to infect more *E. coli*. 4) For the same number of phage particles incubated with the same number of bacteria, this second round of infection is much more successful (as judged by the number of plaques produced) than infection at stage 1.

∏ In view of what we described as the basis of restriction-modification in the introduction to this chapter, how do you think the phenomenon summarised in Figure 9.1 happens?

protection by restriction enzymes is not infallible

In the first round of infection, the bacteriophage injects its DNA genome into a bacterial cell. In most cases, this DNA is immediately digested by the restriction enzyme(s) present in the bacterium. However, in a small percentage of the cells, the first event is that the infective DNA becomes modified by the methylase enzyme(s) and protected against restriction. Hence the viral DNA is able to function within the host to allow viral replication etc. As the DNA in the resulting particles has become suitably modified to escape restriction in this host, re-infection of fresh bacteria will occur with a high efficiency.

One question which this raises is how is newly synthesised DNA protected from restriction? The answer to this lies in the nature of the recognition sites for restriction enzymes. Methylases most commonly modify adenosine nucleotides as shown in Figure 9.2, and, to a lesser extent, cytidine nucleotides.

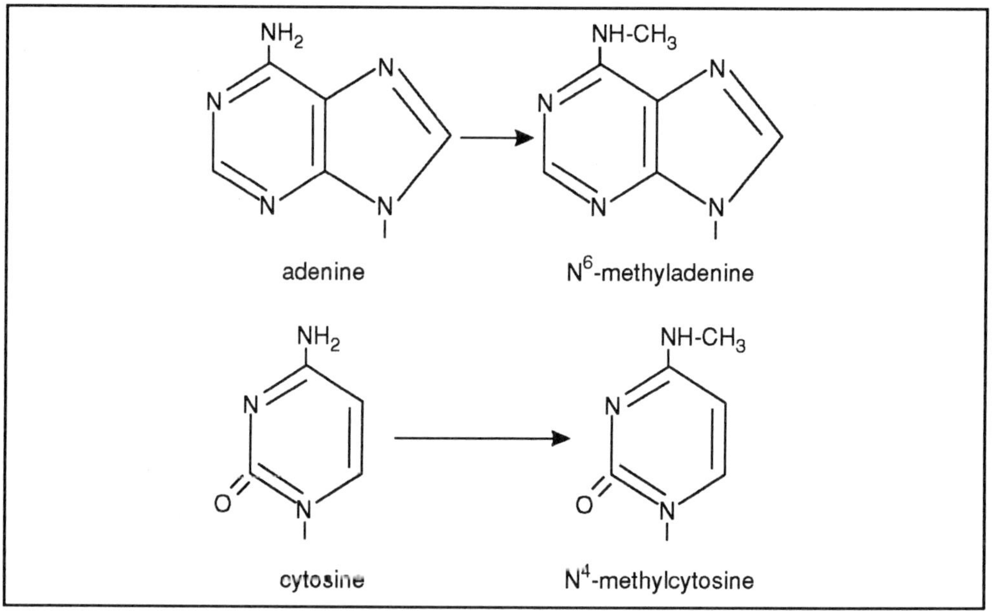

Figure 9.2 Modifications catalysed by prokaryotic methylases.

sites of methylation

Generally, the recognition sequence contains at least one adenosine or cytidine nucleotide in each strand, for example in *E. coli* K the recognition sequence is:

```
            *
   5' AACNNNNNNGTGC 3'        (N is any nucleotide)
   3' TTGNNNNNNCACG 5'
                *
```

In this case, the adenosine nucleotides marked with an asterisk are methylated. On semi-conservative replication of DNA containing this sequence, two hemi-methylated daughter DNA molecules will be produced.

```
         *
   ----AACNNNNNNGTGC----         ----AACNNNNNNGTGC----
   ----TTGNNNNNNCACG----   and   ----TTGNNNNNNCACG----
                                                *
```

Thus each molecule is methylated on one strand only. This is sufficient to protect the DNA from restriction, and the methylase will modify the other strand prior to the next round of replication.

Use of enzymes in molecular biology and biotechnology 251

SAQ 9.1

What would you expect the effects to be of:

1) Plating phage λ on *E.coli* K?

2) Re-plating from plaques produced in a) on further *E.coli* K?

3) Plating phage produced in 2) on *E.coli* C?

4) Replating phage from plaques in 3) on *E.coli* K?

Explain your answers.

9.3 Restriction-modification system nomenclature

basis of names of restriction enzymes

In order to simplify matters, a clear nomenclature has been worked out for specifying restriction endonucleases and their corresponding methylases. This is important as each endonuclease/methylase pair recognises a highly specific DNA sequence or group of related sequences under optimal conditions. If we consider the endonucleases first, these are described in terms of the bacterial species and strain from which they originate. If a particular species produced more than one endonuclease, each is then give a roman numeral, that being identified first being called I, that second, II etc. Hence the third endonuclease to be identified in *Haemophilus influenzae* Rd is called:

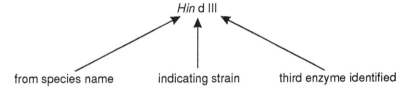

Similarly, the two restriction enzymes isolated from *Bacillus globigii* are called *Bgl* I and *Bgl* II.

Π What would you expect to call restriction enzymes isolated from *Moraxella bovis*, *Providencia stuurlii* and *Escherichia coli* RY13?

The first two are fairly obvious - *Mbo* I and *Pst* I. The last one is less obvious - this enzyme is known as *Eco* RI. *Eco* RII comes from a different strain, *E. coli* R245.

Nomenclature used for the corresponding methylases is also simple, eg the methylase protecting against restriction by *Eco* RI is called M.*Eco* RI.

Π What would you expect the methylases protecting *Bam* HI, *Hae* III and *Ava* II restriction sites to be called?

Yes, it is fairly obvious - M.*Bam* HI, M.*Hae* III and M.*Ava* II.

Before we go on, we should point out that although this nomenclature allows us to identify the organism from which individual enzymes originate, it tells us nothing *per see* about the recognition sites or the class of restriction system involved. Consultation

9.4 Class I restriction-modification systems

Class I enzymes: restriction site is distant from recognition site

The earliest identified and characterised restriction enzyme was, as mentioned in Section 9.2, that from *E. coli* K (*Eco* K). This belongs to the first class of restriction enzymes, Class I. Although these enzymes recognise specific DNA sequences, a distinguishing feature of them is that they cut the DNA at non-specific sites some distance from these (between 1000 and 7000bp). The endonuclease and methylase activities reside in the same enzymes, which are large (at least 400 000 Daltons) heterotrimeric molecules requiring magnesium ions, S-adenosyl methionine (SAM) and ATP as cofactors. All three cofactors are required for the endonuclease activity, whereas only SAM is required for methylase action (as the methyl donor). In some cases (eg *Eco* B) methylase and endonuclease activities may be associated with particular subunits, but in others (eg *Eco* K methylase) this is not so. The mode of action of Class I enzymes is that a molecule interacts with SAM and binds to unmodified DNA at the recognition sequence. The function of SAM is uncertain - it may act as an allosteric effector. The enzyme then tracks along the DNA in an ATP-dependent fashion. After 1000 - 7000 bp, the enzyme cuts *one strand* of the DNA at a random site. It subsequently makes a gap in the cut strand of approximately 75 nucleotides by releasing short oligonucleotides. A *second* enzyme is required to cut the other DNA strand, and there is no clear evidence that the enzymes are in fact catalytic in terms of nuclease activity.

cofactor requirements (Mg^{2+}, SAM and ATP)

It is possible that Class I endonucleases have two sites for DNA binding, which would explain the observation of a DNA loop as a reaction intermediate in the action of *Eco* B and *Eco* K. To understand this let us look at the recognition sequences of *Eco* B and *Eco* K:

Eco B: 5' TGANNNNNNNNTGCT 3'
3' ACTNNNNNNNNACGA 5'

Eco K: 5' AACNNNNNNGTGC 3'
3' TTGNNNNNNCACG 5'

Π What is immediately apparent about these sequences, and what might it tell us about the action of the enzymes?

Yes, both contain two clearly identified sequences separated by a variable region. It is possible that each enzyme has two binding domains, one recognising each of the constant parts of the recognition sequence. This model envisages part of the enzyme remaining anchored to one of the recognition sites whilst the other domain translocates along the DNA duplex. At the cleavage site a nick is introduced into either one of the strands and about 75 nucleotides are removed. The enzyme stays bound to the DNA following this, and hydrolyses ATP. This implies that the recognition and cleavage sites are held close together (see Figure 9.3) allowing a second enzyme to cleave the other strand at the correct site. We should emphasise that this is purely a model for the action of Type I nucleases. Enzymes in this class are highly complex and many questions (thankfully outside the scope of this discussion) remain unanswered about their action. Since the site of cleavage is non-specific, these enzymes are of negligible use for biotechnologists.

Use of enzymes in molecular biology and biotechnology

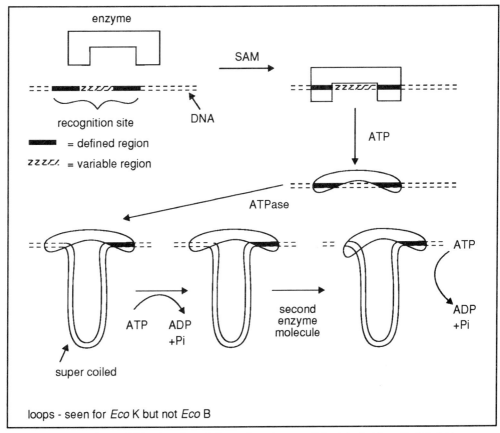

Figure 9.3 A two-site model for class I restriction endonuclease activity. Adapted from Endlich & Linn (1981) The Enzymes vol. XIV pp137-156. Academic press, Inc. SAM, but not ATP, is required for recognition. ATP is hydrolysed following enzyme-DNA binding, and this continues after cleavage of the DNA. NB Translocation has unique polarity for *Eco* B, but not for *Eco* K ie the cut site is always upstream of the recognition site of *Eco* B, but may be on either side of that for *Eco* K.

The fact that non-specific cutting occurs with endonucleases of this class would appear to indicate a clear role in prevention of the incorporation of foreign DNA into cells which contain them. Another role for restriction enzymes has also been postulated, which we will return to very briefly after a consideration of the other two classes of restriction - modification system.

9.5 Class II restriction-modification systems

endonuclease and methylase activities are on separate enzymes

Enzyme systems of this class recognise short nucleotide sequences, and restriction occurs within or close to these. The endonucleases and methylases are separate enzymes, the former having a simple structure and a molecular mass range of 20 200 - 100 000 Daltons. The only requirement for nuclease activity is magnesium ions.

Class II enzymes: recognition sites are short and often palindromes

The recognition sequences of class II endonucleases are interesting in that they are short (usually up to 8bp long) and often have rotational symmetry (that is, the sequence is the same in the 5'-3' direction on both strands of the DNA). Such sequences are often termed 'palindromic' sequences. Some examples of Class II enzymes and their recognition sequences are shown in Table 9.1.

Note that the cleavage positions are also marked, and the fragments resulting from cleavage are shown.

Enzyme	Recognition sequence	Fragments produced on cleavage	
Eco RI	5' G*AATTC 3' 3' CTTAA*G 5'	-G 3' -CTTAA 5'	5'AATTC- 3'G-
Hin dIII	5' A*AGCTT 3' 3' TTCGA*A 5'	-A 3' -TTCGA 5'	5'AGCTT- 3'A-
Bam HI	5' G*GATCC 3' 3' CCTAG*G 5'	-G 3' -CCTAG 5'	5'GATCC- 3'G-
Hae III	5' GG*CC 3' 3' CC*GG 5'	-GG 3' -CC 5'	5'CC- 3'GG-
Pst I	5' CTGCA*G 3' 3' G*ACGTC 5'	-CTGCA 3' -G 5'	5'G- 3'ACGTC-
Bst EII	5' G*GTNACC 3' 3' CCANTG*G 5'	-G 3' -CCANTG 5'	5'GTNACC- 3'G-
Bgl I	5' GCCNNNN*NGGC 5' 3' CGGN*NNNNCCG 5'	-GCCNNNN 3' -CGGN 5'	5'NGGC- 3'NNNNCCG-

* indicates position of endonuclease cleavage
N is any nucleotide

Table 9.1 Examples of some class II restriction enzymes.

∏ Looking at Table 9.1, can you suggest how Class II restriction enzymes may be further sub-divided?

If we consider the types of fragments formed following nuclease action, these fall into three groups:

- fragments which have a 5' overhang, for example *Eco* RI, *Hin* dIII and *Bam* HI restriction fragments;

- fragments with no overhang that is, those with flush or blunt ends such as *Hae* III restriction fragments;

Use of enzymes in molecular biology and biotechnology

usefulness of sticky (cohesive) ends

- fragments with a 3' overhang, for example *Pst* I restriction fragments.

The production of fragments with short overhanging ends (also called cohesive or sticky ends) is useful for production of recombinant molecules as we shall see later. The overhanging ends on two fragments generated by *Eco* RI cleavage are complementary and may anneal together, hence the names cohesive ends/sticky ends.

Generally, due to their specificity of recognition and cut sites, Class II enzymes are the most widely used endonucleases. Use of a particular enzyme under optimal conditions will give a series of defined, reproducible fragments from a given starting piece of DNA. Note that the specificity of restriction enzyme action frequently depends on the buffer and ionic strength of the medium: specificity may drop (and cleavage becomes more random) as conditions are changed from the optimal. Thus it is very important that the 'correct' conditions are used for each enzyme.

Π What property of Class II endonucleases would you consider important if you wanted to generate either a few large fragments, or many small fragments, from a large piece of DNA?

The answer to this is the length of the recognition sequence. For certain purposes you may wish to obtain large pieces of DNA, hence an enzyme with a long recognition sequence is appropriate as long recognition sequences will occur less frequently in the DNA. A short recognition sequence (eg 4 bases) is likely to occur more frequently than a longer sequence: an enzyme recognising the shorter sequence can be expected to give more, smaller, fragments.

Π What would be the average length of restriction fragments produced by an endonuclease with a four base pair recognition sequence and another with an eight base pair one?

As DNA is made up of 4 nucleotides, a particular 4bp sequence will occur on average every 4^4bp (that is once per 256bp) and an 8bp one every 4^8bp (that is once every 64500bp or 64.5kb). The wide range of class II enzymes recognise a variety of sequences of different lengths, making them a versatile and effective tool in the analysis and use of DNA. We shall consider specific examples of their use in Sections 9.8 - 9.11.

9.6 Class III restriction-modification systems

Originally grouped together with Class I systems, these are of intermediate complexity between I and II. They consist of separate endonuclease and methylase enzymes, but these have a subunit in common. Generally, each endonuclease has two subunits with approximately 250 000 D molecular mass.

Π Class I and III endonucleases require the same cofactors; what are they?

Mg^{2+}, SAM and ATP. ATP is a strict requirement for Class III enzymes, but they do not exhibit measurable ATPase activity, unlike members of Class I. The methylases only require SAM as methyl donor.

Class III enzymes: restriction site is close to but separate from recognition site

The cleavage site is also intermediate between those of the other two classes, occurring 24-26bp to the 3' side of the recognition site and having a degeneracy of one or two bases. The recognition sites themselves do not show rotational symmetry, for example:

Eco P1 : 5' AGACC 3'
Eco P15 : 5' CAGCAG 3'

Π One unsolved problem with the Type III system is that only adenosine residues are methylated. Why should this be a problem?

Looking at the recognition sequences of *Eco* P1 and *Eco* P15 you should see that adenosine residues are only observed in one strand. This would appear to cause difficulties when we consider what happens on DNA replication. One daughter molecule will be methylated (having originated from the parental methylated strand) but the other will not. How this is overcome is uncertain. It is thought that possibly the action of the methylase is in some way linked to the replication processes.

SAQ 9.2

To consolidate your understanding of different classes of restriction-modification systems, draw a table comparing the following properties in each class:

1) Size.

2) Complexity/association of endonuclease and methylase.

3) Type of recognition sequence.

4) Cleavage site.

5) Methylation site.

6) Cofactor requirement.

7) Usefulness in recombinant DNA work.

and any other points you consider important.

The fact that Classes II and III restriction involve fairly specific cleavages has led to the suggestion that these enzymes may, at some time, have been important not solely for protection of the host genome, but for site-specific recombination. Similarly Class I enzymes may have been evolved for non-specific recombination. We will not pursue these ideas here, but geneticists may like to bear this in mind.

9.7 Joining DNA molecules together: DNA ligase

differences between E. coli and T4 ligases

As indicated previously, to produce recombinant molecules, we must be able to join different pieces of DNA together. This can be achieved using isolated preparations of the enzyme DNA ligase from *E. coli* or from the bacteriophage T4. *In vivo*, the enzyme catalyses the repair of single strand breaks in duplex DNA molecules in addition to playing a role in normal DNA synthesis. The reaction involves the formation of a phosphodiester bond between the 3' OH and 5' phosphate groups of adjacent nucleotides, and requires an energy input. The bacterial enzyme uses NAD^+ (nicotinamide adenine dinucleotide) as the energy source, whereas the bacteriophage enzyme makes use of ATP. Both require double stranded DNA molecules as substrates, and will not link single stranded DNA. In addition, the bacterial ligase will not join two blunt-ended DNA molecules, since it requires some overlap, whereas T4 ligase will link blunt-ended fragments. This point is important when considering an experimental strategy; it is no use using *E. coli* DNA ligase to add linkers on to blunt-ended cDNA for example.

mechanism of action of ligases

The mechanisms of action of the two types of ligase are very similar (see Figure 9.4), involving three stages: formation of an activated enzyme-AMP complex, transfer of the activated AMP to the 5' phosphate group at the nick, and nucleophilic attack by the 3' OH group on the activated 5' phosphate. In the figure, the mechanisms are illustrated using sealing of a nick in a single strand of a DNA molecule. For joining two duplexes together the same reactions occur, but twice - once on each strand.

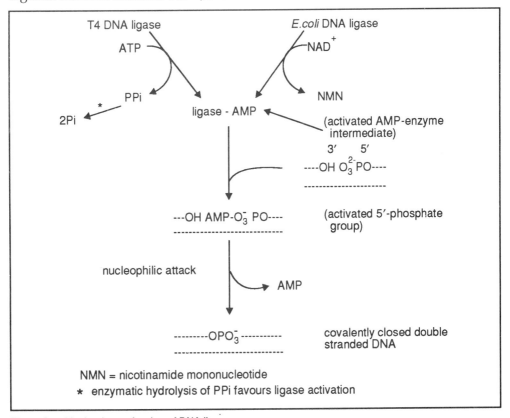

Figure 9.4 Mechanisms of action of DNA ligases.

∏ The activated enzyme complex involves a covalent phosphoamide bond between AMP and a lysyl residue - what would this look like?

The bond will be between the primary amino group at the end of the lysyl side chain and the phosphate group of the AMP as follows:

```
                    H
                    |
enzyme — lysine — N — H
                    |
               ⁻O — P = O         phosphoamide bond
                    |
                    O
                    |
                  ribose
                    |
                  adenine
```

Ligases work optimally at a temperature of 37°C for the repair of nicks. However for cloning procedures, where the joining of both strands of DNA are required this is usually too high.

∏ If you wished to join two DNA fragments which had been prepared by *Eco* RI digestion, which ligase would you choose and why would 37°C be too high a temperature at which to carry out a ligation reaction?

Firstly, we must consider what the ends of our fragments would be like. Using *Eco* RI, fragments with 5' overhangs as shown below will be produced.

```
   5'          3'              5'          3'
  AATT---------------          AATT---------------
       --------------- TTAA         --------------- TTAA
   3'          5'              3'          5'
```

3' ends will terminate in OH groups and 5' ends in phosphate groups. Ligation requires the sticky ends of the fragments to base pair as below:

```
         ------- A A T T -------
         ------- T T A A -------
```

temperature used is a compromise

As the overlap is only 4bp long, and involves all A-T base pairs, the interaction is not strong, (since AT base pairs only possess two hydrogen bonds) and is likely to be disrupted fairly easily at the relatively high temperature of 37°C. It is for this reason that ligation reactions are commonly carried out at between 4°C and 20°C over long (4-16 hour) time periods. The precise temperature used is a compromise between that allowing a good enzyme rate and that allowing annealing of the ends of fragments. For joining sticky-ended fragments such as those generated by *Eco* R1, either *E. coli* or T4 ligase may be used.

9.8 Use of restriction enzymes and DNA ligase in cloning

Cloning is a term used to describe the biological amplification of a piece of DNA. This involves introducing the DNA into a host organism in a form such that it is retained by it, and is passed on to successive generations. This places a requirement for joining the DNA to be cloned to vector DNA that will be accepted and maintained in the host.

Detailed discussion of DNA cloning is the subject of other books in this series, and these should be consulted for further information (see for example 'Techniques for Engineering Genes'). A particular problem arises from the presence in eukaryotic genes of non-coding regions (introns) and the resolution of this by the production of copy DNA (cDNA) is discussed in the next chapter. The most commonly used cloning vectors are plasmids, cosmids and bacteriophages. Common features which all cloning vectors must have include:

- an origin of replication that is recognised by the host cell;

- a marker gene to allow selection of organisms which have received the vector. These are usually genes whose expression confers resistance against an antibiotic. A bacterium transformed by such a plasmid would become resistant to, for example, ampicillin, if the plasmid DNA is correctly expressed;

- one or more unique Class II restriction endonuclease recognition sites into which DNA for cloning may be inserted.

commonly used plasmids

Plasmids are small circular pieces of DNA capable of being maintained in bacteria independently of the genome. The commonly used cloning vectors have evolved by modification of native small plasmids such as Col E1 and RSF 2124 found in wild type populations. Examples of commonly used plasmids include pBR322 and pUC19 (see Figure 9.5). pBR322 was one of the first plasmids to be developed in the laboratory.

To clone a piece of DNA using this plasmid the following steps must be taken:

- the plasmid is linearised by incubation with one of the restriction enzymes that has a single recognition sequence within the plasmid DNA and which produces DNA with sticky ends;

- DNA to be cloned is prepared such that it has sticky ends complementary to those produced on the linearised plasmid in the previous step;

- linearised plasmid and DNA to be cloned are incubated with T4 DNA ligase and ATP or *E. coli* ligase and NAD^+ at an appropriate temperature overnight. This ligation is much more efficient when sticky ended DNA fragments are used rather than blunt-ended ones;

- the solution is used to transform bacteria (ie the DNA is introduced into the bacteria) which are then grown on agar plates in the presence of ampicillin and/or tetracyclin.

random cloning of genomic DNA

This procedure raises several important questions. Firstly, how do we prepare the piece of DNA which we want to clone, such that it has the correct sticky ends? The answer to this depends on the nature of the DNA to be cloned. If we just want to clone random pieces of genomic DNA the best way is to digest it with the same restriction enzyme with which we have linearised the plasmid, or with one that gives the same sticky ends. Various restriction enzymes are known which recognise different DNA sequences but which produce the same sticky ends eg *Bam* HI and *Sau* 3A.

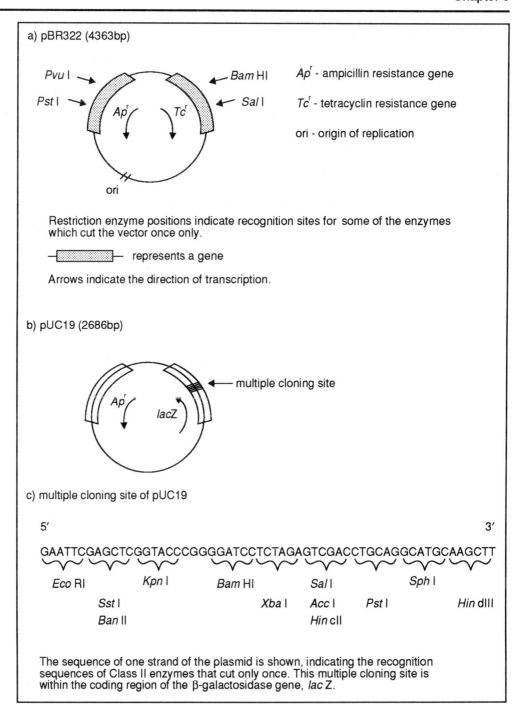

Figure 9.5 Bacterial plasmids pBR322 and pUC19 and the multiple cloning site of pUC19.

⊓ The recognition site for *Bam* HI is 5'GGATCC3' and for *Sau* 3A is 5'GATC3'. Where might you expect these enzymes to cut these sequences?

Use of enzymes in molecular biology and biotechnology

In fact, both enzymes cut to produce a GATC overhangs. This is an interesting example to use, as partial *Sau* 3A digestion is often used to produce random genomic DNA fragments for cloning, and many vectors contain unique *Bam* HI sites.

use of linkers Another way of producing DNA with suitable ends for cloning is to ligate specific 'linkers' on to the ends of blunt ended DNA. Linkers are short double-stranded oligonucleotides which contain within them the recognition site(s) for one or more restriction enzymes. For example an *Eco* RI linker may have the sequence:

```
5' CCGAATTCGG 3'
3' GGCTTAAGCC 5'
```

This contains within it the GAATTC recognition sequence for *Eco* RI.

Linkers are ligated to blunt-ended DNA using T4 DNA ligase, the linkers being present in vast molar excess of the DNA to be cloned. The reason for this is that we do not want the DNA fragments to join together tandemly, but we want each end of each fragment to have at least one linker joined on to it.

The next step is to digest the ligated material with *Eco* RI to produce the required ends (Figure 9.6). Following separation from excess linkers by column chromatography or gel electrophoresis, these fragments are now ready for cloning.

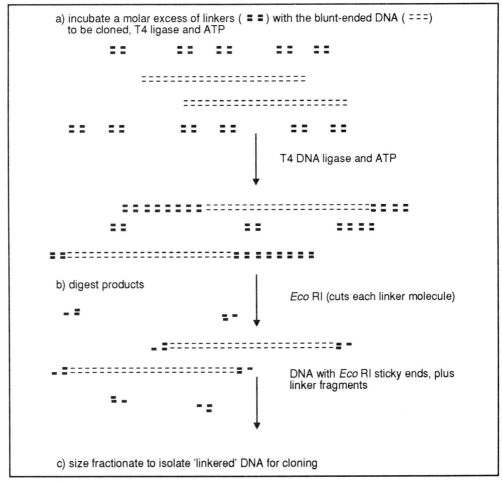

Figure 9.6 Preparation of sticky ends by linker ligation and restriction enzyme digestion.

⁇ What precaution should be taken to protect the DNA fragments in this example from digestion at any internal *Eco* RI sites?

The answer, of course, is to block any internal sites. To do this we treat the DNA with the appropriate methylase enzyme (M.*Eco* R1) prior to ligation with unmethylated linkers. Still other ways exist of modifying the ends of DNA molecules to facilitate cloning, and some of these will be mentioned in Chapter 10.

selection of bacteria transformed with recombinant DNA

An important point about the cloning process is that we need to be able to distinguish those bacteria which have taken up recombinant DNA from those that have taken up only vector and those which have not taken up any DNA. First of all then, we should consider what the products are from the ligation reaction containing the linearised plasmid and the DNA of interest. The reaction is carried out at a moderate DNA concentration such that production of recombinant molecules is favoured, and with equimolar amounts of plasmid and DNA to be cloned. Various products will form in this mixture as shown below:

1) Re-circularised plasmid.

2) Circularised plasmid containing DNA for cloning (recombinant molecules).

3) Tandemly joined pieces of DNA (either plasmids or DNA for cloning or combinations of these).

1) and 2) may both be stably maintained in bacteria into which they are introduced, but it is only the recombinant molecules in 2) that we are interested in. Pieces of DNA without origins of replication will not be maintained in bacteria into which they are introduced, and plasmid dimers tend to be unstable and subject to recombination and elimination from bacterial cells.

use of insertional inactivation of antibiotic resistance

To distinguish between the uptake of recombinant DNA molecules and native vectors, use is made of cloning sites in selectable markers of plasmids. Hence for cloning into pBR322 it is useful to use one of the restriction sites in either the ampicillin or tetracyclin resistance genes. For example, cloning into the *Bam* HI site (see Figure 9.5). Following transformation of bacteria with the ligation products, bacteria are plated on to agar plates containing ampicillin. Only bacteria which contain the vector will grow (as the bacterial strain used does not carry the resistance gene in its genome). If the colonies are replica-plated onto tetracyclin-containing plates (Figure 9.7), only those colonies containing the unmodified vector will grow. This is because for colonies containing the recombinant vector, the tetracyclin resistance gene is interrupted by the inserted DNA and hence cannot be correctly expressed. By comparing the two plates, colonies containing the cloned DNA may be identified for future study.

Use of enzymes in molecular biology and biotechnology

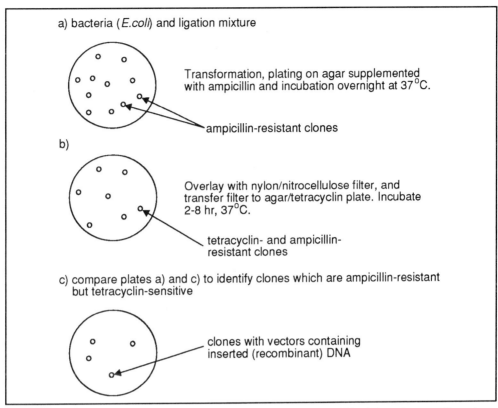

Figure 9.7 Replica plating in screening for clones carrying recombinant DNA. a) Bacteria treated in such a way as to be competent to accept foreign DNA are transformed with DNA in the ligation mixture. Cells which have obtained plasmids from the ligation mix are selected using ampicillin. b) Overlaying the plates with a filter transfers a fraction, but not all, of each colony to the filter. When this is transferred to a plate supplemented with tetracyclin, only those colonies of bacteria containing plasmids with a functional tetracyclin resistance gene will grow. c) As highlighted in the diagram, comparison of the two plates indicates those clones on the first one which contain an interrupted (and thus non-functional) Tc^r gene. These are the clones containing recombinant plasmids.

choice of bacterial strain as host

The third major point about the cloning procedure involves the choice of bacterial strain. If the DNA introduced into the cells in the transformation process is not to be degraded immediately, mutated bacterial strains which lack endogenous restriction-modification systems must be used. The most commonly used bacterial hosts for cloning work are modified strains of *E. coli* K12. In these strains the endogenous *Eco* B and *Eco* K activities have been eliminated. If eukaryotic genomic DNA is to be cloned it is also important to use strains deficient in the recently discovered McrB system, which restricts methylated DNA if methylation occurs at Cs in 5' GC 3' dinucleotide sequences. Such methylation may occur in plant and vertebrate DNA, and the presence of this restriction system would lead to under-representation of methylated regions in random libraries produced in hosts containing it. We will return to a discussion of eukaryotic DNA methylation in Section 9.11.

McrB system

Cloning into cosmid and bacteriophage vectors involves the same considerations, and you are advised to consult the BIOTOL text 'Strategies for Engineering Genes' for further information.

SAQ 9.3

β-Galactosidase (the product of the *lac* Z gene) is the enzyme used by bacteria to break down lactose. The *lac* Z gene may be induced artificially by a chemical analogue called isopropyl-β-D-galactoside (IPTG). Additionally, β-galactosidase will break down the chromogenic substrate 5-bromo-4-chloro-3-indolyl-β-D-galactoside (X-gal) to yield a dark blue product. Using this information, consult Figure 9.5 and explain how bacteria transformed with recombinant pUC19 may be selected in a cloning experiment.

9.9 Use of restriction enzymes in diagnosis of genetic disorders

Many common disorders occur due to inheritance of mutated genes from one generation to the next. Such disorders range from very mild (for example, colour blindness) to very severe (for example, brittle bone disease). It is useful to be able to diagnose inheritance of affected genes at the DNA level. This information is of use, for example, in prenatal diagnosis, giving a firm basis for decisions involving elective abortion.

One technique for identification of genes responsible for particular disorders is to follow inheritance of different forms of the genes through affected families and compare these to inheritance of the disease phenotypes. This relies on the fact that everyone's precise DNA base sequence is unique. This sequence diversity results in the production of different fragments on digestion of DNA from different individuals with restriction enzymes. This is illustrated in a hypothetical example in Figure 9.8 where the effects of sequence differences in a small region of a single chromosome are indicated.

Use of enzymes in molecular biology and biotechnology

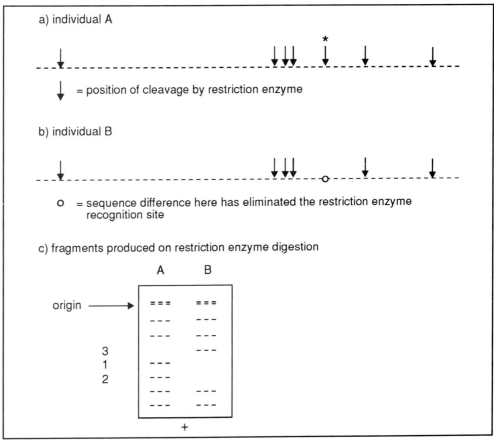

Figure 9.8 Effects of sequence diversity on restriction fragment length. DNA fragments from individuals A and B following restriction enzyme digestion are separated on the basis of size by agarose gel electrophoresis. Smaller fragments run fastest through the gel. Bands are visualised by including ethidium bromide (an intercalating dye) in the running buffer, and 1 and 2 are fragments produced due to the presence of the restriction site marked with an asterix in 1). 3 is the fragment produced on abolition of this recognition sequence. It has a length equal to that of 1 plus 2.

Π Obviously the situation in reality is much more complex and not so easy to visualise. Why do you think this is?

As diploid organisms, we all have pairs of chromosomes, one inherited from our father and one from our mother. These are themselves not identical in terms of precise sequence. Also, the human genome is very large, so that restriction enzymes will produce literally millions of fragments upon digestion of genomic DNA. This means that it is not simple to identify differences in one particular region. To achieve this we need to have available DNA probes which cover or are close to the regions we wish to study. DNA probes are regions of DNA which have been isolated by cloning and characterised in terms of base sequence and chromosomal localisation ie the region of the genome from which they originate. These probes may be 'labelled' by using either radioactive phosphorus, fluorescent dyes or biotin etc. Such labelled probes may be used to detect differences in sequences in the homologous regions of the genome by studying the lengths of restriction fragments to which they hybridise in DNA from different individuals. The presence of different restriction fragments is known as

use of DNA probes

restriction fragment length polymorphism, RFLP

restriction fragment length polymorphism (RFLP)). The process used to identify RFLPs is illustrated in Figure 9.9.

a) Genomic DNA from a number of unrelated individuals is incubated with a restriction endonuclease for 6hr at the optimal temperature.

b) Fragments produced are separated by agarose gel electrophoresis:

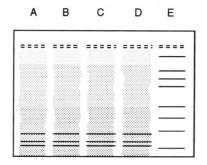

fragments appear as a smear with several discrete bands (due to high repetitive 'satellite' DNA) visible

A, B, C, D = genomic DNA samples. E = marker DNA fragment of known lengths.

c) DNA is transferred to a solid support (eg nitrocellulose or nylon membrane) following denaturation by alkali treatment and neutralisation. This transfer is known as Southern blotting:

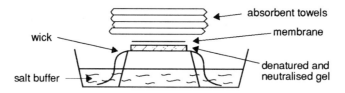

d) DNA is fixed to the membrane by UV irradiation or baking, and the membrane is incubated with a radioactively labelled denatured DNA probe overnight. Following incubation, excess probe is washed away and the membrane is exposed to X-ray film. The film is developed and may appear as shown below on the left if A, B, C and D have the genotypes shown on the right.

(* = restriction site):

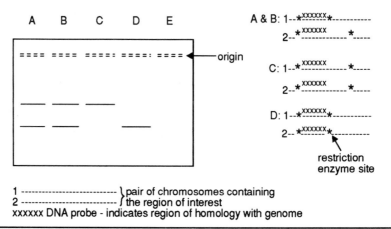

1 --------------------------- } pair of chromosomes containing
2 --------------------------- } the region of interest
xxxxxx DNA probe - indicates region of homology with genome

Figure 9.9 Identification of RFLPs associated with specific DNA probes.

This involves separating fragments produced on restriction enzyme digestion by electrophoresis, followed by denaturing them and transferring them to a solid support. The single stranded fragments are fixed to the supporting filter and allowed to hybridise to single stranded probe DNA. Excess probe is washed away, and fragments to which probe has bound are visualised by a technique appropriate to the label employed.

Using the example shown in Figure 9.9, it could be that A and B are the carrier parents of affected child C who has a severe recessive genetic disorder (ie the child has a defective gene on both homologous chromosomes). This disorder is associated with a gene labelled by the DNA probe used. D is a foetus carried by the mother and is being tested for its status regarding the disease.

∏ Is D likely to be affected by the disorder?

use of linkage analysis with inherited disease

No, because if C is affected the disorder must be associated with the larger restriction fragment in this family. As foetus D has only received chromosomes containing the additional restriction enzyme site from its parents it will be unaffected, not even a carrier. This type of analysis is known as linkage analysis - where RFLPs track through a pedigree along with a disease phenotype.

In attempts to identify RFLPs for which a probe is available, Type II restriction enzymes are tested at random on panels of DNA from unrelated individuals of the same ethnic group. Some RFLPs are only observed in certain populations due to founder effects. There may be many RFLPs associated with a given region or very few. As not all RFLPs are necessarily informative in linkage studies, the more that are available the better.

9.10 Genetic fingerprinting

analysis of hypervariable regions of minisatellite DNA

This technique, developed by Alec Jeffries at Leicester University, UK, relies on the use of restriction enzymes to excise repeated sequences dispersed throughout the genome. Eukaryotic DNA contains large amounts of repeated sequences, some of which are known as minisatellite DNA or hypervariable regions (HVRs) and are made up of short 'core' sequences reiterated variable numbers of times in tandem (Figure 9.10). The precise sequence of each tandem copy of the core is subject to variability due to accumulation of mutations, and between loci the repeated sequences may vary markedly.

DNA fingerprinting

Minisatellite regions of DNA evolve relatively rapidly due, it is thought, to recombination at meiosis and mitosis, or slippage during replication. These events result in differences in lengths of repeated regions at similar points (alleles) on homologous chromosomes (Figure 9.10). This diversity shows up as a series of RFLPs when DNA is cut with suitable enzymes and is the basis of DNA fingerprinting.

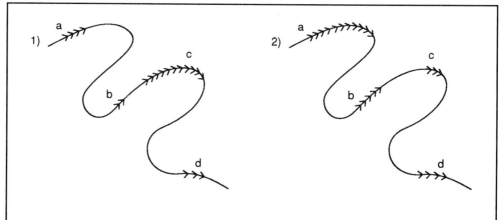

1 and 2 are a pair of homologous chromosomes from a particular individual. One will have been inherited from each parent.

  ⟶⟩⟩⟩⟩⟩⟩⟩⟩— = minisatellite sequence

  ⟶ corresponds to a single copy of the core sequence which is repeated in each minisatellite.

a, b, c, d = hypervariable regions (HVRs) within the chromosomes where mini--satellite sequences are observed. The number of tandem repeats at each locus may vary markedly between chromosomes. This variability is the basis of genetic fingerprinting.

Figure 9.10 Minisatellite DNA (HVRs).

stringency of hybridisation is affected by temperature and salt concentration

Minisatellites appear to exist as families related by the similarities or homologies of their core regions. Indeed cloned HVRs may be used as DNA probes to detect alleles from related loci on Southern blots using suitable hybridisation conditions. The stringency of hybridisation conditions is determined mainly by temperature and salt concentration of the mixture. Decreased temperature and increased ionic strength give a lower stringency, allowing the probes to anneal to DNA to which it is not completely complementary. Conversely, increased temperature and decreased salt levels increase stringency of hybridisation.

⏸ Would high or low stringency conditions be more appropriate for detecting families of HVRs with a single cloned region as probe?

The answer should be obvious - low stringency conditions are more appropriate. This will allow hybridisation of the probe to related (but non-identical) HVRs dispersed throughout the genome. The pattern of regions identified is the genetic fingerprint, and is specific to each individual. The pairs of alleles in the fingerprint are inherited in a normal Mendelian way, one arising from maternal and one from paternal genetic information (Figure 9.11).

Use of enzymes in molecular biology and biotechnology 269

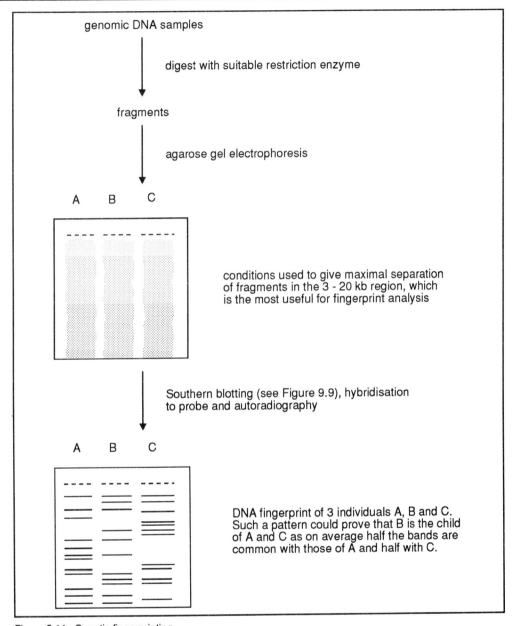

Figure 9.11 Genetic fingerprinting.

choice of restriction enzymes for DNA fingerprinting

We have so far failed to address one important question - which restriction enzymes may best be used for this kind of analysis? Well, the requirement is that the enzyme should cut frequently, but not within the repeated sequence.

∏ Which of the Class II enzymes would be good candidates for frequent cutting?

Those which have a short recognition sequence. Enzymes commonly used for fingerprinting and their recognition sites are shown in Table 9.2.

Enzyme	Recognition sequence
Hin fI	5'-GA*NTC-3' 3'-CTNA*G-5'
Alu I	5'-AG*CT-3' 3'-TC*GA-5'
Hae III	5'-GG*CC-3' 3'-CC*GG-5'
Mbo I	5'-*GATC-3' 3'-CTAG*-5'

* indicates position of cleavage

Table 9.2 Restriction enzymes commonly used in genetic fingerprinting.

Π To what uses other than for the determination of family relationships may genetic fingerprinting be put?

uses of genetic fingerprinting

You should have managed to prepare quite a list! Some of the obvious uses include forensic analysis in cases of rape, assault and other crimes, and determination of whether twins are dizygotic (fraternal, thus having individual fingerprints) or monozygotic (identical, arising from a single fertilised egg and hence having identical fingerprints). The latter application may be important for various medical procedures including organ or tissue transplantation. Mention of this is interesting at this stage, as fingerprinting may be used to monitor success or otherwise of bone marrow transplants between unrelated individuals. Samples of the marrow and/or white blood cells may be obtained at various times following transplantation and their genetic origin tested as an indication of how well the transplanted tissue has colonised the patient's tissue and how active it is.

Genetic fingerprinting can also be used as a quality assurance measure, for example as a method to prove the identity of micro-organisms used in industrial processes.

Many other applications for this technique are known and more detailed descriptions of these may be found in other volumes of this series.

9.11 Detection of transcriptionally active genes

It appears that in certain groups of eukaryotes, particularly mammals and plants, DNA may be subjected to methylation. This methylation, unlike the major type observed in prokaryotes (Section 9.2), involves cytidine nucleotides. The base is modified to 5-methyl cytosine:

Use of enzymes in molecular biology and biotechnology

methylation of cytidine nucleotides in eukaryotic DNA

In animals up to 5% of the cytosines are methylated although the significance of this is uncertain. Even higher levels are observed in plant DNA. The DNA from insects, however, shows no methylation. Only particular cytosines are modified; in animals it is only C residues in the sequence ...CpG..., and the cytosines in both strands are methylated. In plants methylation occurs in the sequence ...CpNpG... where N is any base.

Methylation is maintained when DNA replicates by a specific methylase that modifies hemi-methylated DNA (Figure 9.12).

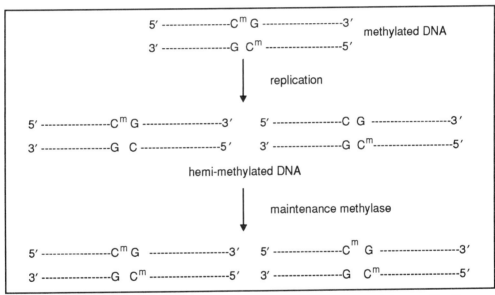

Figure 9.12 Maintenance of methylation in eukaryotic DNA. The maintenance methylase modifies C residues within the methylation sequence on newly synthesised DNA strands. It will not initiate modification of CG sequences not previously methylated, and thus differs from bacterial methylases.

methylation appears to affect gene expression

Selection of sites for initial methylation occurs during embryonic differentiation by unknown mechanisms. However, evidence is now becoming available to indicate that methylation, particularly in regions close to the 5' end of genes, may be responsible for their tissue-specific inactivation. Hence in general, methylated genes are less likely to be expressed. This property has been made use of in identification of genes within genomic DNA. If there are no sequence data available for a particular gene (eg one that when mutated causes a particular disorder) but its approximate chromosomal location has been identified by genetic linkage studies, mapping of the area involved for differentially methylated regions allows the possible gene loci to be identified for detailed studies. By differentially methylated regions, we are referring to locations within the region of interest which are methylated in DNA extracted from tissues where the gene is not expressed and unmethylated in samples from tissues where it is expressed.

use of restriction enzymes to identify genes

A convenient way of monitoring the presence or absence of methylation is to use a pair of restriction enzymes which recognise the same sequence, one being able to digest DNA regardless of the methylation state of the recognition sequence, but the other only recognising either methylated or unmethylated DNA. Such a pair of restriction enzymes are *Msp* I and *Hpa* II. Both recognise the sequence 5'-CCGG'3' but only the former cleaves the DNA when methylated at the second C residue. Both enzymes cleave the unmethylated sequence, the cleavage site being between the two C residues. If a probe was available covering a region near to the 5' end of a candidate gene, analysis such as that indicted in Figure 9.13 could be used to determine the likelihood of the gene being responsible, for example, for a disorder involving the lungs but not the heart. Hence in the example shown in Figure 9.13, the restriction fragment profiles from heart DNA indicate methylation near to the probe regions, whereas this is absent in lung DNA. This may indicate the nearby presence of the gene being sought. Such strategies helped, for example, in the identification of the cystic fibrosis gene before the protein encoded by it had been isolated.

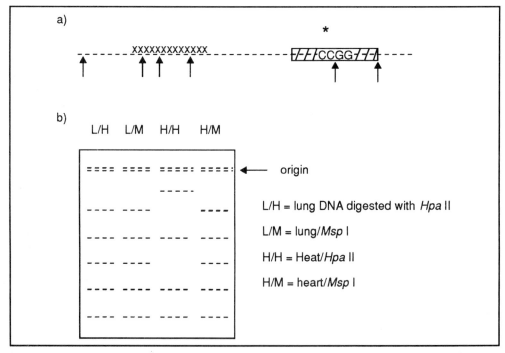

Figure 9.13 Use of *Hpa* II and *Msp* I to identify a candidate gene for a disorder involving the lungs, but not the heart. a) Shows the region of interest on a chromosome (-------). The hatched box indicates the approximate position of the gene under consideration. xxxxxxx = available DNA probe homologous to a nearby sequence. ↑ = recognition site for *Msp* I. * indicates a C residue in recognition site, close to the 5' end of the gene, which may or may not be methylated. b) Samples of DNA from the lungs and heart are digested with *Hpa* II and *Msp* I, separated by agarose gel electrophoresis, Southern blotted and hybridised to labelled probe (see Figure 9.9). An autoradiograph of the filter may appear as shown. The results indicate that the gene in question is methylated in the heart but not in the lungs, and hence may be the disease gene sought.

9.12 Identification and purification of restriction enzymes

detection of restriction enzymes

Generally, the presence of restriction-modification systems within a given bacterial strain may be indicated using phage infection experiments such as described in Section 9.2.

Alternatively, the presence of restriction enzymes may be detected in crude cell extracts by comparing activity on foreign versus host DNA. Various assays have been used based on this ability to recognise foreign DNA and digest it into discrete reproducible fragments. The formation of these fragments can be monitored using various methods, the best of which is agarose gel electrophoresis.

As we have seen, some organisms contain more than one restriction-modification system, each system recognising a different sequence. These different specificities can be identified as purification proceeds, by testing the samples produced for their ability to digest standard DNA after each step.

Π What kind of methods would you expect to be of use in purification of restriction enzymes?

The answer to this can be as simple or as complicated as you care to make it. In general, restriction enzymes are soluble proteins and therefore the commonly used techniques of ion exchange chromatography, ammonium sulphate precipitation, gel filtration etc are suitable. The precise series of steps will vary depending upon the enzyme. (Details of considerations for enzyme purification are given in the BIOTOL text 'Principles of Enzymology for Technological Applications').

The recognition sequences for restriction enzymes may be identified using DNAs of known sequence as substrate, monitoring the size of the fragments produced and looking at the effects of mutating the sequence around the site where cleavage occurs.

Mutations may be introduced at specific points (ie by changing individual bases) using site-directed mutagenesis. This will be described in detail in the next chapter.

SAQ 9.4

Sketch out a scheme showing the kind of experiments you might carry out to determine the recognition site of a Class II restriction enzyme, using a piece of DNA of known sequence as substrate, and assuming that the restriction enzyme only cuts it once.

Summary and objectives

Restriction enzymes are of great importance in medicine (RFLP and linkage studies), in industry (for cloning commercially important genes) and elsewhere within society (use of genetic fingerprinting). Of the three classes known, only type II enzymes are important for these uses, as only these enzymes cut at specific DNA sequences. We have described their properties and uses. DNA ligases are used to join fragments of DNA produced by the use of restriction enzymes: the properties and uses of bacteriophage T4 and *E.coli* DNA ligases have also reviewed. Useful features of commonly used vectors for cloning were described and the main steps in gene cloning summarised. Use of restriction enzymes for diagnosis of inherited disease, genetic fingerprinting and the detection of transcriptionally-active genes have been discussed.

Now that you have completed this chapter you should be able to:

- describe the concept and nomenclature of bacterial restriction-modification systems;

- distinguish between Class I, II and III restriction-modification systems;

- discuss the merits of using T4 or *E.coli* DNA ligase in ligation reactions, and the conditions required for both;

- describe the use of restriction enzymes for cloning into vectors such as pBR322 and pUC19, and discuss methods for recombinant selection;

- discuss the concept of RFLPs and their use in tracing disease genes through pedigrees;

- describe the basis and some uses of genetic fingerprinting;

- discuss the use of restriction enzymes in identification of transcriptionally active genes;

- understand how restriction enzymes may be identified and their recognition sequences determined.

Use of enzymes in molecular biology and biotechnology: enzymes other than restriction endonucleases

10.1 Introduction	276
10.2 Reverse transcriptase (RT)	277
10.3 Use of calf intestinal alkaline phosphatase (CIP) in DNA cloning	280
10.4 Modification of DNA for cloning using terminal deoxynucleotidyl transferase	282
10.5 *Taq* polymerase and the polymerase chain reaction (PCR)	284
10.6 Chemical modification of enzymes for experimental use	290
10.7 Site-directed mutagenesis: The basis of protein engineering	295
Summary and objectives	298

Use of enzymes in molecular biology and biotechnology: enzymes other than restriction endonucleases

10.1 Introduction

In the previous chapter, the important contributions of restriction enzymes to recombinant DNA work and diagnosis were discussed. In this chapter, other important enzymes including *Taq* DNA polymerase, terminal transferase, Klenow fragment of DNA polymerase I and reverse transcriptase will be considered. In addition, ways in which enzymes have been chemically and biologically altered to optimise their experimental properties will be illustrated.

presence of introns in eukaryotic genes

The discovery in the late 1970's that eukaryotic genes contain discontinuous coding regions (Figure 10.1) was, at the time, surprising and disconcerting. The presence of non-coding intervening sequences, or introns, makes the genes very long and not easy to study in terms of identifying the coding regions. However, these introns are removed from initial RNA transcripts of the genes to yield the mature forms with continuous uninterrupted coding regions. Subsequently it was discovered that use can be made of enzymes found in retroviruses, called reverse transcriptases, which are RNA-dependent DNA polymerases (ie make DNA copies of RNA) to make DNA copies of mature RNA. Discovery and use of reverse transcriptase has revolutionised studies of eukaryotic genes and is the subject of Section 10.2.

In Section 10.3 we describe the use of alkaline phosphatase in DNA cloning and in Section 10.4 we describe the use of deoxynucleotidyl transferase to modify DNA for cloning.

In addition to the restriction endonucleases, many other prokaryotic enzymes of use for *in vitro* manipulation of DNA have been described. One of the most recent and, arguably, most important of these is the thermostable DNA polymerase of *Thermus aquaticus* (*Taq* polymerase) which has proved invaluable in the amplification of DNA. This is discussed in detail in Section 10.5.

In addition to enzymes in their native state (such as those described above and others discussed in Sections 10.2-10.5), several enzymes have been modified *in vitro* by chemical or biological means to optimise activity for exploitation in experiments. Examples include T7 DNA polymerase which has been chemically modified for use in DNA sequencing (Section 10.6.2) and *E. coli* DNA polymerase (Section 10.6.1), for DNA probe preparation. Methods available for site-specific mutation of enzymes are varied, and two common ones are discussed in Section 10.7.

Some of the methods included here (eg DNA sequencing, site-directed mutagenesis) have been described elsewhere in the BIOTOL series. Their inclusion in this chapter is for the purpose of describing the enzymes involved and the theories behind the techniques, and should provide useful revision.

Use of enzymes in molecular biology and biotechnology

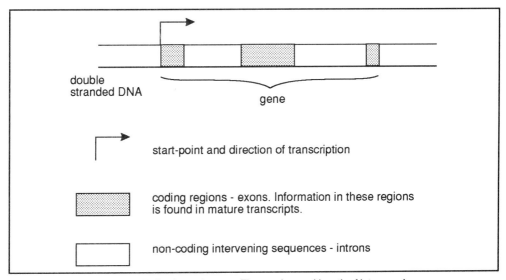

Figure 10.1 Structure of a typical eukaryotic gene. The number and length of intron and exon sequences vary markedly from gene to gene. Bacterial genes are uninterrupted, not containing introns.

10.2 Reverse transcriptase (RT)

This enzyme plays a vital role in the life cycle of retroviruses. These organisms have RNA genomes, which, following infection of suitable hosts, are 'copied', or reverse transcribed, by the enzyme to produce double stranded complementary DNA (cDNA). This DNA is introduced into the host genome to produce a 'provirus', and transcribed to direct synthesis of many new phage particles. Reverse transcriptases are able to copy RNA as they contain multiple enzyme activities within a single molecule.

Π Considering the reactions required to create double stranded DNA, what do you think these activities are?

properties of reverse transcriptase

There are at least three activities. Firstly, RNA-dependent DNA polymerase activity for copying the RNA. Secondly, a ribonuclease (RNase H)-type activity to degrade the RNA; and, finally, a DNA-dependent DNA polymerase activity to copy the single strand to form a duplex.

Unlike bacterial DNA polymerase, RT does not have a 3'-5' exonuclease activity. This activity is important in rapidly polymerising systems in that it provides for proof-reading. Its omission from the viral enzyme means that mutations may accumulate in the genome relatively rapidly, explaining the rapid evolution of retroviruses.

Generally, the two polymerase activities of RT are inseparable, containing a common active site, and the enzyme is a zinc metalloenzyme, requiring a primer for extension. The RNAse H activity is that of an exoribonuclease, requiring free RNA ends upon which to act, and liberates short oligonucleotides. This contrasts with 'normal' cellular RNAse H which is an endonuclease (Figure 10.2). The RNAse H activity of RT resides on the same polypeptide as the two polymerase activities, but has a separate active site.

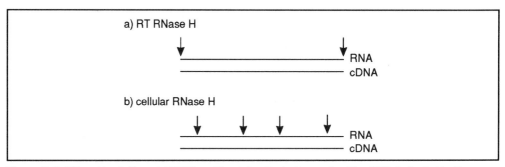

Figure 10.2 RNase H activities. a) RT RNase H activity attacks free ends of the RNA (indicated by the arrows) in an RNA-cDNA hybrid. b) Cellular RNase H attacks within the RNA molecule of RNA-cDNA hybrids.

structure of reverse transcriptase

The two most commonly used RTs for recombinant DNA work are those from avian myoblastosis virus (AMV) and murine leukaemia virus (MuLV). AMV-RT has a molecular mass of 170 000 Daltons, and consists of two subunits - α and β. The α subunit alone contains all three enzyme activities, but appears to be stabilised by the β subunit in terms of its template interaction. MuLV-RT on the other hand comprises a single subunit of molecular mass 84 000 Daltons.

use of reverse transcriptase to produce cDNA

The mechanism by which RT works *in vivo* to produce DNA for integration into the host genome is complex, but the enzyme's RNA-dependent DNA polymerase activity is made use of in simple experiments to copy RNA *in vitro*. As we have seen in the introduction to this chapter, most eukaryotic genes are interrupted by non-coding regions. If we wish to study just the coding information we only want that present in the mature mRNA. This can be obtained in DNA form by copying mature mRNA, using RT. Hence the major use of this enzyme has been in synthesising cDNA copies of eukaryotic mRNA for cloning and for direct amplification (see Section 10.5). A useful feature of many eukaryotic mRNAs is that they contain a stretch of several hundred adenine nucleotides at the 3' end, known as the polyA tail. Hence, incubation with primer oligo(dT)$_{12-18}$ (that is an oligonucleotide of deoxythymidine nucleotides between 12 and 18 residues long) and RT allows them to be copied easily (Figure 10.3). Various methods have been used to convert this single stranded cDNA to the double-stranded form (Figure 10.4) but the most effective way is that illustrated in Figure 10.4b.

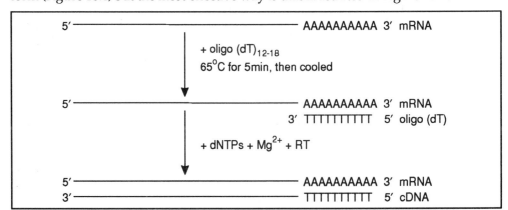

Figure 10.3 Synthesis of single-stranded cDNA. Primer (oligo(dT)) and mRNA are heated together for destruction of any secondary structure in the RNA. On cooling, the primer anneals with the polyA tail. RT subsequently copies the mRNA to produce a complementary cDNA strand by RNA-directed DNA polymerase activity.

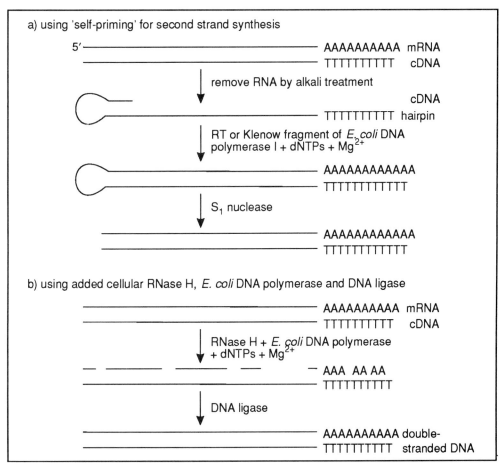

Figure 10.4 Preparation of double-stranded cDNA. a) using 'self-priming' for second strand synthesis. b) using added cellular RNase H, *E. coli* DNA polymerase and DNA ligase. a) RNA is removed from the hybrid product of first strand cDNA synthesis by alkaline hydrolysis, leaving single-stranded cDNA. This folds into a loop at the 3' end by basepairing. The loop acts as a primer for extension by either RT or Klenow fragment (Section 10.6.1) or both enzymes, to produce the second strand. The unpaired hairpin loop region is removed by treatment with an enzyme, S$_1$ nuclease, which preferentially degrades single-stranded DNA. This leaves double-stranded cDNA for cloning. b) Incubation of the mRNA-cDNA hybrid with cellular RNAse H and *E. coli* DNA polymerase I results in replacement of all of the RNA sequences by DNA. The RNAse H degrades the mRNA gradually at multiple points, the remaining RNA acting as primers for the polymerase. Eventually all the RNA is replaced, and addition of DNA ligase joins the fragments together.

Π Why is the method illustrated in Figure 10.4b more popular than that in Figure 10.4a?

Comparing the two, it should be obvious that more information (in terms of DNA sequence corresponding to the 5' end of the mRNA) is retained when using RNase H and *E. coli* DNA polymerase to copy first strand cDNA rather than allowing hairpin formation followed by removing the unpaired sequence with S$_1$ nuclease. Also, S$_1$ nuclease is liable to attack double stranded DNA if the experimental conditions are not maintained very carefully, or if too much enzyme is used, causing further loss of sequences.

For those mRNA species which have no polyA tail, first strand cDNA synthesis may be primed in various ways including:

- randomly, using a mixture of all possible deoxyhexanucleotides. This is most useful if no sequence data are available relating to the RNA sought, or if representation of all RNA species present is required.

- specifically, using an oligodeoxynucleotide primer of complementary sequence to part of the RNA sought.

The disadvantage, again, is that the production of full-length cDNA is unlikely, as these primers will bind part way along rather than at the end of the mRNA template.

10.3 Use of calf intestinal alkaline phosphatase (CIP) in DNA cloning

The way in which DNA is commonly cloned, involves annealing complementary sticky-ended molecules followed by covalent bond formation by DNA ligase.

Π Which chemical groups are required at the 3' and 5' ends of DNA fragments for covalent linkage by DNA ligase to occur?

As we saw in Chapter 9, 3'-OH and 5' phosphate groups are required. In a typical experiment, both DNA strands may be covalently linked (Figure 10.5a). However, this is not necessary: covalent joining of one strand only at a given site is sufficient to produce a molecule stable enough for introduction into the host (Figure 10.5b). The other strand will be repaired inside the host. This property is made use of in cloning experiments to reduce the high background level of non-recombinant clones produced when vectors close on themselves without introduction of the desired DNA into the cloning site.

reduction in number of non-recombinant clones

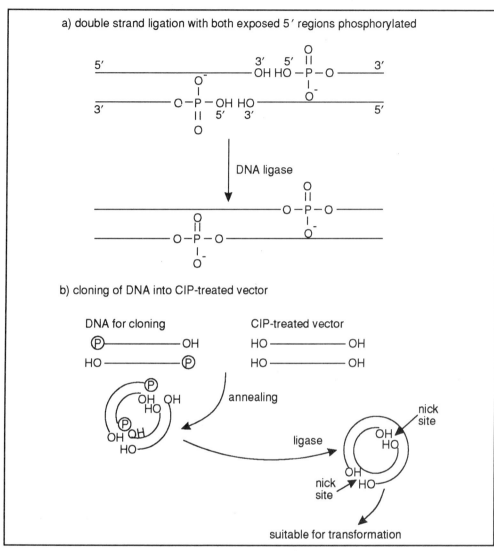

Figure 10.5 DNA ligation. a) double strand ligation with both exposed 5' regions phosphorylated. Both strands are covalently joined by DNA ligase. b) DNA ligase can only covalently join the DNA where 5' phosphate (P) and 3' hydroxyl (OH) groups occur together. Hence in this example, vector molecules cannot be recircularised without an inserted piece of DNA being present. The product of the ligation is a double-stranded DNA with two nicks in it which the enzyme cannot join. This is sufficiently stable to be introduced into a host organism, which itself contains enzymes to repair at these nick sites.

The procedure involves removal of the 5' phosphate groups from either the DNA to be cloned or the linearised vector by CIP. Thus, when the two populations of DNA are mixed and ligated, it will not be possible for two molecules of the CIP-treated DNA to be joined together covalently by ligase.

Π Do you think it is better to phosphatase-treat the vector or the DNA to be cloned?

The answer is the vector. Untreated vector could recircularise and will be maintained inside the host after transformation giving a high background on plating.

To prevent cloning re-circulated plasmid (or other vector), CIP treatment of the plasmid should be carried out. This is demonstrated in Figure 10.5b.

∏ What potential problem remains?

The remaining problem is that more than one piece of DNA for cloning may be introduced into the vector, as there is nothing to stop two phosphorylated DNA fragments from being ligated. In fact, this is relatively easy to overcome, but outside the range of this discussion (see for example the BIOTOL text 'Techniques for Engineering Genes').

10.4 Modification of DNA for cloning using terminal deoxynucleotidyl transferase

use of homopolymer tails in gene cloning

We saw in the preceding chapter that one way of modifying blunt-ended DNA for cloning is by ligation of specific linkers, followed by digestion with the appropriate restriction enzyme. Another commonly used method is to add a homopolymer tail (ie a stretch of identical nucleotides like the polyA tail of mRNA) to each end of the opposite strands of the DNA to be cloned. The linearised vector is tailed with a homopolymer tail made of complementary nucleotide for annealing and cloning.

properties of terminal transferase

Tailing is achieved by making use of an enzyme isolated from plants or calf thymus called terminal deoxynucleotidyl transferase, or terminal transferase (TdT). This is a small cobalt-requiring protein containing an 8000 and a 26 000 Dalton subunits, which acts as a 5'→3' DNA polymerase. It requires a free 3' hydroxyl group but no template. It works best on DNA with a 3' overhang (such as produced by *Pst* I action), although under appropriate conditions it adds nucleotides to blunt-ended and 3' recessed DNA molecules. *In vitro*, terminal transferase adds purine nucleotides to DNA at an optimal rate when Mg^{2+}, rather than cobalt, is included in the reaction buffer. Conversely, addition of pyrimidine nucleotides is favoured in the presence of cobalt ions.

For cloning, the lengths of tails added by terminal transferase must be carefully controlled. The length of tails produced depends on several factors.

∏ What may these factors be?

factors influencing length of tails

As in any chemical reaction, the amount of catalyst (in this case the enzyme), and the concentration of substrates (the chosen deoxynucleotide and the DNA) are vital. As the enzyme is processive (adding nucleotides one at a time to each terminal), a high concentration of enzyme is needed. This ensures that all termini are extended equally, rather than a few being extended extensively and others not at all, and removes the DNA concentration as a factor determining optimal reaction conditions. The other major parameters will now be the type and concentration of nucleotide, the ionic conditions, and the temperature. Optimal conditions must be determined empirically for each nucleotide.

Tailing with dGTP is self-regulating as TdT stops after adding approximately 30 guanine nucleotides on to each 3′ terminus of substrate DNA. Tailing with other nucleotides requires careful adjustment of reaction times and conditions. Under optimal conditions 15-40 nucleotide dG or dC tails may be added and 30-80 nucleotide dA or dT tails. cDNA is usually tailed with dG as this is reliable and reproducible and hence does not waste valuable sample material in the optimisation of experimental conditions.

Incubation of dG-tailed cDNA with dC-tailed vector results in covalently associated species which can be introduced into hosts (Figure 10.6) and hence cloned.

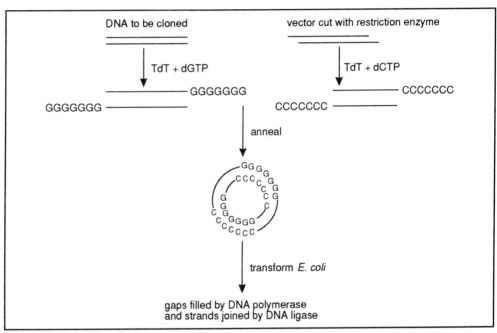

Figure 10.6 Cloning by homopolymer tailing. DNA for cloning and restriction enzyme-cut plasmid are tailed with complementary nucleotides using terminal transferase (TdT). The purified tailed DNAs are mixed and allowed to anneal. This preparation is introduced into *E. coli*, where gaps are filled and DNA ligase joins the strands covalently.

Π Why are these species so stable?

GC pairing stabilisation

The regions whereby the vector and inserted DNA associate are very stable as all the basepairing involves G≡C bonds - the strongest type (since they have three hydrogen bonds). This overlaping region is approximately 30bp long and the co-operativity of the hydrogen bonding involved is great. Once inside the host, any gaps are filled and each strand is covalently sealed by DNA ligase (Figure 10.6).

Π Will CIP-treatment of the tailed vector be required?

No, of course not. As each end of the vector has the same homopolymer tail associated with it, these will not anneal to one another; thus vector recircularisation will not occur.

TdT may also be used to aid amplification of 5' and 3' ends of specific cDNA molecules. This will be discussed later in Section 10.5.

SAQ 10.1

1) Using the information given in Chapter 9 and Section 10.4, draw flow diagrams to compare linker-ligation and homopolymer-tailing methods of preparing cDNA libraries in pUC19. Indicate all the enzymes required.

2) Think of a way that homopolymer-tailing may be used to produce a primer for cDNA synthesis that allows a short-cut in cDNA cloning.

10.5 *Taq* polymerase and the polymerase chain reaction (PCR)

Classical methods for amplifying DNA involve cloning with its associated problems (see above). In the early 1980's, however, a method for direct amplification of DNA fragments was described. This involves the use of a pair of short (15-30 nucleotide) oligonucleotides complementary to regions bounding the fragment of DNA to be amplified (Figure 10.7). The DNA to be amplified is rendered single-stranded by heating at 94°C for a short time. Cooling the solution in the presence of a molar excess of the oligonucleotides allows annealing to occur between them and the DNA to be amplified. On addition of the Klenow fragment (Section 10.6), these short primers are extended in the 5'→3' direction. This cycle of reactions is repeated over and over, the quantity of the DNA sequences between the primers doubling after each cycle (assuming 100% efficiency). Although the products of the first cycle are longer than the target region, after a few rounds of amplification the major product is the sequence that is sought.

use of oligonucleotides bordering the DNA sequence of interest

Π Sketch out the effect of three subsequent rounds of amplification after that shown in Figure 10.7 to convince yourself of this. Remember that the position at which the primers anneal represent the starting points for polymerisation and longer fragments will rapidly be diluted out.

This method, although potentially useful, did not catch on quickly as it had a major draw-back. This is that fresh enzyme has to be added for each round of amplification, as the denaturation step not only produced single-stranded DNA but also destroyed the enzyme activity. This made the procedure very time-consuming and also expensive. These problems were overcome in the late 1980's by the discovery and marketing of a thermostable DNA polymerase (*Taq* polymerase), by the Cetus Corporation. This enzyme is produced by *Thermus aquaticus*, an organism which lives in hot springs. The enzyme is relatively stable at the high temperatures required for DNA denaturation, and has optimal activity at approximately 72°C. This high temperature is ideal for primer extension, as it overcomes problems of mis-priming which often occur at low temperatures. Mis-priming occurs when oligonucleotide primers anneal to sequences to which they are not completely complementary. This creates problems because if mis-priming occurs in an early amplification cycle, large amounts of unwanted non-specific products from other DNA loci are formed during the series of amplifications in addition to the DNA sequence sought.

properties of Taq polymerase

Use of enzymes in molecular biology and biotechnology 285

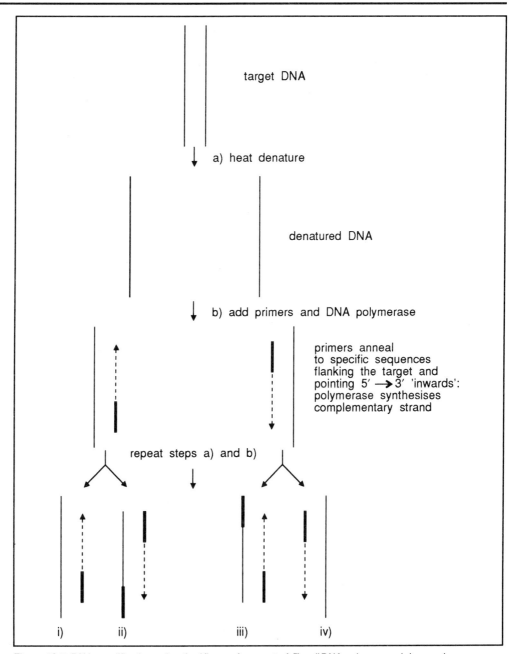

Figure 10.7 DNA amplification using the Klenow fragment of E. coli DNA polymerase I. Long primer extensions (strands i and iv are produced on the original template and increase arithmetically (20 copies in 20 cycles). Primer terminated copies (strands ii and iii) increase rapidly (10^6 copies of each template in 20 cycles).

polymerase chain reaction, PCR

A typical series of amplification cycles using *Taq* polymerase (known as a polymerase chain reaction or PCR), involves denaturation at 94-96°C for up to two minutes, followed by annealing of primers to denatured template at 40-65°C for up to two minutes, and extension for a variable time (depending on target product size) at 65-72°C. Commonly, between 15 and 40 cycles are carried out. Various automated, programmable heating blocks have been produced to allow this cycling to be carried out quickly and efficiently, and enzyme is added once at the beginning of the programme. The precise denaturing, annealing and extension temperatures, and times at each temperature, are governed by the lengths and GC contents of the target DNA and oligonucleotide primers used.

choice of conditions for PCR

∏ Starting with one molecule of target DNA, how many copies will be made under optimal cycling conditions after 10, 25 and 40 cycles?

potential yields from PCR

The answers are approximately one thousand (2^{10}), 33½ million (2^{25}) and a million million (2^{40}) copies! This shows how powerful the PCR technique is. We can start off with single cells (containing pg quantities of DNA) and amplify stretches of DNA to such an extent that they are visible as ethidium bromide stained bands on agarose gels. The largest segment of DNA amplified to date is approximately 10kb. Generally, smaller fragments are amplified most efficiently.

One potential variable in different PCR reactions is the Mg^{2+} concentration. *Taq* polymerase has a requirement for sub-millimolar concentrations of Mg^{2+}, but the precise free concentration in a given reaction is affected by the dNTP and oligonucleotide primer concentrations, as these all chelate Mg^{2+}. Some pairs of primers appear to have considerable effects on free [Mg^{2+}], and often it is necessary to carry out a Mg^{2+} titration curve to determine optimal amplification conditions. A common starting point is to use buffer containing 1.5 mmol l^{-1} MgCl$_2$ (total) and 200 µmol l^{-1} of each dNTP, with 0.25 µmol l^{-1} each primer.

selection of primers

The oligonucleotide primers used in PCRs need to be chosen carefully. They should not be self-complementary or, within a pair, complementary to one another, particularly at the 3' end. Any significant complementarity will result in preferential amplification of 'primer-dimers', rather than the longer target DNA. In addition, secondary structure due to internal complementarity must be avoided for efficient amplification.

Ideally, the base composition of the primers should reflect that of the target DNA, with a fairly random sequence, avoiding polypurine, polypyrimidine or repetitive sequences. To obtain good annealing at the 3' end of the primer (essential for extension by *Taq* polymerase), sequences ending in a C or a G are often used. To facilitate cloning of amplified DNA, primers are sometimes made which have a 3' end complementary to the target, and a Class II restriction enzyme recognition sequence at the 5' end. Following amplification, the PCR products may be digested with the restriction enzyme for cloning into a suitably cut vector. The fact that the 5' end of the primer is not complementary to the original target DNA is not a problem if a sufficient length at the 3' end is complementary, and appropriate annealing temperatures in PCR are used.

∏ How might PCR products be cloned in pUC19 in one given orientation only?

Use of enzymes in molecular biology and biotechnology

If you look back to Chapter 9, pUC19 contains a multiple cloning site. Primers may be designed such that one primer sequence contains the recognition sequence of one of these restriction enzyme sites at its 5' end while the other primer contains the recognition site of another restriction enzyme. Following restriction of the PCR product with both enzymes, each end of the amplified material will be different. This material may then be ligated in the orientation specified by these ends into the vector cut with the same enzymes (Figure 10.8). This approach may be used for cloning into any vector with a multiple cloning site, and is of particular use if expression of the amplified material is sought.

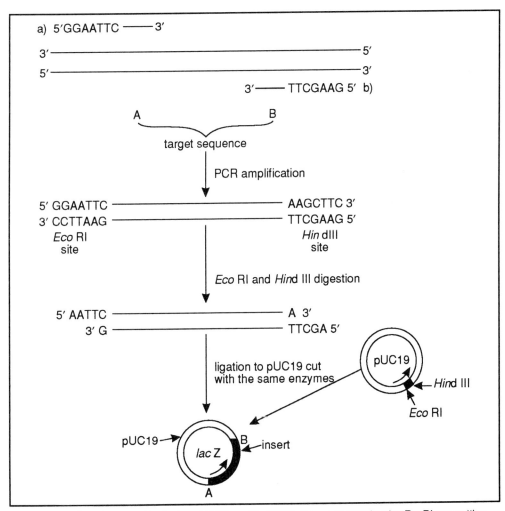

Figure 10.8 Orientation-specific cloning of PCR amplified DNA. Primer a) contains the *Eco* RI recognition site at its 5' end, and primer b) that of *Hin*d III. The 3' ends of both primers are unique, specifying the boundaries of the DNA to be amplified. Following amplification, digestion with *Eco* RI and *Hin*d III orientates the DNA such that it can be cloned into pUC19 cut with the same enzymes, in one orientation only. This orientation is illustrated with respect to the *lac* Z gene in this vector (→) - see also Figure 9.5.

As we have seen in Section 10.2, a problem often associated with cloned cDNA is that sequences corresponding to the 5' end of the mRNA are lost. This information may be obtained using a 'semi-specific' PCR. This involves the use of TdT, and is illustrated in Figure 10.9. First strand cDNA synthesis is carried out as normal, but rather than producing a second strand, the 3' end of the cDNA is extended using TdT and dATP. Extended first strand cDNA is recovered and the second strand is synthesised using an oligo(dT) primer (primer a) to which is attached a multiple cloning site (specifying three restriction enzyme recognition sites and about 18 nucleotides long). Double stranded cDNA is recovered, and subjected to PCR using fresh primers. One of these is complementary to a known sequence within the mRNA to be amplified (primer b), and the other is the multiple cloning site part of the second strand cDNA primer (primer c).

rapid amplification of cDNA ends

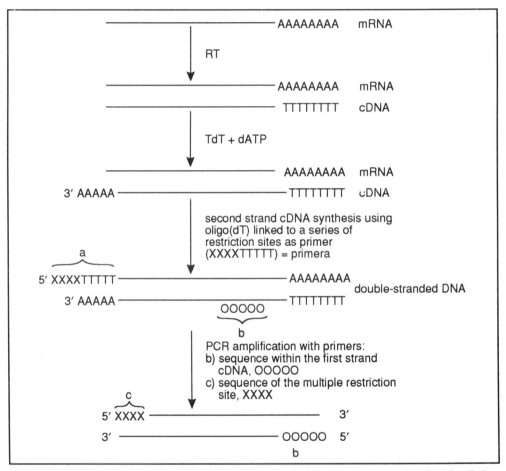

Figure 10.9 Amplification of the 5' end of cDNAs by PCR: The RACE protocol. First strand cDNA is tailed with poly(dA), and the second strand synthesised, after removal of the mRNA, using a primer consisting of oligo(dT) and a DNA sequence containing recognition sites of restriction enzymes at its 5' end, (primer a). This DNA is amplified in a PCR using one primer specific to the particular sequence sought (primer b) and another corresponding to the multiple restriction enzyme sequence (primer c). See text for further explanation.

Π Why do you think it is not a good idea just to use an oligo(dT) sequence as one of the PCR primers?

Use of enzymes in molecular biology and biotechnology

As mentioned earlier, polypyrimidine sequences should be avoided, and primers should have a fairly random sequence with about 50% Gs and Cs. Hence a multiple cloning site is attached to the second strand of the cDNA during its synthesis, to act as a priming site for PCR. Using oligo(dT) (or any other nucleotide homopolymer) would result in poor PCR efficiency, with difficulty in determining suitable cycling conditions to give a specific product.

RACE protocol This method of amplification is known as the RACE protocol (rapid amplification of cDNA ends) and was first described by Michael Frohman and colleagues.

Π How might this procedure be adapted to amplify the 3' ends of specific mRNAs?

This is useful if only a small amount of cDNA, mRNA or protein amino acid sequence is known and a longer cDNA sequence is required for further study. Basically, an oligonucleotide corresponding to the known sequence is synthesised, (primer b), and used along with primers a and c described above (see Figure 10.9).

First strand DNA is synthesised using primer a. This material is purified and used as a PCR template with primers b and c. Thus the product formed will depend on the sequence of b and will correspond to the 3' end of the required mRNA.

PCR has many other applications, including the amplification and study of ancient DNA samples and amplification of forensic material for DNA fingerprint analysis.

Π What are the potential problems of PCR amplification of small amounts of starting material?

precautions in using PCR One obvious problem is that of contamination. If only a few copies of starting material are present in the test sample, contamination by even one similar molecule will dramatically affect the results. Hence, to overcome this, great care must be taken to use clean reagents (molecular biology grade: RNA-, DNA-free), apparatus etc. For this reason, PCR reactions are often set up in tissue culture cabinets using dedicated pipettes, solutions and disposable plastic-ware.

Another precaution is to include negative and positive control samples in each reaction series.

SAQ 10.2 An important use of PCR is in the identification of genetic disease in foetuses. How might PCR be adapted for use in:

1) direct studies of inheritance of genetic disorders for which gene loci and associated RFLPs have been identified and sequenced?

2) rapid sexing of foetuses which are at risk of X-chromosome linked disorders for which no gene markers are available?

Some disorders (eg Muscular Dystrophy) are genetically heterogeneous. Any one case may result from one of a variety of mutations in a very large gene. An additional problem with such disorders is that the large size of the gene involved (in this case the 2Mb dystrophin gene), means that using RFLPs as an aid to diagnosis is not reliable as recombinational events can occur between the ends of the gene during meiosis. To overcome this, methods have been developed to amplify numerous regions of such genes where mutation 'hot-spots' occur, simultaneously. This is known as multiplex PCR, and has been applied to the detection of exon deletions in the dystrophin gene. In each reaction tube, genomic DNA is mixed with several pairs of primers. Each pair of primers allows amplification of different complete exons by PCR. Absence of any of the expected amplified bands indicates a mutation in the gene. This is a feasible diagnostic approach to this disorder in males because it is linked to the X chromosome. This is because males only have one copy of the gene, so the presence or absence of amplified bands on a gel indicates unequivocally the presence or absence of that exon. Because females contain 2 X chromosomes, the presence of a normal dystrophin gene on one chromosome means that a normal profile of amplified exons will be produced which will mask the absence or changes to the exons derived from the other (mutant) chromosome. Thus it is more difficult to unequivocally prove the presence of an aberrant gene in females. Such an approach would be less useful in studying autosomal dominant disorders.

use of multiplex PCR in analysis of genetics of Muscular Dystrophy

10.6 Chemical modification of enzymes for experimental use

Various enzymes have activities which may be exploited for experimental purposes, but may also have undesired traits. Modification of the enzymes may remove the latter while preserving the former. Two examples are discussed.

10.6.1 The Klenow fragment

E. coli contains three DNA polymerases - I, II and III. The first two are involved in DNA repair processes and the last is the enzyme used for DNA replication *in vivo*. All three contain the following activities:

- $5' \rightarrow 3'$ elongation of DNA strands complementary to a template strand;

- $3' \rightarrow 5'$ exonuclease activity. This is used to 'proof-read' DNA strands as they are synthesised. Incorrectly incorporated nucleotides are removed by this activity;

- $5' \rightarrow 3'$ exonuclease activity. This is a prominent feature of *E. coli* DNA polymerase I, and is considered to be involved in excision-repair mechanisms of damaged DNA molecules.

If we wish to use DNA polymerase *in vitro* to produce DNA strands, the first two activities are most useful, but the 5'-3' exonuclease activity is an unwelcome problem. Fortunately the structure of *E. coli* DNA pol I is such that we can treat it in a way to retain the desired activities whilst removing the unwanted one.

Klenow fragment of DNA polymerase I

Pol I comprises a single polypeptide of 109kD which is folded into two domains joined by a protease-sensitive linking region. Cleavage at this linking region liberates two fragments - a large one of 76kD, known as the Klenow fragment, and a small one of 36kD. The Klenow fragment contains the polymerase and $3' \rightarrow 5'$ exonuclease activities and is easily separated from the smaller fragment containing the $5' \rightarrow 3'$ exonuclease activity. In fact, amino acid sequence data have enabled the cloning of the Klenow fragment for its rapid production and isolation free of any contaminating 3'-5' exonuclease activity.

The Klenow fragment has had many applications in molecular biology and biotechnology including:

- the production of labelled DNA probes;
- DNA sequencing.

Production of labelled DNA probes using the Klenow fragment

The Klenow fragment has been used to incorporate radiolabelled, biotinylated and fluorescently-labelled nucleotides into DNA for use as probes for hybridisation. A method of probe production in common usage is depicted in Figure 10.10. DNA for which a probe is required is denatured by heating and random hexanucleotides, dNTPs, Mg^{2+} and Klenow fragment are added. The hexanucleotides hybridise to complementary strands of the template DNA, and are extended by the enzyme. Use of three normal and one labelled nucleotides ensures incorporation of the label into the newly synthesised oligonucleotides. Denatured reaction mixtures serve as probes in DNA or RNA hybridisation experiments. It is not generally necessary to separate incorporated from unincorporated label although some workers prefer to do so.

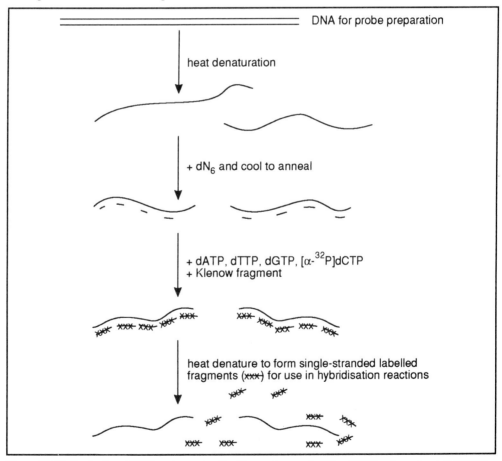

Figure 10.10 Production of DNA probes using the Klenow fragment of *E. coli* DNA polymerase I. Following heat denaturation, a mixture containing all possible hexanucleotide sequences is added and annealed to the DNA. These random primers are extended by the Klenow fragment, one of the nucleotides used being labelled (in this case, radioactively labelled dCTP has been chosen). The probes produced on denaturation of the reaction products may be used to monitor the presence of the same or related sequences on, for example, Southern blots, by hybridisation. dN_6 = deoxy-hexanucleotides.

| SAQ 10.3 | The Klenow fragment is often used to produce DNA probes. In addition, this enzyme is used to prepare radioactively labelled markers (for example, the restriction fragments produced on *Hin*d III digestion of DNA). What would be the requirements for the reaction involved? Hint! - think about what the ends of the fragments produced on *Hin*d III digestion look like. |

Use of the Klenow fragment in DNA sequencing

The most commonly used method for DNA sequencing is the dideoxy chain termination method developed and described by Sanger. A dideoxynucleotide (ddNTP) has the following structure:

$$^{-}O-\overset{\overset{O}{\|}}{\underset{\underset{O^{-}}{|}}{P}}-O-\overset{\overset{O}{\|}}{\underset{\underset{O^{-}}{|}}{P}}-O-\overset{\overset{O}{\|}}{\underset{\underset{O^{-}}{|}}{P}}-O-H_2C-\text{sugar}-\text{BASE}$$

∏ How does this differ from a normal deoxynucleotide?

In addition to having no hydroxyl group at the 2′ carbon atom, it does not have one at the 3′ carbon atom of the sugar ring either. These nucleotides are more correctly known as 2′, 3′ dideoxynucleotides.

∏ What effect will incorporation of such a nucleotide into a growing DNA strand have?

As the sugar has no 3′ hydroxyl group onto which an enzyme may attach a further nucleotide, elongation of the strand is terminated.

∏ Consider how this property of dideoxynucleotides is made use of to determine the sequence of a piece of DNA.

procedure for DNA sequencing using dideoxy-nucleotides

Firstly, single stranded DNA to be sequenced is prepared (usually by cloning it into a single stranded DNA vector), and a complementary oligonucleotide primer is annealed to its 3′ end. Aliquots of this are introduced into 4 separate tubes, each of which contain all four dNTPs (one of them radioactively labelled) and Klenow fragment. However, each tube contains a different dideoxynucleotide (ddATP, ddGTP, ddCTP or ddTTP). The enzyme will randomly incorporate either the deoxy- or dideoxynucleotide at any position where that nucleotide is specified by basepairing. Where the dideoxynucleotide is incorporated, elongation will terminate for that copy of the template. The ratio of deoxy to dideoxy nucleotides will determine whether a larger number of longer or shorter fragments are produced. Gel electrophoresis of denatured samples from each tube followed by autoradiography will allow the DNA sequence to be determined as illustrated in Figure 10.11.

∏ What will be the sequence of DNA giving the results observed in Figure 10.11?

Use of enzymes in molecular biology and biotechnology

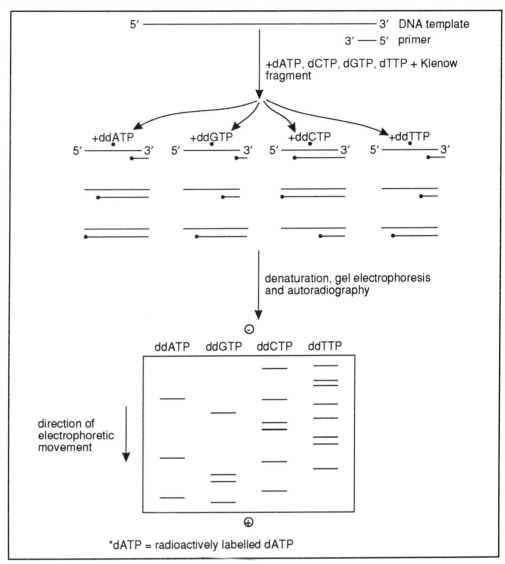

Figure 10.11 DNA sequencing by the dideoxy chain termination method. Single-stranded DNA template is produced and annealed to a short primer. This is extended in the 5'-3' direction by the enzyme in each of the four tubes. Each tube contains a different ddNTP, which will be incorporated randomly into the growing chain in place of the corresponding nucleotide. Incorporation of a ddNTP terminates elongation of that particular fragment, and on average the 4 tubes will contain fragments of a series of lengths dictated by the positions of the dideoxynucleotides. These are separated by gel electrophoresis, and their positions determined by autoradiography. The presence of a band in any lane indicates the position where the corresponding nucleotide is found in that sequence.

To answer this, you have to think about how DNA fragments run on electrophoresis. They run towards the anode (+), and the smallest fragments run fastest. Hence, the fragments terminated nearest to the primer will be those at the bottom of the gel. By 'reading' the gel, in terms of where bands are found, starting from the bottom and working up, the sequence of the strand produced, in the 5'→ 3' direction, is deduced. In the example illustrated in Figure 10.11 this is:

GACGGTCATTCCTGTCATTCT

The corresponding sequence of the template molecule in the 5'→3' direction will be:

AGAATGACAGGAATGACCGTC

(Remember the fact that the two strands run anti-parallel)

10.6.2 Modified T7 DNA polymerase

problems with the Klenow fragment

As we have seen, the Klenow fragment of *E. coli* DNA polymerase may be used for DNA sequencing. Unfortunately, this enzyme has a low 'processivity', that is it dissociates from its template easily, usually before about 10 nucleotides have been incorporated. Additionally, it has significant 3'→5' exonuclease activity and discriminates markedly against dideoxynucleotides, and other nucleotide analogues.

role of thioredoxin

To overcome these problems, a system based on the DNA polymerase of bacteriophage T7 was developed in the late 1980s. This system involves a chemically modified form of T7 polymerase (84kD) in a 1:1 stoichiometry with *E. coli* thioredoxin (12kD). The polymerase contains the catalytic activity, but high processivity is only observed in the presence of the thioredoxin. The latter protein confers increased processivity by binding the polymerase to the primer-template complex.

The T7 polymerase component needs to be chemically modified as it contains a highly significant 3'→5' exonuclease activity. Oxidation of the domain containing the active exonuclease site is used to selectively inactive this activity. This oxidation is brought about by highly localised production of free radicals.

Sequenase

The resulting protein complex is marketed under the name 'Sequenase' and is the preferred choice in many laboratories for efficient DNA sequencing by the dideoxy chain termination method. In addition to having high processivity and lacking 3'→5' exonuclease activity, Sequenase incorporates nucleotide analogues (such as dideoxynucleotides, azo-nucleotides etc) very efficiently. Good rates of nucleotide analogue incorporation are of use for obtaining sequence data from regions of DNA where repetitive or highly GC rich regions are found.

Π Why should repetitive sequences be hard to determine?

Such sequences are difficult to determine due to unusual or extensive secondary structures forming in the template, but use of modified nucleotides in the sequencing reactions can overcome these problems.

Use of enzymes in molecular biology and biotechnology

SAQ 10.4

DNA sequencing has been carried out for some time, using the Klenow fragment of DNA polymerase I (Section 10.6.1) or modified T7 DNA polymerase (Section 10.6.2). More recently, similar methods have been developed for RNA sequencing. These involves using an enzyme to copy the RNA, which will use dideoxynucleotides as substrates.

1) Which enzyme would be used?

2) What primers would be required?

3) This enzyme often terminates chain elongation randomly, (ie it exhibits a low processivity), which will result in bands appearing at the same mobility in several lanes giving problems in reading the sequence. How might such problems be overcome?

10.7 Site-directed mutagenesis: The basis of protein engineering

Our ability to study mechanisms of action of proteins, particularly enzymes, has been dramatically improved with the advent of recombinant DNA techniques. The best way to determine the importance of a particular amino acid residue within an enzyme to the enzyme's function is to see what happens if we change it. This is now relatively easy, provided a cDNA corresponding to that enzyme is available. Let us consider the situation where such a cDNA is available, cloned into a vector (such as a plasmid) from which it may be transcribed and the resulting mRNA translated. Such a vector is known as an expression vector, having a strong promoter (which the host cell RNA polymerase will recognise) just upstream of the cloning site, and a transcription termination signal downstream of this site. Host cells into which the recombinant vector is cloned will express the normal enzyme.

use of an expression vector

∏ How would you mutate this sequence?

The way to mutate the sequence is to prepare a single stranded form of the recombinant vector. Meanwhile, an oligonucleotide, approximately 15-20 nucleotides long, is prepared covering the sequence coding for the region where a change is required. This oligonucleotide is complementary to the normal sequence in the single-stranded vector except at one position where a change is introduced. The oligonucleotide acts as a primer for extension by the Klenow fragment, and the whole sequence is copied (see Figure 10.12).

∏ What enzyme must we add to complete the new strand?

DNA ligase must be used to close the circular plasmid.

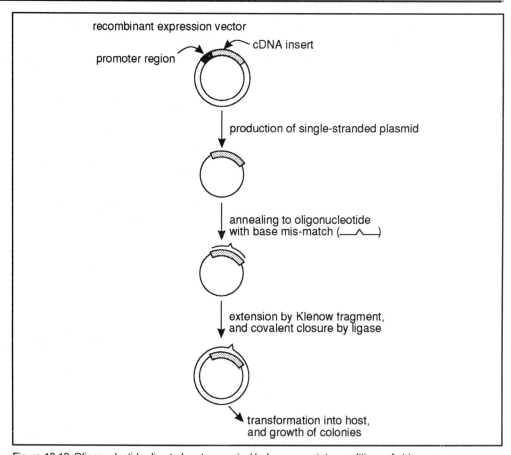

Figure 10.12 Oligonucleotide-directed mutagenesis. Under appropriate conditions of stringency, incompletely complementary oligonucleotides will hybridise to DNA. This is used as shown above to produce hybrid plasmid molecules. Semiconservative replication of the plasmid to give rise to two populations: normal and mutant. These will be passed on randomly on cell division such that two populations of bacteria will result, one containing and expressing the wild-type cDNA sequence, and the other the mutated sequence. The cDNA is arranged in the plasmid such that it may be efficiently transcribed from a strong promoter.

The double-stranded vector containing one normal and one mutated strand can then be introduced into its host, and on replication two types of clones will be produced - one bearing the native enzyme cDNA sequence, and the other the mutated sequence. Sequencing should allow identification of the mutant-bearing organisms, and large scale culture will provide suitable amounts of the enzyme for analysis and comparison to the normal form.

∏ The PCR can be used to generate large amounts of site-specifically mutated cDNA. How do you think this is achieved?

use of mismatched primer

If the mutation to be introduced is near to the end of a cDNA sequence, direct amplification of a single strand of this cDNA can be used to incorporate a mutation. One of the primers used for PCR can cover that region, including the appropriate change (see Figure 10.13). The other primer corresponds to the other end of the cDNA sequence. Direct amplification produces large amounts of altered cDNA which can be cloned into an expression vector and introduced into a host without the necessity for checking each resulting clone for the presence of the mutation.

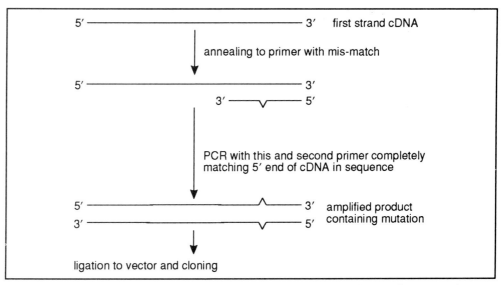

Figure 10.13 PCR and site-directed mutagenesis. PCR-based incorporation of mutated sequences is easily achieved using a pair of primers to amplify single-stranded cDNA. The first primer contains the sequence mis-match and effectively primes second strand synthesis. The second primer is completely identical to the 5' end of the original strand. PCR amplification results in a population of molecules which virtually all contain the mutation in both their polynucleotide strands.

A variety of vectors may be used, to allow expression in either prokaryotic or eukaryotic cells. Should the encoded protein require post-translational modification, a host which will perform these reactions must be used.

Summary and objectives

A wide variety of enzymes produced by micro-organisms have been identified, isolated and, if required, modified, for experimental use in the manipulation of DNA. The restriction enzymes have been discussed in Chapter 9, whereas some of the many others have been introduced in this chapter.

We described the use of reverse transcriptase to make cDNA copies of RNA *in vitro*. Modifications to cloned cDNA sequences may readily be made *in vitro*, and the effects on the function of the protein products studied. Additionally, reverse transcriptase has been used for direct RNA sequencing.

We also discussed the thermostable DNA polymerase from *Thermus aquaticus*, *Taq* polymerase. This enzyme allows efficient and rapid amplification of DNA by PCR and has been put to numerous uses including the study of ancient DNA samples, amplification of the ends of mRNA sequences and use in prenatal diagnosis. Two further DNA polymerases are widely used, but both in modified form - these are the Klenow fragment of *E. coli* DNA polymerase I, and Sequenase. The former is used extensively in DNA probe preparation, and the latter for DNA sequencing.

Alkaline phosphatase and terminal deoxynucleotidyl transferase are used commonly for DNA cloning. Both of these enzymes modify the ends of double-stranded DNA molecules, although in different ways, such that formation of recombinant DNA molecules is favoured in ligation reactions.

Now that you have completed this chapter, you should be able to:

- describe the use of phosphatases and terminal deoxynucleotide transferase in DNA cloning;
- understand the use of reverse transcriptase in formation of cDNA, and in RNA sequencing;
- describe the use of *Taq* polymerase in the amplification of DNA by the PCR;
- discuss some of the applications of PCR in cloning and diagnosis;
- understand the considerations involved in chemical modification of *E. coli* DNA polymerase I and T7 DNA polymerase for molecular biological applications;
- discuss the use of the Klenow fragment in formation of DNA probes;
- describe the use of Sequenase and other polymerases in DNA sequencing;
- discuss methods whereby site-specific mutations may be introduced into genes to modify activities of the encoded enzymes.

The application of protein and genetic engineering to industrial enzymes

11.1 Introduction	300
11.2 The objectives of protein and genetic engineers	300
11.3 The range of techniques available for modifying protein structure	301
11.4 Chemical approaches to protein engineering	303
11.5 Mutational approaches to protein engineering	305
11.6 Industrial applications of protein engineering	313
11.7 Hybrid enzymes and structure-function analysis	320
11.8 Modifying enzymes to simplify purification	320
11.9 Improving yield by genetic engineering	322
Summary and objectives	325

The application of protein and genetic engineering to industrial enzymes

11.1 Introduction

the evolution of enzymes vs their industrial use

In previous chapters, we have shown how enzymes are used in industry. These have included the uses of enzymes on a large scale in solution and in immobilised forms to bring about chemical changes, their uses in analysis and their application in the manipulation of genetic material. The structures of enzymes have, however, evolved to perform functions within living organisms and these may not be ideal for their use *in vitro*. Many enzymes have, for example, evolved in such a manner that they may be regulated *in vivo* by metabolite activation or inhibition. Similarly, many enzymes have evolved so that they associate with particular cellular structures (eg membranes) or other molecules in order to fulfil their biological functions more efficiently. Many of these features are not needed or are even undesirable in the industrial application of enzymes. The association of enzymes with other cellular constituents may, for example, complicate their extraction and purification; there is, of course, no evolutionary pressure for organisms to produce enzymes that are easily extracted and purified!

Of further concern to the industrial enzymologist is the inherent instability of enzymes. Although enzymes may show satisfactory stability in the milieu within cells, they often denature rapidly within the artificial environments of *in vitro* systems. For these reasons, much effort has been invested in producing enzymes with modified structures and properties which, although making them less well adapted to their biological functions, are more suitable for their industrial applications.

protein/genetic engineering

The processes by which this is achieved are referred to as protein engineering. Engineering, in this context, means the application of scientific principles to achieve practical objectives. We shall learn that, for the most part, this involves the use of the techniques of genetic engineering using the enzymes discussed in the two previous chapters. We will also learn that the techniques of genetic engineering not only enable us to modify the structure and, therefore, the properties of enzymes but also allows us to modify the yield of enzymes.

We begin this chapter by examining the objectives of protein engineering. We then examine the techniques involved in modifying enzyme structure to achieve these objectives. We will use specific examples to illustrate the achievements of protein engineering. Towards the end of this chapter, we will also include a brief discussion of the strategies used to improve the yield of enzymes and to facilitate their purification.

11.2 The objectives of protein and genetic engineers

In industry there are many different reasons why it may be desirable to modify the structure and yield of enzymes. The overall aims are to produce enzymes that show more desirable properties and to achieve greater economic return. Within these overall aims we can identify a number of specific objectives.

Π Before reading on, see if you can make a list of generalised objectives that may be set for the application of protein and genetic engineering to industrial enzymes.

The objectives are, to some extent, dependent upon the nature of the organism producing the enzyme, the properties of the naturally produced enzyme and the industrial process that is being undertaken. The sorts of objectives we hoped you would cite are:

- to increase the stability of enzymes *in vitro*;

- to increase the catalytic efficiency of enzymes;

- to modify (change) the substrate specificity of enzymes;

- to alter the pH dependence of enzymes;

- to make isolation and purification of enzymes simpler;

- to increase knowledge of the structure - activity relationships of enzyme to enable predictions to be made about the consequences of changing the structures of enzymes;

- to transfer the ability to make desirable enzymes to more suitable (easier to cultivate; safer to use) organisms;

- to increase the yield of the desired enzymes by increasing the expression of the relevant structural genes.

The first six of these objectives, although achievable by genetic engineering techniques, are essentially the objectives of protein engineering as they involve changes to the structure of enzymes. The final two objectives are the objectives of genetic engineering since they do not involve structural changes to proteins, but entail re-organisation of the genes coding for these proteins. It should, however, become self-evident as you work through this chapter that, in this area, protein and genetic engineering interact intimately in industrial enzymology.

We will begin our discussion of the techniques used in this area from the perspective of protein engineering.

11.3 The range of techniques available for modifying protein structure

To change the functions and/or properties of an enzyme by protein engineering, we mainly need to change its three dimensional structure. The techniques used to achieve this can be broadly divided into genetic-based and chemical-based techniques (see Figure 11.1).

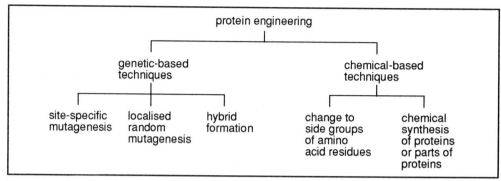

Figure 11.1 Overview of the techniques available for engineering enzymes (see text for details).

changes in amino acid composition and the physical environment

The three dimensional structure of an enzyme is dependent upon its amino acid sequence (that is, its primary structure) and the environment (pH, ionic strength, temperature etc) in which the enzyme is placed. Therefore, changes made in the amino acid sequence of an enzyme will be reflected by changes in the three dimensional structure of the enzyme and, subsequently, in its properties. This is the basis of the main strategy used by protein engineers.

three approaches to changing the primary sequence

Changes in the primary sequence are made by mutating the corresponding gene which codes for the enzyme. Such mutations can be made by using three types of strategies. One approach is to use site-specific mutagenesis in which the techniques of genetic engineering are employed to bring about specified changes to the nucleotide sequence (and thus the amino acids in the gene product) in specific regions of the gene. The second approach is to produce mutations in a localised region of the gene by random mutagenic procedures in the hope that at least one of the mutants will display the desired characteristics. The third approach is to form hybrid genes (chimeras) in which two genes are fused together to produce a novel protein. We will examine these genetic-based techniques in greater detail in later sections of this chapter.

Π There is, however, an alternative to the 'genetic' strategy for modifying the structures of enzymes. Can you think what it might be?

modifying the functional groups of amino acid residues

Instead of changing the enzyme that is being synthesised by an organism, we can first produce the 'natural' enzyme and then modify its structure by chemical means. For example we may change the functional group on the side-chains of certain amino acid residues within the protein. In some ways, the immobilisation of enzymes using linking agents described in earlier chapters are examples of such a post-enzyme synthesis treatment. As you have already learnt, such changes can greatly influence the properties (for example, stability, K_M, pH optimum) of enzymes. We should anticipate, therefore, that chemical modification of enzymes may be a method of achieving desired changes to enzymes.

11.4 Chemical approaches to protein engineering

It has become technically possible to make proteins from their constituent amino acids using chemical synthesis. Thus, in principle, we can produce proteins of more-or-less any amino acid sequence we want. We can even substitute amino acids with structural analogues. Thus, in theory, we are able to design an enzyme. In practice however, total synthesis of enzymes is prohibitively expensive and our knowledge of structure:activity relationships too rudimentary to make this a viable option for the commercial production of enzymes. (Note that this approach has been used to produce short peptides of commercial value for example to convert porcine insulin into human insulin).

total synthesis not economically viable

An alternative to total synthesis is to use partial synthesis (semi-synthesis). In this approach, the protein is first cleaved into shorter peptides. One, or more, of these peptides can be either modified or replaced by an alternative peptide and the protein reconstructed. Again, this approach has not been successfully applied to enzymes used industrially. Nevertheless the semi-synthesis of proteins has potential applications in research and medicine. Using this approach it is possible, for example, to examine the consequences of amino acid substitution/modification on enzyme activity. In medicine, this approach may also enable us to 'tag' enzymes with peptides which facilitate their binding to particular target cells.

partial synthesis

Side-chain modification of proteins is by far the most widely applied chemical method of modifying proteins *in vitro*. Prior to the development of site-specific mutagenesis, the chemical modification of side-chains was the principal technique used for the identification of active-site amino acid residues. This aspect of enzymology is covered in the partner BIOTOL text 'Principles of Enzymology for Technological Applications' and will not be greatly enlarged upon here. However, the chemical modification of amino acids residues in an enzyme also modifies its activity. For example, chemically changing the serine residue involved in the catalytic activity of the enzyme subtilisin into a cysteine residue resulted in a major change in catalytic activity.

chemical modification of amino acid side-chains

Π There are many limitations of the chemical modification of enzymes. See how many you can list.

The main limitations are:

- the process can only be undertaken with amino acid residue that have reactive side-chains and only certain alterations can be made;

- only amino acid residues at the surface of the protein are accessible to the chemical modifier;

- there is little specificity. All amino acids of one type (eg histidine) at the surface will be chemically modified. It is difficult to achieve modifications of a single residue.

Despite these disadvantages, there are some advantages of chemical modifications. It does, for example, enable us to attach compounds, such as co-enzymes, to the enzyme. Using this approach, Kaiser and Lawrence have been able to change the thiolprotease activity of the enzyme papain into an oxidoreductase activity. Such modifications have, as yet, found little practical application in industry. However, all is not lost for the chemical modification strategy. Chemical modification has been applied with some success in conjunction with genetic approaches. We will describe a specific example which, although not an enzyme, illustrates the principle.

The limited supply of insulin from natural sources (eg porcine pancreas) has led to its production using recombinant DNA technology. In this approach, genes for chimeric proteins have been generated with the overall structures shown in Figure 11.2. In these, a portion of the insulin gene coding for chain A or chain B of insulin (chain A is illustrated in Figure 11.2) or the pro-insulin gene was joined with another gene (for example *tryp* E or β-*gal*).

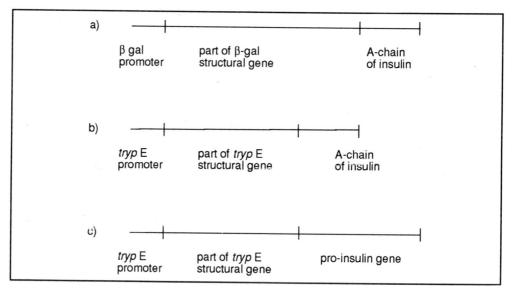

Figure 11.2 Chimeric gene constructs used in the production of insulin using recombinant DNA technology. In a) the expression of the chimeric gene is under the control of the β-*gal* promoter. In b) and c) the expression of the chimeric genes is under the control of the *tryp* E promoter. Note that the insulin gene portions were constructed from cDNA produced by using insulin mRNA and reverse transcriptase.

The gene combination shown in Figure 11.2 produce the proteins shown below:

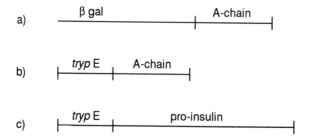

The key to the success of this process is in the method used to separate the insulin components from the other components.

To achieve this, genetic engineers inserted the codon for methionine (AUG) between the β-*gal* or *tryp* E and the insulin gene portions. The proteins that are produced from such genetic constructs are of the type:

These proteins can be isolated and chemically treated with cyanogen bromide (CNBr). Cyanogen bromide causes the cleavage of the protein at the methionine site. Thus the insulin portions can be isolated. This process is illustrated in Figure 11.3 and shows how chemical methods can be combined with genetic approaches to produce desired peptides.

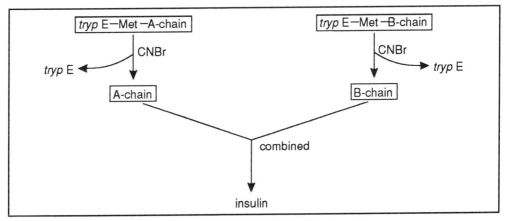

Figure 11.3 Simplified representation of insulin production from chimeric proteins.

SAQ 11.1

Why is the CNBr approach used to separate insulin from chimeric gene products unlikely to enable us to separate a desired enzyme from chimeric proteins of the type illustrated below?

tryp E ─── ├ Met ─┤ desired enzyme ├ ?

11.5 Mutational approaches to protein engineering

In Section 11.3, we indicated that the genetic approaches to protein engineering can be divided into three broad categories. These are:

- site-specific (also called site-directed) mutagenesis;

- localised random mutagenesis;

- hybrid formation.

In this section we will examine the two approaches which use mutagenesis as the basis of changing the structures of enzymes.

11.5.1 Selection of method

The mutagenic approaches to protein engineering depend to a large extent on how much we know of the three dimensional structure of the enzyme and how this structure is related to the properties and functions of the enzyme. If we have thorough knowledge of the structure of the enzyme and of the relationship between this structure and the activity of the enzyme, usually a site-directed mutagenesis approach can be used. If on the other hand, our knowledge of the structure: activity relationship is very limited, it is more usual to use a localised random mutagenesis approach.

Based on this basic division, we can identify four types of protein engineering based on mutagenesis. These are illustrated in Figure 11.4 along with the strategy used to select them.

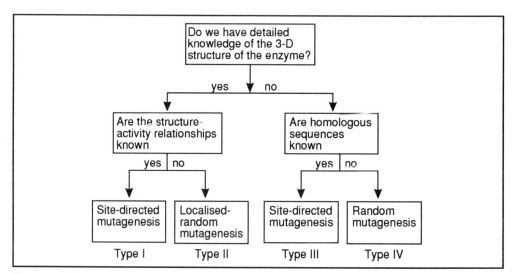

Figure 11.4 The four types of mutagenesis used in engineering proteins and the criteria used in their selection (see text for details).

In all four types of protein engineering based on mutagenesis, the gene coding for the enzyme has to be cloned. This may be done either by screening restriction enzyme fragments of genomic DNA or by using reverse transcriptase and the appropriate mRNA. Many of the techniques used to isolate and clone genes were described in the two previous chapters. If, however, you wish to explore these techniques further, we recommend the BIOTOL text 'Techniques for Engineering Genes'.

Let us assume that we have been able to clone the appropriate gene or cDNA. Which of the four types of protein engineering illustrated in Figure 11.4 should we use?

If we have a lot of information of the 3 dimensional structure of the enzyme and a detailed knowledge of the structure:activity relationship for the enzyme, this means we know exactly the role and contribution of important amino acid residues in the functioning of the enzyme. In this case, we can perform site directed mutagenesis. Using this technique, we can mutate one amino acid residue into any other. In this way, we can obtain a mutant protein with an altered function. This we have called type I protein engineering by mutagenesis. In the type I approach we can carefully select the appropriate amino acid to be replaced and make some predictions as to the likely consequences of replacing this amino acid by others.

type I approach: site-directed mutagenesis

The application of protein and genetic engineering to industrial enzymes

type II approach: localised random mutagenesis

If we do have knowledge of 3D-structure of the enzyme, but our knowledge of the structure:function relationship is limited in that we do not know the specific contribution of amino acid residues to the function of the enzyme, we adopt type II protein engineering by mutagenesis. In this case a few amino acid residues which are suspected to play a role in, for example, catalysis or stability may be mutated randomly by a process called localised or directed random mutagenesis. For four positions, this means $20 \times 20 \times 20 \times 20 = 160\,000$ mutant proteins (remember that nature has provided us with 20 different amino acids) may be produced. Since this number of mutants is too large to handle, a selection test which selects for the desired property is included and usually reduces the number of mutants we need to handle by orders of magnitude.

type III approach: mutagenesis using homologues

If we have little or no knowledge of the enzyme's 3D structure, we can still do some protein engineering. In type III protein engineering by mutagenesis, we use information from other enzymes which are homologous to our enzyme of interest. By analysing the homology between the amino acid sequences, an alignment can be made such that it becomes possible to identify which regions are conserved and which are variable. It may then become possible to deduce the role of certain amino acids. This then enables us to carry out site-directed mutagenesis to confirm their role. We can also carry out mutations which can change the properties of the enzyme in a desired way. It is also possible in this case to randomly mutate the gene in a limited region (see type II described above).

type IV approach: random mutagenesis

If we have only limited knowledge of the amino acid sequence of the enzyme and have little knowledge of its 3D-structure or of structure-activity relationships and no homologous sequences are known, we have to subject the whole gene to random mutagenesis. This we have called type IV protein engineering by mutagenesis. The key to the success of this method is to devise suitable selection tests that enable us to identify and isolate those mutations displaying the desirable characteristics. It is the development of suitable selection methods that hamper most attempts to carry out these mutagenic procedures.

11.5.2 Protein engineering by the mutagenesis cycle

Protein engineering by mutagenesis can be represented as a cyclical activity as shown in Figure 11.5. We can identify several stages.

Use Figure 11.5 to help you follow the description given.

We begin with the cloned gene (right hand side of Figure 11.5). If this is placed in an organism in which it is expressed, we can extract and purify the enzyme and using conventional approaches (X-ray crystallography, NMR chemistry), we can determine the structure and function of the enzyme. By function we mean what reactions it catalyses, its catalytic efficiency and the effects of environmental factors (pH, temperature, ionic strength, substrate concentration etc) on its catalytic activity. In this way, we are usually able to establish structure:function (structure:activity) relationships. (Note that this topic is covered in the BIOTOL text 'Principles of Enzymology for Technological Applications'). In establishing these relationships we assign the functional attributes of the enzyme to its 3D-structure. For example, the measured substrate specificity of the enzyme may be related with the ability of the substrate(s) to fit (dock) in the active site.

analysis of structure-activity relationships

use of computer aided graphics

Here of course computer aided graphic systems are invaluable in allowing us to examine the complex 3D-structures of enzymes. These systems contain very large files of the atomic co-ordinates of the protein. These are read by the computer and interpreted into a structure which is usually displayed on a high resolution, coloured

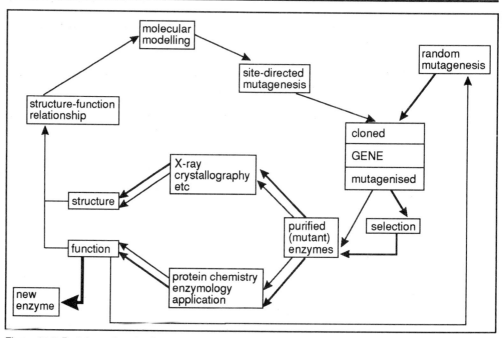

Figure 11.5 Protein engineering by mutagenesis cycles (see text for a description).

display. The computer system enables us to rotate the molecule and magnify particular parts of its structure thereby enabling us to understand how, for example, the substrate fits into the active site and which amino acid residues are important in holding the substrate at the active site. This approach can also be used to understand the features of the molecule which contribute to its stability/instability.

modelling

Using this information, we can propose mutations that will change the properties of the enzyme. This is usually done using models. Thus, using the type of molecular imaging described above, we feed in co-ordinates of a new amino acid residue to replace an existing amino acid residue within the enzyme and examine the likely consequences on, for example, substrate binding or stability. The models can also be combined with energy calculations which predict the likely structural consequences of producing mutated proteins. The models used to calculate the energy content of the protein molecule (so-called force-fields) are still in the process of development and are, therefore, not widely used as predictive devices. At the moment, the reverse is true, protein engineering serves as a tool to test the validity of the force-field models.

Π Now refer back to Figure 11.5. The figure does in fact show two quite distinct cycles. One of these cycles passes from the box containing the word 'function' to 'random mutagenesis'. This cycle passes through cloned gene, selection, purified enzymes, protein chemistry to function. It does not encompass structural analysis or structure-function relationships. Is this typical of type I, II, III or IV protein engineering by mutagenesis?

The answer is type IV. You may have been tempted to include type II operations as well but remember that in type II operations, although we may not have a detailed knowledge of the structure-activity relationships, we must have some knowledge of these in order to direct our mutagenesis to the desired location in the gene. Thus type I, II and III all require some knowledge about the structure of the enzyme.

The application of protein and genetic engineering to industrial enzymes

We will now turn our attention to the actual processes of producing desirable mutations.

SAQ 11.2

1) Which type of protein engineering by mutagenesis would you employ to modify the active site of lysozyme?

2) An enzyme has just been isolated from a bacterium which catalyses the reaction.

[reaction scheme: a naphthalene derivative with HO and CH₂–CHO substituents is converted, using NADP⁺ → NADPH + H⁺, to the corresponding naphthalene derivative with HO and CH₂–COOH substituents]

The enzyme appears to be carried on a plasmid and can be readily cloned into a suitable expression vector in *Escherichia coli*. The amino acid sequence of the enzyme has only been partially elucidated and little is known about its 3D-structure. The enzyme is fairly stable at 25°C, but unstable at 37°C.

a) Explain the generalised approach that could be used to isolate mutants that produce a version of the enzyme that was more stable at 37°C.

b) Explain how you could screen for such mutants.

11.5.3 Site-directed and localised random mutagenesis

In the chapters devoted to discussion of the enzymes used in molecular biology we described many of the techniques used to clone and recombine DNA fragments which underpin genetic engineering. We also briefly mentioned site-directed mutagenesis. These techniques are described in detail in the BIOTOL text 'Techniques for Engineering Genes' so we will not treat this aspect in detail here. We will however refresh your memory by giving a broad overview of the techniques involved to provide a context in which we can discuss the achievements of these mutagenic techniques in engineering proteins.

The technique of site-directed mutagenesis, is, in fact, elegantly simple. The technique is outlined in Figure 11.6 and described in the following paragraphs.

use of a single stranded vector

Lets see how a DNA sequence encoding an Asp residue is changed into a sequence encoding a Ser residue. If one wants to change a residue in a certain protein, the DNA fragment encoding that protein is firstly cloned into a single stranded DNA vector, for instance phage M13 vector. This single stranded DNA is subsequently converted into a double stranded DNA. As a primer for this DNA polymerisation, however, a synthetic DNA oligonucleotide with a mutation at the position of the 'to be changed' triplet is hybridised to the target site. This mutation is incorporated into the complementary strand (Figure 11.6a-b).

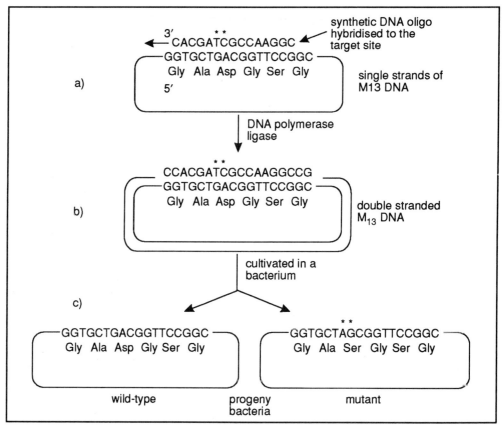

Figure 11.6 Site-directed mutagenesis. The stars indicate the points of mutation. Note single stranded M_{13} DNA is generated in a single stranded DNA vector (see text).

This double-stranded heteroduplex molecule is transformed into, for example, *E. coli* and propagated.

After growth, two types of progeny can be found. One is carrying the wild type Asp triplet and the other is carrying the mutant Ser triplet (Figure 11.6c).

The capability to make DNA synthetically has been the breakthrough, which has opened the way to site-directed mutagenesis. These days most molecular biology laboratories have equipment which can synthesize DNA of a specific sequence automatically.

Initially the efficiency of site-directed mutagenesis as outlined in Figure 11.6 was low. Several improvements of the technique are shown in Figure 11.7.

The application of protein and genetic engineering to industrial enzymes 311

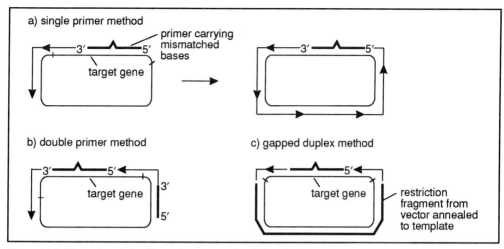

Figure 11.7 Strategies of site-directed mutagenesis. a) The simplest approach to M13 mutagenesis is 'single priming', where a mutagenic primer is annealed to the single-stranded template, extended briefly with Klenow fragment and used to transfect an *E. coli* host. After extending 'all-the-way-around' the template the ends of the new strand are ligated. b) In the 'double-priming' technique a second primer 5' to the mutagenic primer is used. c) In the 'gapped duplex' technique, mutagenesis is carried out in a single-stranded region formed by annealing the template with a restriction fragment from the vector (see text for further details).

gapped duplex approach

A frequently used technique is the gapped duplex approach. With this technique, the single strand DNA fragment carrying the DNA insert with the 'to be changed' codon is heated together with linearised M13 DNA without an insert. After cooling heteroduplex molecules are formed, which are double stranded in M13 sequences and single stranded in the insert sequences. This single stranded insert is subsequently annealed with a mutagenic synthetic oligonucleotide and filled in with DNA polymerase. Because of the small size of the insert as compared to the whole vector molecule the efficiency of completing the heteroduplex is much higher resulting in a higher mutation frequency.

use of expression vectors

After the desired mutation has been constructed the DNA fragment can be expressed either directly from the M13 vector or after transfer into a specialised expression vector. These expression vectors are designed in such a way that the mutated gene is inserted adjacent to a promoter which allows transcription of the inserted gene.

advantages and disadvantages of site-directed mutagenesis

The advantage of site-directed mutagenesis stem from the fact that we can select both the amino acid(s) we want to change and the amino acid we want to replace it with. The disadvantage is, of course, that we have to synthesise the primer DNA containing nucleotides of known sequence. If we do not control this process very carefully we will of course produce randomised changes in amino acids in the target region (a form of localised random mutations).

localised random mutagenesis

The more usual approach to introduce localised random mutations is to use site-directed mutagenesis to insert restriction enzyme site(s) into the cloned gene adjacent to the area of interest. The gene is then cloned into a vector, produced in large quantities and isolated. It is subsequently 'opened' using a restriction enzyme (see Figure 11.8). The nucleotides flanking the restriction enzyme sites are removed using an appropriate exonuclease. Short, mutagenic, nucleotide sequences (usually 10-25 basepairs in length) are inserted in the gap and subsequently the DNA is re-ligated inserted into an expression vector. This is then used to transform bacteria (or an appropriate host).

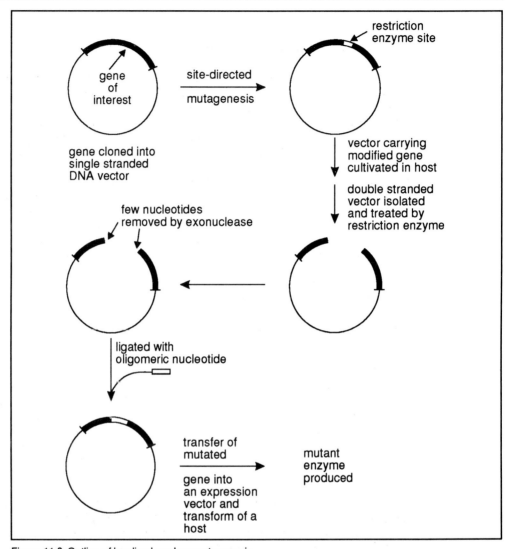

Figure 11.8 Outline of localised random mutagenesis.

use of PCR to introduce mutations

These are not the only basic strategies that are available to change the amino acid sequences of enzymes. In the previous chapter, we also described how a modification of the polymerase chain reaction (PCR) can be used to produce mutated sequences of nucleotides. These techniques enable us to generate large libraries of mutant proteins. In some ways these techniques enable us to use an empirical approach to mutating proteins. Instead of carrying out detailed structure:activity studies and analysis and using the results of these to make a deliberate change in the amino acid sequence, we can produce a large number of mutants quite simply and then analyse the consequences of these for enzyme stability and/or specificity.

We will leave the mutational techniques of protein engineering here. If you wish to find out more about these techniques, we recommend the BIOTOL text 'Techniques for Engineering Genes'. We will now turn our attention to the industrial applications of these techniques.

11.6 Industrial applications of protein engineering

Protein engineering is very widely used. In the academic environment, it is mainly used to study structure:function relationships. It has almost replaced chemical modification as a tool to deduce the role of particular amino acid residues in enzyme activity. Also, it is used to unravel the reaction mechanism of enzyme catalysed reactions. The most extensive study in this respect has been on the enzyme tyrosyl tRNA-synthetase and has been carried out by one of the pioneers in protein engineering Alan Fersht.

As an illustration of the power of this approach, we shall briefly consider some data described by Fersht and co-workers for tyrosyl-tRNA synthetase from *Bacillus stearothermophilus*. They constructed a 'family' of enzymes which differed at position 51. This enabled the contribution of each amino acid residue at this position to substrate binding (via effects on K_M) and catalysis (via effects on k_{cat}) to be assessed. It was possible to create active enzymes in which effects on k_{cat} were relatively slight, whilst those on K_M for ATP were substantial (Table 11.1). The effect on the specificity constant (k_{cat}/K_M) was dramatic. One welcome result of this type of analysis is the possibility of producing enzymes which are optimised for particular substrate concentrations.

Enzyme	k_{cat} (s^{-1})	K_M (mmol l^{-1})	k_{cat}/K_M (l mol^{-1} s^{-1})
wild-type	4.7	2.5	1860
Ala at position 51	4.0	1.25	3200
Cys at position 51	2.9	0.29	8920
Pro at position 51	1.8	0.019	95800

Table 11.1 Effect of substitutions at position 51 on kinetic parameters of tyrosyl-tRNA synthetase of *Bacillus stearothermopilus*. The data relates to ATP as substrate; wild type enzyme has a threonine residue at position 51. Data from Fersht, A.R. and Leatherbarrow R.J., in Protein Engineering (Eds Oxender, D.L. and Fox, C.F.) p275, Alan R. Liss, Inc., 1987.

The commercial value of protein engineering has also been recognised by several companies. It is recognised that industrial enzymes and pharmaceutical proteins could be improved, that is made more suited for their applications under (rather) unnatural conditions. Therefore, many patent applications have been produced and granted during the last decade. The first patent was granted to Genentech Inc. for their modified subtilisin (a protease from *Bacillus amyloliquefaciens*).

We will now give an indepth review of the proteases used in laundry detergents as an example of how protein engineering has been used to improve the performance of industrial enzymes.

11.6.1 Detergent proteases

Proteases are enzymes belonging to the class of hydrolases. Their catalytic activity resides in their ability to hydrolyse (cleave) peptide bonds. Depending on the catalytic mechanism, proteases can be divided into four major classes.

serine proteases — Serine proteases have a serine residue in the active site which is reversibly acylated during the catalytic cycle. From this class, the serine proteases from *Bacillus* species are used in laundry detergents. Well known members of this class also comprise the pancreatic proteases trypsin, chymotrypsin and elastase but they are also abundant in other tissues and species.

metalloproteases

Metalloproteases need a metal ion (usually Zn^{2+}) for their activity. Therefore, they cannot be applied in laundry detergents since these contain a lot of cation-sequestering agents (see below). Thermolysin, one member of the metalloproteases, is successfully used for the synthesis (in fact the reverse reaction of hydrolysis) of the dipeptide, aspartame which is used as an artificial sweetner.

aspartyl proteases

Aspartyl proteases have aspartic acid residues playing an essential role in the active site. Their pH optimum is usually on the acidic side (in contrast with the serine- and metalloproteases) and therefore they are not suited for use in laundry detergents (the pH of such is 7-9 for liquid detergents and 9-10 for powder detergents). A well known member is chymosin, the milk clotting enzyme from the calf stomach which can now be produced via recombinant DNA technology.

cysteine proteases

Cysteine proteases are similar to serine proteases in that the cysteine is also reversibly acylated during the catalytic cycle. However, the pH-optimum is on the acidic side. Examples are papain and bromelain.

stability of Bacillal proteases

The application of proteases in laundry detergents, aiming at the removal of proteinaceous stains, has been studied for a long time. Initially (around 1930), trypsin or a pancreatic extract was used, but these proteases could not withstand the surfactants used in the detergents. Later, in the 1950s, the proteases from Bacilli were discovered and these appeared to be remarkably stable against surfactants and even worked up to pH values 11-12! Despite the good properties of these proteases, they have a few weak points, for example, their sensitivy towards oxidation by bleaching agents and their tendency to digest themselves (after all they are protein molecules). Furthermore, their specific activity tends to be rather low.

We will explain how protein engineering may help to solve these problems.

11.6.2 Oxidation resistant proteases

The composition of a typical powder laundry detergent is given in Table 11.2.

Group	Ingredient	Content (w/w)
Surfactants	anionic	2-10%
	non-ionic	0.5-6%
	soap	1-5%
Sequestering agents	polyphosphate or zeolites or others	30-50%
Bleaching agents	sodium perborate activator (TAED)	20-30%
Enzymes	proteases	0.3-0.6%
Sodium sulphate		to make up to 100%

Table 11.2 Composition of a heavy duty powder detergent.

Π Which of the components in the heavy duty detergent described in Table 11.2 is (are) likely to inactivate the proteases?

We expect that you will have mentioned surfactants. These will, of course, interact with hydrophobic (internal) portions of enzymes and make them more compatible with the aqueous milieu. In other words, we can anticipate that surfactants will modify the structure and, therefore, the activity of enzymes in solution. Perhaps more important are the bleaching agents. These not only remove coloured stains by oxidation, but also have the capacity to oxidise proteases (and other enzymes) thereby inactivating them. Table 11.3 gives details of some of the bleaching systems commonly used in detergents. Also included in this table are details of the active component generated by these bleaching systems.

Bleaching system	Active oxygen
Sodium perborate (NaBO$_3$)	H$_2$O$_2$
sodium perborate plus tetra acetyl ethylene diamine (NaBO$_3$/TAED)	Peracetic acid
Diperoxydodecanoic-di-acid (DPDDA)	Peroxy-acid

Table 11.3 Some bleaching agents commonly used in detergents.

subtilisin methionine 222 oxidised by bleach

Scientists at the laboratories of Procter and Gamble (a major laundry detergent manufacturer), discovered in 1969 that in one of the proteases used in detergents (subtilisin derived from *Bacillus licheniformis*), methionine at amino acid position 222, next to the active site serine at position 221, was oxidised by a bleach (hydrogen peroxide), thereby severely inactivating the enzyme. However, at that time, genetic engineering techniques had not yet been developed. Thus, although the oxidation process had been established, it could not be remedied. The development of site directed mutagenesis some 13 years later allowed for the development of oxidation resistant enzymes. Genentech, for example, applied it with success to a subtilisin from *Bacillus amyloliquefaciens* which has a methionine at a corresponding position.

However, it became evident that protein engineering using site-directed mutagenesis has its drawbacks. It was possible to create an enzyme which was resistant towards bleaches, but the mutant enzymes were not as active as the wild type enzyme. The best subtilisin mutant appeared to be M222A (ie methionine at position 222 replaced by alanine - using the one letter code for amino acids). This mutant still had a specific activity of 80% of that of the wild type when measured with a model substrate. It was at least much better than the oxidised wild type enzyme (25% residual activity with the same substrate).

By using model substrate and washing experiments, the enzyme manufacturer Gist-brocades/International Biosynthetics, engineered oxidation resistancy into their high alkaline protease from *Bacillus alcaligenes* and maintained full washing performance. In this case, a methionine in the active site was replaced by non-oxidisable amino acids. Washing experiments showed that M216S (ie the serine mutant) and M216Q (ie the glutamine mutant) had a washing performance of 100% relative to that of the wild type enzyme (see Table 11.4).

Enzyme	Casein	SAAPFpNA k_{cat} (s^{-1})	K_M (mmol l^{-1})	Wash performance (%)
Wild type	100	100	1.1	100
M216A	31	7	1.1	50
M216E	6	2	1.3	4
M216H	12	< 1	1.5	35
M216L	17	17	1.2	22
M216N	19	11	1.8	74
M216P	9	5	1.1	13
M216Q	33	5	1.7	100
M216S	39	7	1.3	100
M216C	37	33	1.2	50
M216K	7	4	0.6	18
M216T	22	10	1.2	38
M216W	6	2	1.0	13
M216I	10	2	1.3	25
M216G	25	25		

Table 11.4 Specific activities of oxidation resistant high alkaline protease mutants. Casein and SAAPFpNA are used as model substrates. Results are reported as % relative to those of the wild type enzyme. Note that the designation of the mutants, shows which amino acid replaces methionine (M) at position 216 in the protease.

Π From the data presented in Table 11.4, which amino acid, when it replaces methionine at position 216 produces an enzyme with the lowest washing performance? (You may need to use a biochemical reference source giving details of the abbreviations used to represent the amino acids).

The lowest washing performance is given by enzyme M216E (E represents glutamic acid). In other words when the methionine at position 216 is replaced by glutamic acid, the washing performance is very poor (4%) compared with the wild type enzyme.

Π From the data presented in Table 11.4, is the activity of the enzyme measured using model substrates a good indicator of wash performance?

The answer is no. You will see that in many instances the activity, as measured using model substrates, has little correlation with wash performance. Look, for example at enzyme M216Q, M216S, M216C.

Figure 11.9 shows some data relating the remaining activity of the alkaline proteases remaining in bleach-containing detergent after storage at 30°C and 80% relative humidity.

The application of protein and genetic engineering to industrial enzymes

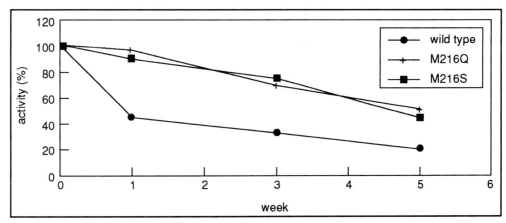

Figure 11.9 Storage stability of the high alkaline protease and its oxidation resistance mutants in bleach-containing detergent. Activity was measured with casein. Bleach: NaBO$_3$/TAED. Storage conditions: 30°C, 80% relative humidity.

Π What can be concluded about the replacement of methionine at position 216 by glutamine or serine on the shelf-life of these detergents?

It should be obvious that the storage stability of the mutant enzyme forms is greater (less susceptible to oxidation) than that of the original wild-type enzyme.

SAQ 11.3

1) Explain why the activity of high alkaline proteases as measured using casein need not be the same as that measured using other protein substrates.

2) Is the amino acid at position 216 in alkaline protease strongly involved in the binding of the substrate to the enzyme? (Give reasons for your answer).

11.6.3 Proteases with a higher wash performance

In the previous section, we learnt that by identifying a specific amino acid residue that made the enzyme susceptible to oxidation, site-directed mutagenesis led to the production of a more stable enzyme. Gist-brocades/International Biosynthetics have also used a directed random mutagenesis (Type II - Figure 11.4) to produce proteases with a higher washing performance (200% relative to wild type).

From the 3D-structure of the wild type enzyme certain amino acid residues were recognised as being involved in the binding of proteinaceous substrates. A protein inhibitor molecule (a substrate analogue) was modelled into the active site, and from the complex the interactions could be studied.

The amino acid residues in close contact with the substrate analogue were identified. It was assumed that these amino acid residues also interacted with the normal substrate (proteinaceous stains).

use of random localised mutagenesis

Of these, three amino acid residues, at position 126, 127 and 128 were chosen to be mutated. Since protein stains are very complex and of unknown 3D-structure, there was no clue as to what to mutate these residues into. Therefore, the amino acid residues were mutated randomly. As has been described above, in this case a selection test has to be included since 20 x 20 x 20 = 8000 mutants were to be expected. Several hundreds of mutant cultures were subjected to mini-wash tests, and only those mutants scoring better than wild type, were purified to homogeneity and sequenced. This effort resulted in several mutants with a higher specific washing performance (activity per unit mass of enzyme preparation Table 11.5).

Protease	Washing performance (%)	Specific activity (%)
Wild type	100	100
S126V, P127E, S128K	250	43
S126L, P127N, S128V	165	81
S126L, P127Q, S218A	200	62
S126L, P127M	250	69
S126M, P127A, S128G	200	115
S126Y, P127G, S128L	80	69
S126N, P127H, S128L	90	115
S126H, P127Y	113	73
S126R, P127S, S128P	130	81
S126F, P127Q	225	
S126G, P127Q, S128L	75	95
S126F, P127L, S128T	150	

Table 11.5 Directed-random mutants of the high alkaline protease. Results are reported relative to those of the wild type. Specific activities were measured using artificial substrates.

This example shows that both site-directed mutagenesis and localised (directed)-random mutagenesis are capable of generating industrial enzymes with interesting (improved) properties. Such approaches are, of course, not confined only to the proteases used in commercial laundry detergents.

constraints on genetically engineering enzymes

In the commercial world, the application of these techniques to modify enzymes is mainly constrained by the balance between the cost of carrying out the research and the cost-benefit accrued from producing enzymes of superior quality. Also important in this calculation is the unpredictable nature of protein engineering. Because the limitation in our knowledge of how enzymes work, it is not always possible to predict accurately the consequences of producing mutant forms of enzymes.

11.6.4 Producing more thermostable enzymes

Despite these limitations, much effort is being directed towards using mutagenesis to generate new forms of enzymes. These efforts have been especially directed towards more thermostable enzymes and some general rules have been identified that find application in converting thermolabile enzymes into more thermostable forms.

Comparison of thermostable forms of enzymes, isolated from thermophilic organisms, with those of their mesophilic counterparts, shows that these are only 20-30 kJ mol^{-1} more stable. This suggests that the thermolabile form of enzymes may need only to be modified slightly to improve their thermostability.

The application of protein and genetic engineering to industrial enzymes

∏ From your knowledge of protein structure, see if you can suggest ways in which the thermostability of an enzyme might be improved.

changes to hydrophobic cores of enzymes

We know that the 3D-structure of an enzyme depends on interactions between the side-chains of amino acid residues in the enzyme and that the core of the enzyme usually contains hydrophobic residues. We also know that the active sites of enzymes are on their outer surface. Thus we might anticipate that if we replaced amino acid residues in the core of the enzyme by more hydrophobic residues, this would increase the hydrophobicity of the core of the enzyme and improve its thermostability. Also important are the number of hydrogen bonds which maintain the 3D-structure of the enzyme. If, therefore, we replace internal amino acid residues to produce greater hydrogen bonding, we might increase the thermostability of the enzyme. It is however important that the secondary and tertiary structure of the enzyme is conserved in order to maintain activity. This often restricts the changes that can be introduced to external amino acid residues.

more hydrogen bonds may improve stability

Experience shows, however, that increased thermostability may be achieved using the following substitutions:

- lysine → glutamine or arginine;
- valine → threonine;
- serine → asparagine;
- aspartate → glutamate;
- isoleucine → threonine;
- asparagine → aspartate.

Internally, replacement of glycine or serine by alanine may also increase thermostability.

An alternative approach to protein stabilisation involves the introduction of an individually much stronger covalent bond in the form of a disulphide bridge. This involves replacement of one or more amino acid residues by cysteines, such that the side chains of two cysteine residues are sufficiently close to form a disulphide bridge (by oxidation). It is usually important that such substitutions are isosteric (ie do not not result in steric distortion). As an example of this approach, Perry and Wetzel replaced the isoleucine at positions 3 in T4 lysozyme by cysteine: from molecular graphic modelling it was expected that a disulphide bridge could form between Cys 3 and Cys 97. A disulphide bridge was formed on oxidation: the half-life of the enzyme at $67°C$ increased from 11 minutes to more than 6 hours.

SAQ 11.4 What are the likely consequences of increasing the thermostability of an enzyme on its catalytic efficiency?

11.7 Hybrid enzymes and structure-function analysis

Hybrid enzymes may be created using genetic engineering techniques. For this, gene portions coding for two different enzymes, are ligated together to form a hybrid (chimeric) gene. We can represent this process in the following way.

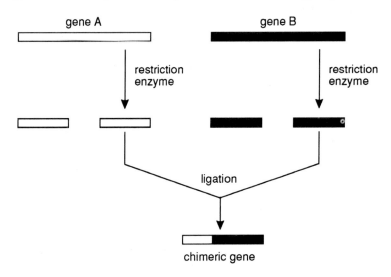

In this way, we can produce a new protein.

For the most part, this approach has been used to gain insight into the relationship between protein structure and function.

Let us examine this strategy in a little more detail. Closely related proteins can often have quite different functional properties despite having very similar amino acid sequences. By producing hybrid proteins with different crossover points (ie the relative positions of joining of the two genes), it becomes possible to relate functional properties to particular regions of the protein.

hybrids help to target mutagenesis

Once sites of functional importance have been identified by this procedure, it becomes possible to modify these by site-directed or localised-random mutations as described in earlier sections. Hybrid enzymes are, therefore, extremely useful in helping to identify target regions for mutagenesis.

11.8 Modifying enzymes to simplify purification

At the beginning of this chapter, we indicate that another potential objective of the protein engineer could be to simplify the purification of an enzyme.

∏ Before reading on, make a list of the properties of enzymes that are used in their purification.

The sorts of properties that are used to separate enzymes from other cell constituents are:

- solubility (eg ammonium sulphate precipitation);
- size (eg gel filtration);
- charge (eg ion exchange chromatography);
- pI (isoelectric point - used in isoelectric focusing).

Changing any one of these might aid purification. The simplest to change is perhaps the charge on the enzyme.

∏ Consider the following two forms of an enzyme.

native enzyme   NH₂ [   ] COOH

modified enzyme NH₂ [   ]–Arg–(Arg)$_n$–Arg–COOH

If the pI value of the native enzymes is pH7, what will be the pI value of the modified enzyme? (We would not expect you to give a precise numerical value).

changing the pI of an enzyme

The presence of many arginine residues would raise the pI value. At pH7, these would all carry a net positive charge. The pKa value for the side-chains of arginine is 12.5. Thus the presence of the many arginine residues would raise the pI of the enzyme to a value approaching 10.5 or even higher. This means that below about pH 10.5, the modified enzyme will carry a net positive charge and will be retained by cation exchange resins. Above pH 10.5 the modified enzyme would carry little or even a net negative charge and would be eluted from the cation exchange resin.

By adding arginine residues to the C (or N) terminal of the enzyme we have, in effect, produced a protein with an unusually high pI value. Thus we could now take our cell extract, at a fairly high pH value (eg pH 9) and pass it through a cation exchange column. Most of the proteins present would be above their pI values and pass through the column. Our modified enzyme would be retained. We could subsequently elute our enzyme by raising the pH to a value above pH 10.5.

∏ List reasons which would limit the success of this approach.

There are several reasons which would limit the success of this strategy. Firstly, the enzyme must be stable to the high pH values used to elute it from the column. Secondly, the addition of arginine residues to the enzyme should not impair its function. Thirdly, if the extract contains other proteins with similarly high pI values, these will also be eluted with our enzyme of interest.

use of signal peptides

⬚ Can you think of another strategy for modifying an enzyme which would simplify its purification? (Begin by thinking about the relative difficulty of extracting and purifying intracellular and extracellular enzymes).

Generally, extracellular enzymes are easier to purify than intracellular enzymes; they can, after all, be separated from other cellular constituents by harvesting the cells by centrifugation. Whether a protein is retained within the cell or exported depends upon the protein being produced with a signal peptide attached to its N-terminal. This peptide is usually removed by the host cell as the protein is secreted. Thus another strategy that may be used is to combine the nucleotide sequence coding for a signal peptide with the structural gene for the enzyme to form the following combination:

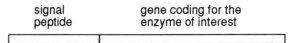

Although this approach has been demonstrated as feasible in some cases, the large size of some enzymes means that success is not always guaranteed. This is especially so for enzymes which contain a number of peptide chains. In these cases, it is the individual peptides that tend to be secreted, not the active enzyme.

A further limitation is that signal peptides tend to be host cell specific. In other words using sequences which code for signal peptides in one organism may be of little value if used in an alternative host. Nevertheless, hijacking the export system of organisms, to export desired enzymes does, in some instances, offer opportunities to reduce the costs of enzyme purification. It does, however, impose the condition that we need to use organisms which naturally secrete enzymes.

11.9 Improving yield by genetic engineering

The ability to clone a gene coding for an enzyme opens up enormous opportunities to improve the production of enzymes. We have seen already that it offers opportunities:

- to modify the properties of the enzyme to improve its commercial value;

- to simplify the purification of enzymes by for example adding poly Arg tails or through the addition of signal peptides;

- to provide a route to obtaining information regarding structure: activity relationships.

transfer of genes to new hosts

Also implicit in our discussion is that the ability to clone a gene means that we can select the host cell type we can use to produce the enzyme. Thus in principle we can take a gene (or cDNA prepared from mRNA) from one organism and use it in an organism (cell type) with more desirable properties (for example easier to cultivate, less harmful). There are, of course, some limitations. For example, some enzymes are 'processed' by the host cell before they become fully active. This may include removal of amino acid sequences in for example the conversion of an apoenzyme into its active form or may include glycosylation etc. In these cases, the new host must also be capable of carrying out these processes if the gene transfer is to lead to the successful production of the enzyme. Also remember that eukaryotic genes contain introns (non-expressed regions) that have to be removed if we wish to use these genes in prokaryotes. This is why we often use cDNA rather than genome DNA with eukaryotic genes. It is also difficult to transfer multi-subunit enzymes composed of more than one type of peptide chain.

> ∏ Let us assume we have isolated a gene for a single subunit enzyme and we can clone this into vectors which can infect the bacterium *Escherichia coli*. How could you maximise the production of this enzyme?

The expression of a gene largely depends upon:

- the strength of the promoter that controls the transcription of the gene;

- the number of copies of the gene present in the cell.

Thus, the two strategies that can be applied are:

- to ligate the gene to a strong promoter. Bacteriophage promoters, for example are often strong promoters and are useful for this purpose. It is also important to create the physiological conditions which maximise the expression of the gene;

- to use high copy number vectors. Some vectors exist in single copies within each host cell. These are stringently controlled. Other vectors are less tightly controlled and are produced in multiple copies. Thus by using these high copy number vectors we will produce more of the gene product. Alternatively a low copy number vector carrying multiple copies of the gene may be used to the same effect. There are however limits to these two approaches. Firstly the size of the insert that can be successfully packaged into a vector is not without limits and this restricts the number of copies of a gene that can be introduced using low copy number vectors. With high copy number vectors, although our enzyme may be produced in substantial amounts, this has an energy cost and the growth rates of the host cells are often reduced.

SAQ 11.5

1) The following gene combination has been produced and inserted into a vector which has then been used to transform the bacterium *Escherichia coli*.

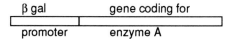

Enzyme A was derived from a related organism *Salmonella typhi*.

Which of the following media would be suitable for cultivating this organism to produce enzyme A?

a) Mineral salts + glucose.

b) Mineral salts + lactose.

c) Mineral salts + glycerol.

d) Nutrient broth.

(If your knowledge of microbial metabolism is limited, read our response carefully).

2) Assume you have selected the appropriate medium in 1), but your strain of *E. coli* carrying the vector fails to produce active enzyme A. Explain the possible reasons for this observation.

Summary and objectives

In this chapter, we have examined the strategies used by protein and genetic engineers to improve the properties and production of enzymes for industrial use. Although still, to some extent, in its infancy, it is recognised that these approaches have enormous potential. The chemical modification of proteins *in vitro*, although historically important, has many limitations. The use of site-directed and localised-random mutagenesis, made possible by advances in molecular biology and genetic engineering, has already led to the successful modification of enzymes. In this chapter, we illustrated this by using the example of the alkaline proteases used in laundry detergents. We also explained how the techniques of genetic engineering have enabled industrial enzymologists to have a wider choice of organisms in which to produce the enzyme of interest and to maximise production of enzymes. These techniques have also opened up opportunities to simplify enzyme purification procedures.

Now that you have completed this chapter, you should be able to:

- list the main objectives of protein engineering in the production and use of enzymes in industry;

- distinguish between the different types of protein engineering by mutagenesis and apply criteria to select an appropriate strategy;

- explain the limitations of using chemical methods to modify enzyme properties;

- explain why differences may arise when measuring enzymes using artificial and 'natural' substrates with particular reference to detergent proteases;

- describe how enzymes may be made more thermostable and explain the likely consequences this increased thermostability may have on catalytic efficiency;

- apply criteria in selecting appropriate promoters and incubation conditions to maximise the production of enzymes cloned into vectors;

- explain why merely joining a promoter to a structural gene does not guarantee the successful expression of the structural gene.

Industrial enzymology using non-aqueous systems

12.1 Introduction	328
12.2 Enzyme reactions in non-aqueous system	328
12.3 Enzyme stabilisation in non-aqueous solvents	330
12.4 The application of biphasic systems	332
12.5 Concluding remarks	336
Summary and objectives	337

Industrial enzymology using non-aqueous systems

12.1 Introduction

The impetus given to industrial enzymology by developments in other branches of biotechnology and the search for cleaner ('greener') technologies is not confined to the developments described in the previous chapters. A particular area of development worthy of comment is the use of enzymes in non-aqueous systems.

12.2 Enzyme reactions in non-aqueous system

possible advantages and disadvantages of using non-aqueous systems

Many substrate and product molecules of interest (eg steroids, lipids, hydrocarbons, etc) are barely soluble in water and carrying out reactions with such materials in aqueous systems presents difficulties. If we could accomplish such reactions in a non-aqueous solvent then we might be able to use much higher concentrations and, at the same time, reduce the chances of microbial contamination. However, enzymes are usually found in an aqueous milieu, the enzymes embedded in membranes forming an exception to this general rule. It is often found that immersion of an enzyme in a non-aqueous solvent causes denaturation and/or inactivation because the interactions holding the enzyme in its particular 3D-configuration are modified. How then can we reconcile these two facts?

One way is to immerse a damp (almost dehydrated) enzyme into the organic solvent. In this way, the enzyme is surrounded by a thin interphase consisting of water. The enzyme is then retained within an aqueous environment and the substrate has only a thin aqueous barrier to diffuse through.

We can represent this in the following way.

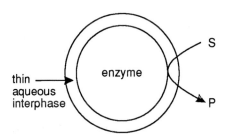

use of surfactants

The formation of such thin interphases may be enhanced by using suitable surfactants (eg cetyltrimethylammonium bromide = CTAB; phosphatidylcholine). These form 'inverted' micelles around the enzyme molecules, thus:

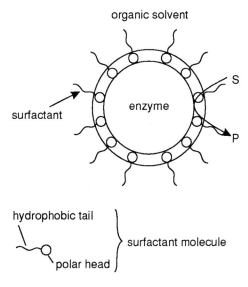

formation of reverse (inverted) micelles

Considerable research has been done on the formation of inverted (or reverse) micelles and various formulations of surfactants and co-surfactants (aids to surfactants such as butanol, hexanol, octanol) have been developed.

As you might anticipate these systems may cause substantial changes to the apparent equilibrium (Keq') constant of reactions.

∏ Suggest a group of enzymes for which the equilibrium constant is likely to be substantially influenced by the reaction occurring in a low water activity (ie concentration) environment.

You might have thought of proteases (which normally hydrolyse peptide bonds), esterases (which normally hydrolyse esters), lipases (normally hydrolyse triglycerides) or carbohydrases (normally hydrolyse carbohydrates, eg starch). All are classified as hydrolases, because all use water as a substrate.

∏ What is the likely impact on the apparent equilibrium constant (Keq') if a reaction involving a hydrolase takes place in an environment in which water is present at 1 mol l^{-1}, as compared to the normal circumstance (where water-based buffer is solvent)?

Assume the reaction is as follows:

Dipeptide + $H_2O \rightleftharpoons$ amino acid$_1$ + amino acid$_2$

therefore $K_{eq} = \dfrac{[aa_1]\,[aa_2]}{[dipeptide]\,[H_2O]}$

What is the 'normal' concentration of water? In molar terms, pure water has a concentration of 55.5 mol l^{-1} (from $\frac{100}{18}$ = 55.5 mol l^{-1}). Thus lowering the concentration of water from 55.5 to 1 mol l^{-1} would change K_{eq} by more than 50 x !

The situation with real bi-phasic systems is more complicated, as a result of partitioning considerations. None-the-less, the potential impact of lowering the water content of the system should be clear. We shall return to this later.

The amount of water attracted to the interphase differs for different enzymes. Hydrophobic enzymes such as lipases attract little water (perhaps less than 30 molecules of water per enzyme molecule). With hydrophillic (water soluble) enzymes the figure is higher (50-500 water molecules per enzyme molecule). These enzymes are nevertheless in a more-or-less anhydrous state.

12.3 Enzyme stabilisation in non-aqueous solvents

Placed as they are in such small volumes of water, the direct effects of such factors as pH are difficult to determine. Nevertheless it appears that enzymes placed in such an environment maintain the configuration/form they had in the aqueous solution from which they were taken. It also appears that enzymes in such environments are more thermostable. For example pancreatic lipase is very quickly completely inactivated in aqueous solution at 100°C whereas 50% of its activity is retained even after 12h incubation at 100°C in 0.02% water in tributyrin.

increased thermostability

Generally, factors which tend to disrupt the water interphase around the enzyme, are more likely to disrupt the structure of the enzyme.

Π We have two organic solvents, butanol and hexadecane. Which of these is most likely to lead to inactivation of the enzyme?

The answer is butanol. The more polar solvent is most able to disrupt the water layer around the enzyme and is thus more likely to disrupt the structure of the enzyme. Hexadecane is very non-polar and has little ability to disrupt the water layer. Thus we might anticipate that enzymes will be more stable in solvents of least polarity. This is in effect what happens in practice.

Usually polarity is measured by the logarithm of the partition coefficient of the liquid (ie the solvent) ('P') between octanol and water. Thus polarity of a solvent is given by:

$$\text{polarity} \equiv \log P = \log \left(\frac{[\text{ concentration of } X_{octanol}]}{[\text{ concentration of } X_{water}]} \right)$$

A typical relationship between activity of an enzyme and log P in a biphasic system in which the enzyme is surrounded by a thin interphase of water within an organic solvent is shown in Figure 12.1.

Industrial enzymology using non-aqueous systems

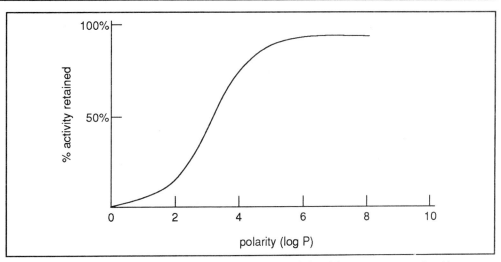

Figure 12.1 Relationship between % activity retained and the polarity of the non-aqueous solvent in biphasic systems (see text for details).

To give you some idea of the solvents which give the polarity values shown in Figure 12.1, we have provided some values in Table 12.1.

Solvent	log P
Hexadecane	8.7
Heptane	4.0
Petroleum ether (60-80)	3.5
Hexane	3.5
Cyclohexane	3.1
Carbon tetrachloride	2.8
Toluene	2.7
Chloroform	2.2
Benzene	2.0
Butyl acetate	1.7
Diethyl ether	0.8
Butanol	0.8
Ethyl acetate	0.7

Table 12.1 Polarity (log P values) of commonly used organic solvents.

SAQ 12.1

1) Would you expect an enzyme to retain more or less of its activity if it was immersed in hexane compared with the level of activity it would retain if immersed in cyclohexane?

2) A bacterium produces an enzyme which will dehydrogenate cortisol to produce prednisolone.

[cortisol structure] →(dehydrogenase)→ [prednisolone structure]

It is proposed to extract this enzyme to use it in a biphasic system (water : hexane) to convert cortisol to prednisolone. What problems do you envisage with this proposal and how would you overcome them?

12.4 The application of biphasic systems

In the previous section we explained that enzymes in biphasic systems may be substantially stabilised and that such systems are potentially useful for delivering water-insoluble substrates or removing water insoluble products to and from enzymes. This however is not the main attraction of these systems.

Consider the reaction.

$$R''-\underset{\underset{H_2C-O-C-R'''}{|}}{\overset{\overset{H_2C-O-C-R'}{|}}{\overset{O}{\|}}}\!\!\!C-O-CH \;+\; H_2O \;\rightleftharpoons\; \underset{\underset{H_2C-O-C-R'''}{|}}{\overset{\overset{H_2C-O-C-R'}{|}}{\overset{O}{\|}}}\!\!\!H-O-CH \;+\; R''\,COOH$$

It is typical of a lipase catalysed reaction. In an aqueous environment the equilibrium is well to the right, favouring hydrolysis. Is this also true in a non-aqueous environment?

The answer is no. Remember that the change in Gibbs function (Gibbs free energy) of a reaction is dependent upon the concentrations of reactants and products.

$$\Delta G = \Delta G^{o'} + 2.303\, RT \log_{10}\left(\frac{[\text{product}_1][\text{product}_2]}{[\text{reactant}_1][\text{reactant}_2]}\right)$$

Industrial enzymology using non-aqueous systems

shift in equilibrium position

In aqueous systems, the concentration of water is about 55 mol l^{-1}. Reducing this to the low levels in non-aqueous systems means that instead of producing net hydrolysis we achieve net synthesis. This principle can be applied to more or less any hydrolytic enzyme.

There are many examples where this principle has been used but perhaps the most successful application has been in the inter-esterification of lipids. Essentially, for this, suitable lipases (for example from *Rhizopus spp* or *Mucor spp*) are used as transacylases.

We can represent the reaction scheme as follows:

$$\begin{bmatrix} FA_1 \\ FA_2 \\ FA_3 \end{bmatrix} + \begin{matrix} FA_4 \\ +H_2O \end{matrix} \rightleftharpoons \begin{bmatrix} FA_1 \\ OH \\ FA_3 \end{bmatrix} + \begin{matrix} FA_2 \\ FA_4 \end{matrix} \rightleftharpoons \begin{bmatrix} FA_1 \\ FA_4 \\ FA_3 \end{bmatrix} + H_2O + FA_2$$

where:

$$FA = \text{fatty acid}; \quad \begin{bmatrix} FA_1 \\ FA_2 \\ FA_3 \end{bmatrix} = \text{triglyceride}$$

transesterification of triglycerides

We can replace a high proportion of the fatty acids in triglycerides in this way. The reasons for doing this is to produce triglycerides with particular properties. For example, the melting temperature of the triglycerides may be changed by substituting with different fatty acids. Generally unsaturated fatty acid residues lower the melting temperature. In this way we can create fats with that 'melt in the month' feel, a useful product for the food industry. Thus cheap feedstock triglycerides can be readily converted to expensive food ingredients. (Note that this example is discussed in greater detail in the BIOTOL text 'Biotechnological Innovations in Food Processing'.

A similar approach has been used to produce isoamyl acetate (a natural aroma) from isoamyl alcohol and ethyl acetate using lipase.

$$(CH_3)_2CHCH_2CH_2OH + (C_2H_5)-O-\overset{\overset{O}{\|}}{C}-CH_3$$
isoamyl alcohol   ethyl acetate

$$\Updownarrow \text{lipase}$$

$$(CH_3)_2CHCH_2CH_2-O-\overset{\overset{O}{\|}}{C}-CH_3 + C_2H_5OH$$
isoamyl acetate   ethanol

∏ See if you can identify two types of hydrolytic enzymes that might be used to produce polymers from monomeric units.

The enzymes we hoped you would identify are proteases and polysaccharide hydrolases (eg amylases, cellulases etc).

synthesis of polymers

Using the same thermodynamic arguments as we used before, these enzyme can, in principle, be used to synthesise polymers. Thus proteases could be used to synthesise peptide bonds and enzymes such as amylase could be used to produce glycosidic links. The main focus of interest has been on producing peptides so we will concentrate on that aspect here.

Π If we wish to sequentially add amino acids in a known order to a peptide using proteases, what do you think most limits the success of this process?

The key to the successful addition of amino acids to a growing peptide in a known order lies in the specificity of the protease, not so much in its catalytic ability.

aspartame synthesis

Most proteases are relatively non-specific and so faced with a mixture of peptide and free amino acids, will join the amino acids in random order. This is not so much a problem when the reaction mixture is simple or particular reactive groups are protected. For example the sweetener L-aspartame can be produced by using the protease thermolysin. Aspartame is a dipeptide of L-aspartic acid linked to the methyl ester of L-phenylalanine (aspartame = α-L-aspartyl-L-phenylalanine methyl ester).

In this instance, the amino group of the aspartate is protected by reacting it with benzyl chloroformate to form the benzyloxycarbonyl (BOC) derivative.

Then, by incubating this with L-phenylalanine methyl ester, BOC-L-aspartame is produced. Thus:

BOC — L — aspartate + L phenylalanine methyl ester

⇅ thermolysin

BOC — L — aspartame

The aspartame may be released by simple hydrogenation. In practice this gives rather a small yield. However it has been found that by using an excess of D-phenylalanine methyl ester, yields giving concentrations greater than 1 mol l^{-1} can be achieved. This is because the excess phenylalanine forms an addition compound with BOC-L-aspartame which precipitates out thus pulling the equilibrium towards BOC-L-aspartame synthesis. The L-aspartame can be recovered from the addition compound simply by adjusting the pH and a hydrogenation step.

Industrial enzymology using non-aqueous systems

Thus:

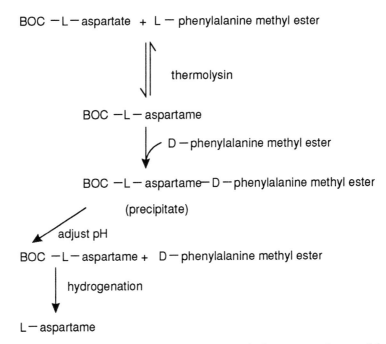

Note that a racemic mixture of phenylalanine methyl ester can be used because the specificity of thermolysin ensures that it is the L-phenylalanine methyl ester which links to BOC-L-aspartate. Particularly note the impact of enzyme specificity in this process: if non-enzymic condensations occurred, the side-chain (β-COOH) carboxyl of BOC-L-aspartate would also react. This does not occur because of the regio-specificity of the thermolysin. Similarly, D, L-phenylalanine methyl ester may be used (which is cheaper to produce than L-phenylalanine methyl ester), as thermolysin is also stereo-specific. This process is also described in detail in the BIOTOL text 'Biotechnological Innovations in Food Processing'.

proteases used to produce synthetic human insulin

Generally, proteases are not frequently used to carry out synthesis. A noticeable exception is the use of trypsin to convert porcine insulin into 'synthetic' human insulin. This requires an alanine on the B chain to be replaced by threonine. To achieve this, trypsin and threonine with a protected carbonyl group are incubated with the porcine insulin. The reaction sequence can be written as:

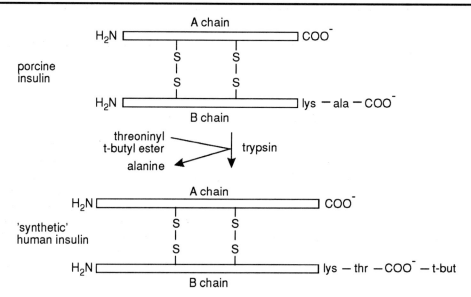

The t-butanol can be easily removed by treatment with trifluoroacetic acid and anisole to release the synthetic human insulin. This process is described in greater detail in the BIOTOL text 'Biotechnological Innovations in Health Care'.

SAQ 12.2

Cocoa butter is an expensive fat used in confectionary. It has a melting point between room-temperature and body temperature. This triglyceride has mainly palmitic acid and stearic acid in the 1 and 3 positions and oleic acid in the 2 position.

Palm oil and olive oil are relatively cheap and have lower melting temperatures because they have a greater proportion of unsaturated fatty acids esterified to the glycerol moiety. Explain how palm oil and olive oil may be converted to a cocoa butter substitute using the enzyme lipase.

12.5 Concluding remarks

In this chapter, we have briefly examined the use of enzymes in non-aqueous systems. The practical application of this use of enzymes is still, industrially, in its infancy. Nevertheless the application of these systems to modify food constituents, especially lipids, and to produce important pharmacological peptides is well established and constitutes important commercial operations. These types of applications, together with the extensive commercial applications of enzymes in aqueous systems and the use of enzymes for analysis and in the manipulation of genetic material is a major growth area. We can anticipate that the technological application of enzymes will become progressively more extensive in a wide variety of business sectors. Specific examples of these applications in various business sectors are described in greater detail in the 'Innovations' series of BIOTOL texts. These include: 'Biotechnological Innovations in Health Care'; 'Biotechnological Innovations in Food Processing'; 'Biotechnological Innovations in Chemical Synthesis' and 'Biotechnological Innovations in Energy and Environmental Management'.

Summary and objectives

In this chapter, we have briefly reviewed the potential of using enzymes in biphasic (organic-water) systems. We explained why enzymes may be stabilised in such systems and why the equilibrium position of reactions may be changed by transferring the reaction from an aqueous to a non-aqueous milieu. This change in equilibrium position offers opportunities to use enzymes hitherto regarded as hydrolases, to perform synthetic or exchange reactions. We used the synthesis of peptides and the modification of triglycerides to illustrate this principle. The equilibrium position may also be manipulated, for example by precipitation of the product.

Now that you have completed this chapter, you should be able to:

- explain why enzymes may be stabilised in biphasic systems;

- explain why using biphasic systems may aid the delivery of water-insoluble substrates to enzymes and thereby enhance reaction rates and product yield;

- select suitable solvents to achieve a good retention of enzyme activity;

- explain how lipases and proteases may be used for modifying lipids and for producing peptides with desirable properties.

Responses to SAQs

Responses to Chapter 1 SAQs

1.1
1) The sequence of amino acids in a protein is referred to as the **primary** structure of the protein.

2) Hydrophobic sidechains of amino acids are predominantly found on the **inside** of proteins.

3) If the pI value (isoelectric point) of a protein is 7, at a pH of 8 the protein will carry a net **negative** charge.

4) Proteins absorb UV light at 280 nm because of the presence of aromatic amino acids (tyrosine, phenylalanine and tryptophan).

1.2
1) False. SDS denatures proteins. Thus, although SDS gel electrophoresis allows separation of proteins (according to size), the proteins (enzymes) are inactive after separation.

2) True. Note that all molecules larger than the exclusion limit of the gel will be eluted in the void volume. Molecules smaller than the exclusion limit will be eluted in order of their molecular masses. (Assumes they have the same general shape).

3) True.

4) True.

1.3 For enzyme A: original specific activity = 45/50 = 0.9 units mg^{-1}.

After heat treatment, specific activity = 2/15 = 0.13 units mg^{-1}. Since specific activity should increase at each step, this is obviously not a useful method as the specific activity has gone down from 0.9 to 0.13. Also the yield of enzyme at this step is 2/45 x 100 = 4.4%. A yield as low as this is totally unacceptable.

For enzyme B: original specific activity = 120/50 = 2.4 units mg^{-1}.

After heat treatment, specific activity = 60/15 = 4.0 units mg^{-1}. This treatment has obviously resulted in some purification since the specific activity has gone up from 2.4 to 4.0 (a 4.0/2.4 = 1.7-fold purification). However, the yield of enzyme is only 50% (60/120 x 100). A loss of 50% of your enzyme in a single step is normally unacceptable (perhaps except when the step results in a dramatic increase in protein purification) and therefore, this would not be a useful method for purifying enzyme B.

For enzyme C: original specific activity = 85/50 = 1.7 units mg^{-1}.

After heat treatment specific activity = 83/15 = 5.5 units mg^{-1}. We can see that the specific activity has increased from 1.7 to 5.5 (a 3.2 fold purification) and also the yield (83/85 x 100 = 98%) is extremely good. This is, therefore a good purification step for enzyme C.

1.4 If you have decided to collect the fraction that precipitates at 55% saturation you obviously understand what you are doing. By adjusting the extract initially to 45% ammonium sulphate and centrifuging we will remove 650 mg of protein but only lose 15 units (~5%) of our enzyme. If we now make the supernatant 55% saturated in ammonium sulphate, most of our enzyme will be in the precipitate that forms, whereas 550 mg of protein will still remain in solution (as well as 25 units of enzyme). We should, therefore, collect the 55% precipitate and re-dissolve this in buffer solution. The original specific activity of the enzyme was 310/1700 = 0.18 units mg^{-1}. The specific activity of the 55% ammonium sulphate fraction is 270/500 = 0.54 units mg^{-1}. We, therefore, have a threefold purification (0.54/0.18 = 3.0) and a yield of 87% (270/310 x 100) which is quite acceptable for this method.

1.5 We would expect five peaks of protein. The first five proteins in this list have molecular masses greater than the exclusion limit (30 000D) of G50 and, therefore, will all elute together in the totally excluded (or void) volume. The molecular masses of the remaining four proteins fall within the fractionation range of G50 (30 000-1500D) and, therefore, will elute from the column in order of decreasing molecular mass, ie trypsin will elute first, followed by soya bean trypsin inhibitor, lysozyme and finally insulin.

Would a better separation have been achieved using a G-75 column?

Yes it would. Phosphorylase and transferrin would still elute in the totally excluded volume, but the remaining seven proteins, being within the exclusion limit (70 000D) of G-75 should separate. However, note that because of the similar molecular masses of ovalbumin and hexokinase, it is unlikely that these two proteins would separate and thus they would almost certainly elute in the same fraction.

1.6 You should have matched cofactors (coenzymes) with reaction types in the following way:

Cofactor/coenzyme	Reaction types
nicotinamide adenine dinucleotide (NAD^+) flavin mononucleotide (FMN) flavin adenine dinucleotide (FAD)	oxidation - reductions involving hydride anion (H^-) transfer
pyridoxal phosphate	transamination, decarboxylation, racemisation
biotin	carboxylation, transcarboxylation
tetrahydrofolate	transfer of one-carbon units
thiamine pyrophosphate	cleavage or formation of C-C bonds adjacent to carbonyl carbon atoms
coenzyme A	transfer of acyl groups

1.7
1) This is false. Enzymes certainly increase the rate of a reaction but do not alter the equilibrium position.

2) True. Enzymatic catalysis is achieved by lowering the activation energy of a reaction; this is accomplished by providing an alternative reaction route. Substrates must bind to their enzyme and form an ES complex for catalysis to occur.

3) False. Binding of 2 groups of the substrate may not be sufficient to guarantee distinction between 2 stereoisomers. With only 2 recognition sites, both stereoisomers could bind.

4) False: at least as a generalisation. Enzymes may be highly specific and only act on a single compound. Others, however, show rather loose specificity (such as hydrolytic enzymes involved in breaking foodstuffs into smaller fragments) and act on a number of related compounds. Thus wide variations in the specificity displayed occur.

5) Correct. A substrate only binds reasonably tightly and correctly positioned for catalysis to occur if there is a good match between the shape and charge distribution of active site and substrate. This is how enzymes discriminate between different molecules.

1.8
1) We hope you resisted any temptation to plot v_0 against $[S_0]$, then attempting to extrapolate the line to get V_{max}. Instead, the data should be transformed such that a straight line will be formed (providing Michaelis-Menten kinetics apply; assume so at this stage). Whilst there are various ways to do this, we have described the Lineweaver-Burk plot. Reciprocals of v_0 and $[S_0]$ have to be calculated, giving $1/v_0$ and $1/[S_0]$, respectively:

	$[S_0]$ mmol l^{-1}	$1/[S_0]$ l mmol^{-1}	v_0 µmol min^{-1}	$1/v_0$ min µmol^{-1}
(A)	0.6	1.67	1.14	0.88
(B)	1.0	1.0	1.70	0.59
(C)	1.5	0.67	2.36	0.42
(D)	3.0	0.33	3.64	0.27
(E)	6.0	0.17	5.00	0.2
(F)	15.0	0.067	6.46	0.15

These are then plotted as follows:

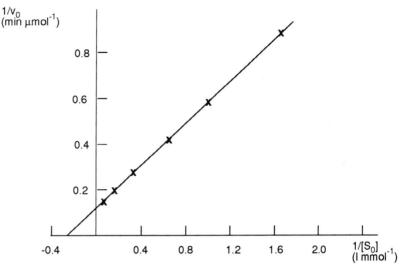

The line of best fit is determined (by eye; better by regression analysis) and K_M and V_{max} obtained as:

a) The intercept on the $1/v_0$ axis $= 1/V_{max}$; in this case it is 0.12; thus:

$$1/V_{max} = 0.12 \text{ and } V_{max} = 8.33 \text{ μmol min}^{-1}$$

b) K_M is determined from the intercept on the $1/[S_0]$ axis; this is -0.255 in this case; thus:

$$-1/K_M = -0.255 \quad K_M = 1/0.255 = 3.9 \text{ mmol l}^{-1}$$

2) The crucial data are obtained from substrate concentrations around the K_M value, particularly in the range 0.2-5.0 x K_M. Assays at very low [S] may cause problems, because they tend to be more prone to error; additionally, when expressed as 1/[S], they give very high values on the x axis. Within the range given above, try to arrange substrate concentrations which will give a good spread of points when plotted as reciprocals. Note how this was done with the numerical data above; whilst it was a highly non-linear distribution in mol l^{-1} terms, they were more evenly spaced as reciprocals.

Suitable values would be:

$[S_0]$, mmol l^{-1}	giving	$1/[S_0]$, l mmol^{-1}
0.8		1.25
1.2		0.83
2.0		0.5
4.0		0.25
10.0		0.1
20.0		0.05

1.9 The correct answer is 4). Non-competitive inhibition means that less active enzyme is available when inhibitor is bound to any enzyme molecules; thus V_{max} is lowered. Since I can bind to either E or ES, there is no effect on substrate binding ie K_M is not affected.

1) would be expected for mixed inhibition, where adverse effects on both substrate binding and catalysis are seen;

2) is an interesting case, in which the inhibitor makes it easier for substrate to bind, whilst none-the-less lowering enzyme activity. This occurs if inhibitor can bind to ES but not to E. This is a form of inhibition which is known as uncompetitive inhibition. It is rare with single substrate enzymes but can occur when more than one substrate is involved;

3) is the result expected for competitive inhibition. In this case, since S can displace the inhibitor, more substrate will be needed to achieve V_{max}, hence the increase in K_M;

5) is consistent with extreme damage to the enzyme, such as denaturation or perhaps irreversible inhibition.

1.10

1) Whilst plotting v_0 against $[S_0]$ may give some indication of the type of inhibition, it is unlikely to distinguish between the possibilities. The data need to be linearised, as in the Lineweaver-Burk plot. Thus $1/v_0$ and $1/[S_0]$ for both inhibited and non-inhibited assays must be calculated, giving:

$[S_0]$	$1/[S_0]$	A (− inhibition)		B (+ inhibition)	
mmol l^{-1}	1 mmol^{-1}	v_0	$1/v_0$	v_0	$1/v_0$
0.02	50.0	31.3	0.032	16.1	0.062
0.03	33.3	41.7	0.024	22.2	0.045
0.04	25.0	47.6	0.021	27.8	0.036
0.06	16.7	57.1	0.0175	35.7	0.028
0.12	8.33	71.4	0.014	52.6	0.019

These are then plotted as $1/v_0$ against $1/[S_0]$, and straight lines can be fitted:

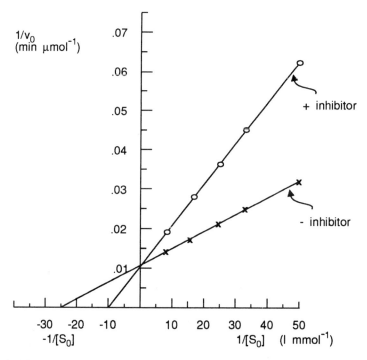

It is apparent that the intercept on the $1/v_0$ axis is unchanged in the presence of inhibitor, V_{max} has not been affected by the presence of the inhibitor (intercept on $1/v_0 = 1/V_{max}$). The intercept on the negative side of the $1/[S_0]$ axis is changed and K_M has been increased (since this intercept = $-1/K_M$). This pattern is characteristic of competitive inhibition. With competitive inhibition, substrate can displace the inhibitor, the original V_{max} can still be obtained. However, a raised $[S_0]$ will be required to achieve this, hence the increase in K_M.

2) All experiments should be repeated several times before conclusions are made. In this case, if reversible competitive inhibition is suspected, then regeneration of activity after careful removal of the inhibitor must be demonstrated. If we cannot regenerate activity, then irreversible inhibition must be suspected. Additional experiments which address this question should be conducted.

1.11

1) Whilst all enzymes are probably best thought of as being flexible, allosteric enzymes undergo conformational changes as a consequence of substrate binding, which is not necessarily the case with non-allosteric enzymes. This statement is thus not true.

2) If an enzyme gives a hyperbolic rate against $[S_0]$ plot and gives a straight line when transformed to a double reciprocal plot (ie one displaying Michaelis-Menten kinetics), it is impossible for a sigmoid plot to also give a straight line after a similar transformation - at least over a wide $[S_0]$ range. Further, K_M is not very helpful with allosteric enzymes, since the rate against $[S_0]$ relationship changes depending on the circumstances.

3) Yes, because allosteric enzymes show a greater change in activity over a given $[S_0]$ range, through co-operative binding of substrate. They are also amenable to activation and inhibition by compounds other than the substrate.

4) True. The requirement for allosteric behaviour is that multiple active sites (for substrate and perhaps also for regulatory molecules) are present. This enables binding of substrate at one site to influence substrate binding at other sites (co-operativity). This is routinely achieved by the possession of several subunits (often 4); however, recent research has identified allosteric behaviour in some single polypeptides.

Responses to Chapter 2 SAQs

2.1

feature	encourage/discourage
1) great diversity of microbes available;	+
2) fast growth rates;	+
3) can be grown in pure culture;	+
4) some microbes can grow in extremes of physical conditions;	+
5) many are unicellular;	+
6) micro-organisms are often associated with disease;	-
7) many micro-organisms carry plasmids;	+
8) post-translational modification of proteins;	-
9) RNA processing (eg removal of introns);	-
10) metabolic regulation.	+

Explanation:

The positive influence of features 1), 2) and 3) should be obvious. 4) The ability of some micro-organisms to grow in extreme conditions (eg high temperatures) indicates that they produce enzymes which are particularly stable. Generally the occurrence of microbes as unicellular organisms 5) is an advantage because they can be readily cloned and all cells are the same. It does, however, make them potentially more difficult to harvest. The negative influence of 6) should be obvious. The presence of plasmids can be an advantage because it means we have a potentially easy route for introducing new genes into the organism. RNA processing (removal of introns) and post-translational modification (8 and 9) are features of eukaryotic cells not carried out by prokaryotic micro-organisms. These features put constraints on transferring genes encoding enzymes from eukaryotes into prokaryotes. Metabolic regulation 10) may be a complicating factor as it usually implies that the organism will limit its production of a particular enzyme depending upon culture conditions. However, it is much easier to either manipulate the culture conditions of micro-organisms or to mutate these organisms in order to stimulate the production of the desired enzyme.

2.2 There was very little difference in the overall yield between the two processes. For the first process, the overall recovery of active enzyme was 59% (ie 100 (.9 x .9 x .9 x .9 x .9) = 100 x .59).

For the second, the overall recovery was 60% (ie 100 x 1 x 1 x 1 x 1 x 0.6).

2.3 The likely location of the enzyme is in the periplasm since osmotic shock releases a substantial amount of the enzyme (55% of that released by sonication) but relatively little of the total protein (5% of that released by sonication). If the enzyme had been intracellular, we would anticipate similar comparative extraction efficiencies of the enzyme and the total protein.

The specific activity of the enzyme extracted by sonication is 500 units/g of total protein whilst that of the enzyme extracted by osmotic shock is 5500 units/g of total protein. In other words is appears that the enzyme prepared by osmotic shock is not so heavily contaminated by other proteins. Instinctively, therefore, we might believe this to be the better approach to producing the enzyme. In practice, however, the matter is not so simple as this. First the osmotic shock only yields 55% as much enzyme and we have to ask the question, 'will the additional costs incurred in purifying the enzyme from sonicated cells be more than offset by the increased yield?' We would also need to consider such issues as, 'is the enzyme produced in dilute form after osmotic shock (that is, in a large volume)?' If so, there are costs incurred in producing a concentrate. Also the relative costs of sonication and osmotic shock need to be considered. These are, of course, commercial considerations. Critical to the selection of strategy is the question, 'can the process be readily scaled-up?' In the examples given in this SAQ, sonication cannot be readily applied on a large scale. If a cell disruption procedure was deemed the best route because of the increased yield, an alternative technique to sonication would need to be found.

2.4 From the data presented in Figure 2.4, you should have concluded that the decanter type is best suited for larger particles and that the chamber type can only handle dilute suspensions.

2.5 1) From the description given about filters especially in Figure 2.5, you should have realised that particles may associate to form 'bridges' across larger apertures. If in doubt refer back to Figure 2.5.

2) From the data given, the best type of centrifuge to use would be the disc stack type (see Figure 2.4). These can cope both with particles of this size and with this feed concentration. The size of particles in the suspensions are at the bottom end of the range that decanters can separate. (We would have to use slow feed rates). Likewise the chamber bowl type is only suitable for low concentrations of feed. With the dense suspension, we would have to keep stopping the device to remove the collected cells.

2.6 There is a lot of scope in the question but we hope you will have included the following factors:

- micro-organism used:
  a) cell wall thickness and composition;
  b) size of the cells;

- location of the desired enzyme:
  a) in the cytoplasm;
  b) in a sub-cellular organelle;
  c) in the periplasmic space;

- design of the bead mill:
  a) bead diameter;
  b) type of impeller;
  c) speed of the impeller;
  d) bead loading;

- operational conditions:
  a) rate of flow of cells through the mill (ie residence time in the mill);
  b) concentration of biomass;
  c) operational temperature (this will tend to rise as energy is dissipated in the mill).

You may have included additional factors but those listed above are the main ones.

2.7 To determine the value of k_b we need to consider the initial rate of product release. From the data given in the SAQ we can see that the rate of release is linear for the first 10 seconds for impeller type A.

Thus, from the graph $[E_t]/[E^{max}] = 0.35$ at $t = 10$.

The $[E^{max}]$ for this impeller is 0.084.

Thus $[E_t] = 0.35 \times 0.084 = .029$.

Substituting $[E_t]$, $[E^{max}]$ and t into $[E_t] = [E^{max}](1-e^{-K_bt})$ gives:

$k_b = 0.043$ min^{-1}.

A similar approach for impeller type B gives a value of k_b of about 0.019 min^{-1}.

2.8 At one level, this appears to be quite a good scheme. There is relatively little loss in activity at each stage and the specific activity (U g^{-1} protein) indicates that there is considerable purification of the enzyme away from the contaminating proteins at each stage. We have tabulated the data:

		yield $\left(\dfrac{\text{total activity}}{\text{initial activity}} \times 100\right)$	specific activity (U /g^{-1} protein)	degree of purification $\left(\dfrac{\text{specific activity}}{\text{initial specific activity}}\right)$
	initial extract	100%	5×10^6	1
step 1	ion exchange chromatography	96%	2.4×10^7	x 4.8
step 2	gel filtration	92%	9.2×10^7	x 18.4
step 3	affinity chromatography	86%	4.3×10^9	x 860

The key question is do we really need to carry out three steps? Since we have an affinity chromatographic method available, would passing the initial extract down such a column lead to a sufficient purification to meet the specification for the product? If so, then why conduct the two other steps? The point we are making is that we should always challenge the scheme in terms of asking if it is the most cost-effective way we can devise in order to achieve our objectives.

In the example given in this SAQ other compounds or proteins in the initial extract might compete with the enzyme for binding sites during affinity chromatography. If this was the case, then using affinity chromatography at step 1 may not be effective. The only way to determine whether this is so would be to conduct suitable experiments to see if affinity chromatography could replace the multiple steps indicated in the question. As we have tried to emphasise throughout, there are no absolute answers.

2.9 Of the techniques listed, only affinity chromatography, hydrophobic interaction chromatography, high-pressure homogenisation and tubular bowl centrifugation are widely-used for large-scale enzyme extraction and purification. Sonication, freeze-pressing, ammonium sulphate precipitation and detergent extraction are commonly used in laboratory-scale systems for the extraction and purification of enzymes, and ultracentrifugation and PAGE (polyacrylamide gel electrophoresis) may be involved for analytical purposes.

Responses to Chapter 3 SAQs

3.1 1) a) 50.1 min. The amount of enzyme added to the incubation mixture would be capable of hydrolysing 100 µmol min^{-1} mg^{-1} x 100 mg l^{-1} = 10 mmol min^{-1} l^{-1}, thus V_{max} = 10 mmol min^{-1} l^{-1}.

We can calculate that 50% hydrolysis starting with $[S_0]$ = 1000 mmol l^{-1}, implies that $[S_t]$ = 500 mmol l^{-1} and as we know K_M = 1 mmol l^{-1} and V_{max} = 10 mmol min^{-1}.

Thus substituting into:

$$V_{max}t = 2.303\, K_M \log \frac{[S_0]}{[S_t]} + ([S_0] - [S_t])$$

$$10t = 2.303 \log 2 + 500$$

$$t = \frac{500 + 0.69}{10} = 50.1 \text{ min.}$$

b) 75.1 min. In this case:

$$10t = 2.303 \log \frac{1000}{250} + 750$$

$$t = 75.1 \text{ min.}$$

c) 90.2 min since:

$$10t = 2.303 \log \frac{1000}{100} + 900$$

$$t = 90.2 \text{ min.}$$

2) We can use the same type of calculation as in 1) except that $K_M = 100$ mmol l^{-1}.

Thus for: a) $t = 56.9$ min; b) $t = 889$ min; c) $t = 113.0$ min.

The contrast between the times calculated in 1) and 2) shows the effect of using enzymes of lower affinity (higher K_M values). The incubation time is prolonged.

3) 200 mg l^{-1}.

Again we can use:

$$V_{max} t\, 2.303 \log \frac{[S_0]}{[S_t]} + ([S_0] - [S_t])$$

We need to achieve 50% hydrolysis in ½ of 56.9 minutes (= 28.5 minutes).

Thus:

$$V_{max}\, 28.5 = 100\, (2.303 \log 2) + 500$$

$$V_{max} = \frac{569}{28.5} = 19.96 \text{ mmol min}^{-1} l^{-1}$$

But the specific activity of the enzyme is 100 μmol min^{-1} mg^{-1} protein.

So we need $\frac{19.96 \times 1000}{100}$ mg of enzyme $l^{-1} = 199.6$ mg $l^{-1} = 200$ mg l^{-1} of enzyme.

You could have taken a short cut to this answer. We pointed out that the rate of an enzyme reaction is proportional to the amount of enzyme present, thus to halve the time, we need to use twice as much enzyme.

3.2

1) Before fermentation. Alcohol is an inhibitor of pectinases and thus adding this enzyme after fermentation would prove fruitless.

2) It could be an advantage. Since glucose isomerase is inhibited by oxygen, the glucose oxidase might remove this oxygen at the expense of a small amount of the substrate being converted to gluconic acid. This latter product would, of course, contaminate the glucose:fructose product but this may not be a problem.

If the subsequent use of the product demanded that no gluconic acid was present, then the presence of glucose oxidase might be a disadvantage. Its activity could be prevented by adding D-arabinose but this again would contaminate the product and may lead to higher purification costs. An alternative might be to further purify the enzyme before use to remove residual glucose oxidase activity.

3) The answer is no. The likelihood is that the maltose released by the enzyme will itself inhibit the activity of the enzyme. Addition of an inhibitor such as ascorbic acid (expensive) or oxalate (poisonous) is clearly not desirable and nothing is gained by trying to inhibit the enzyme.

3.3

We believe homogeniser A is the most suitable. Although it is slightly more expensive both in capital and running costs, it produces small particles. The smaller the particles, the larger the surface:volume ratio. Thus homogeniser A will produce a larger surface of substrate on which the enzyme can work.

If we assume that the particles are spherical we could calculate the actual surface area m^{-3} of meat using the equation:

$$\text{volume of a sphere} = \frac{4}{3} \pi r^3; \text{ area of a sphere} = 4 \pi r^2.$$

Responses to SAQs

(Perhaps you would like to do this for yourself).

We calculated that each sphere of volume 25 mm³ would have a radius of 1.81 mm and the surface area of each sphere would thus be 41.2 mm². But in 1m³ of meat there would be $\frac{10^9}{25}$ spheres = 4×10^7 spheres. Thus the total area of the spheres produced from 1m³ would be 1.648×10^9 mm².

A similar calculation for the spheres produced by homogeniser B gives a radius of 2.73 mm, a surface of 93.7 mm²/sphere.

Since the total number of spheres is $\frac{10^9}{85} = 1.18 \times 10^7$, the total area of the spheres is 1m³ of meat = $93.7 \times 1.18 \times 10^7 = 1.11 \times 10^9$ mm².

Remember that enzyme reaction rates in systems in which the substrate is particulate is often dependent upon the interfacial area between substrate and medium. Thus we anticipate that the relative rates of reaction with the same amount of enzyme present would be 1.648:1.11 (that is 1.48:1) in favour of the product from homogeniser A. This would mean that we could use substantially less enzyme and these savings would more than outweigh the higher running costs of homogeniser A. Furthermore, the smaller particle sizes produced by this homogeniser would mean that there would be less compacted, slow-reacting proteins in these particles. Thus it should be anticipated that maximum hydrolysis would be easier to achieve.

You must, however, realise that this is a somewhat over-simplified situation. In practice many other factors need to be considered. For example we need to know the distribution of particle sizes in order to gauge properly the likely effects on reaction kinetics. Also factors like operational life-span and back-up servicing of the homogenisers need to be taken into account.

3.4

1) There are several options here. One would be to heat the protein extract prior to incubation with trypsin. This might denature the inhibitor and thus trypsin could be used. Alternatively we might use an alternative proteolytic enzyme which was not sensitive to the inhibitor (eg papain).

    You might have considered extracting the inhibitor from the extract but this would be likely to be too expensive.

2) The likely problem you should have spotted is that the long incubation period could allow the growth of micro-organisms. Fruit extracts, although at a low pH, are full of potential nutrients for micro-organisms. The low pH would particularly favour fungi. (Fruit juices usually go 'mouldy' when they are spoilt by micro-organisms). There are three basic solutions to this problem. One would be to use more enzyme and shorten the incubation time. The other would be to pasteurise the crushed fruit and then incubate it under microbiological 'clean' conditions. A third option would be to include an anti-microbial (anti-fungal) agent, preferably one that could be easily removed or one that was acceptable in the food.

3) We would suggest either ferric acetate or ferric chloride. Ferric oxalate, although it is soluble and would supply Fe^{3+}, is toxic and not suitable as a component of a food additive. Its removal may add significantly to downstream processing. Ferric oxide is insoluble. Haemoglobin, contains Fe^{3+} ions, but these are bound into the porphyrin prosthetic group of haemoglobin and would not be available for the enzyme. It is also relatively expensive and not a desirable component to add to our proposed product.

3.5

1) We would choose a pH in the range 4.5-5.5. At these pHs, the enzyme is relatively stable and shows maximum activity. Below pH 4.5 the activity falls very dramatically and the enzyme is unstable (k_1 increases). Above pH 5.5, whilst enzyme activity remains high, the enzyme is unstable.

2) We would choose a pH in the range 7.5-8.5. At these pHs enzyme activity is maximal and the equilibrium favours the formation of B by a factor of more than 100. Below pH7, although the enzyme is still very active, the equilibrium position would imply that the reaction would stop after only a small proportion of A had been converted to B. (Remember that in our figure we were plotting log K_{eq}).

3.6

1) Increase reaction rate by $10^{Z_{his} Z_{arg} \sqrt{2X}} - 10^{Z_{his} Z_{arg} \sqrt{X}}$.

    Remember that reactions involving identical charged groups are increased by increases in ionic strength.

    The relationship we use is:

    $$\log [k] = \log [k_0] + Z_A Z_B \sqrt{I}$$

Thus using this relationship:

$$\log k_{cat\,(x)} = \log k_{cat(0)} + Z_{his} Z_{arg} \sqrt{X}$$

at the lower ionic strength.

At the higher ionic strength:

$$\log k_{cat\,(2x)} = \log k_{cat(0)} + Z_{his} Z_{arg} \sqrt{2X}$$

Thus taking antilogs:

$$k_{cat\,(x)} = k_{cat(0)} + 10^{Z_{his} Z_{arg} \sqrt{X}}$$

and:

$$k_{cat\,(2x)} = k_{cat(0)} + 10^{Z_{his} Z_{arg} \sqrt{2X}}$$

Thus the change in rate will be:

$$k_{cat\,(2x)} - k_{cat\,(x)} = 10^{Z_{his} Z_{arg} \sqrt{2X}} - 10^{S_{his} Z_{arg} \sqrt{X}}$$

2) c) would have most effect. Ionic strength $I = 0.5 \Sigma(cZ^2)$ where c = concentration, Z = charge on the ion.

Thus the ionic strength of a):

$$= 0.5 \times [(0.1 \times 2^2) + (0.1 \times 1^2)] \text{ mol l}^{-1} = 0.3 \text{ mol l}^{-1};$$

similarly for b):

$$I = 0.5 \times [(0.2 \times 1^2) + (0.2 \times 1^2)] = 0.2 \text{ mol l}^{-1};$$

and for c):

$$I = 0.5 \times [(0.1 \times 3^2) + (3 \times 0.1 \times 1^2)] = 0.6 \text{ mol l}^{-1}.$$

Thus solution c) has the biggest influence on the ionic strength of the incubation mixture and thus would have the biggest effect on reaction rate.

3.7 1) It is best to tackle this problem graphically. Since deactivation is first order, then $[E_t] = [E_0]e^{-k_d t}$ or:

$$\log [E_t] = \log [E_0] - 2.303 \, k_d t.$$

Thus a plot of $\log [E_t]$ against t should give a straight line with a slope of $-2.303 \, k_d$. Using the data given in the question.

	Log activity of the enzyme (μmol mg^{-1} protein min^{-1})			
time (min)	40°C	50°C	53°C	58°C
0	2	2	2	2
20	1.995	1.977	1.954	1.771
40	1.988	1.954	1.908	1.538
60	1.986	1.922	1.863	1.310
80	1.982	1.908	1.816	1.009

Note, we have plotted these on a single graph (to save paper). To get more accurate estimates of k_d, these would need to be plotted on separate graphs to achieve more accurate k_d values.

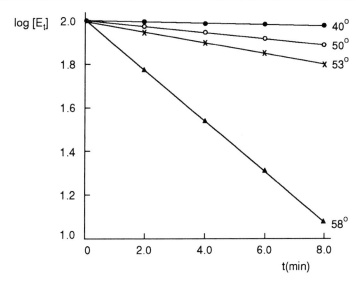

The slopes of the lines (= -2.303 K_d):

40°C = 2.3 × 10^{-4} min^{-1}; 50°C = 11.5 × 10^{-4} min^{-1}; 53°C = 23.0 × 10^{-4} min^{-1}; 58°C = 115 × 0^{-4} min^{-1}.

Thus:

k_d = 1 × 10^{-4} min^{-1} (40°C); 5 × 10^{-4} min^{-1} (50°C); 10 × 10^{-4} min^{-1} (53°C); 50 × 10^{-4} min^{-1} (58°C).

2) $t_{\frac{1}{2}}$ at 50°C = 1386 min; $t_{\frac{1}{2}}$ at 58°C = 138.6 min;

since: $t_{\frac{1}{2}}$ = 0.693/k_d.

3) We use the relationship $[E_t] = E_0 e^{-k_d t}$ and log $[E_t]$ = log $[E_0]$ − 2.303 $k_d t$.

Since $[E_0]$ = 100%, then log $[E_t]$ = 2 − 2.303 $k_d t$.

Thus for a) log $[E_t]$ = 2 − 2.303 × 1 × 10^{-4} × 120 = 2 − 0.02764 = 1.97236;

$[E_t]$ = 93.8%.

By analogy: b) $[E_t]$ = 73%; c) $[E_t]$ = 4.2%; d) $[E_t]$ = 38.5%; e) $[E_t]$ = 0.0072%.

Responses to Chapter 4 SAQs

4.1

1) Tyrosine and histidine. If you got this wrong, re-examine Figure 4.3.

2) All of them. To produce a diazonium intermediate we require a free amine group. The support must, however, be stable to the NaNO$_3$:HCl reagent used to generate the diazonium intermediate from the amine.

3) a) A co-polymer of ethylene and maleic anhydride has many reactive anhydride groups. If enzyme alone was added to the co-polymer, these groups would interact so much with the amine groups on the enzyme, the enzyme would be so rigidly held and structurally modified that it would probably be inactivated. The addition of NH$_2$ - R - NH$_2$ blocks off some of these reactive groups and leads to a looser network. Obviously a balance in the ratio of enzyme to NH$_2$ - R - NH$_2$ has to be maintained. If we add too much NH$_2$ - R - NH$_2$ then no enzyme would be bound because NH$_2$ - R - NH$_2$ would saturate the anhydride sites. NH$_2$ - R - NH$_2$ typically has the structure NH$_2$ - (CH$_2$)$_n$ - NH$_2$ where n = 6 or more.

b) The support would be negatively charged. You will learn why this is important later.

4.2

1) a) Although there is a slight reduction in the specific activity of the enzyme in the later stages of the incubation, this is only slight. Thus we can conclude that we are not losing much additional enzyme activity by using prolonged incubations. On the other hand, the total amount of protein being attached to the support continues to increase. In other words the total amount of enzyme activity (specific activity x total protein) attached to the support continues to rise. This suggests that we can use as long as possible incubation to maximise the binding of active enzyme. Perhaps 12 hours is most appropriate, since after this period, there would presumably be a continued slowing in further binding of enzyme.

b) Although there is a reduction in the specific activity of the enzyme attached to cellulose when compared to the enzyme in solution, we cannot conclude that this is simply a reflection of the enzyme being denatured. It could be more difficult for the substrate to gain appropriate access to the enzyme when attached to the support. The sort of experiment we could do is to couple the enzyme with a variety of supports via diazonium groups and also to cellulose using a range of reactive groups. In this way, we might be able to establish whether it is the reactivity of the diazonium groups or the steric effects of the support which causes a drop in specific activity.

2) a) In determining what the optimum time of incubation is we have first to decide what the objective is. Three possible objectives spring to mind:

- to bind the maximum amount of protein;
- to obtain a bound enzyme with maximal specific activity;
- to obtain a preparation with maximal total activity.

In practice, we are usually not interested in binding maximal amounts of protein which show little or no enzyme activity. Nor are we usually interested in binding very small amounts of enzyme even if it shows very high specific activity. We are, however, usually interested in producing a preparation of bound enzymes which show maximal total activity providing this is achieved cost-effectively.

In the case of the example shown in the question, if we multiply the specific activity of the enzyme by the total amount of protein bound at each time interval, this will give us the total amount of enzyme activity bound to the support at each time interval. This will enable us to identify the optimum incubation time (about 6 hours).

b) Replacement of leucine by glutamate in the co-polymer raises a number of potential problems. First, introducing this amino acid means that new functional groups (- COOH) are introduced in the polymer. These may react with the reagents used in the diazonium coupling process and also may cause further interactions with the enzyme. For example, above pH4 the carboxyl groups will be deprotonated (-COO$^-$) which would attract $^+NH_2$ - groups on the protein. It may also increase the solubility of the co-polymer. One final point is that the negative charge on the - COO$^-$ groups might repel the substrate (RNA) which also carries negative charges.

The aim of this question was to point out that in selecting a carrier we need to consider the presence of functional groups to consider how these may interact with reagents used in the coupling process. They may also influence direct interaction of the carrier with the enzyme as well as influence the behaviour of substrates within the vicinity of the enzyme.

4.3

1) R - COOH would be the most suitable. We would like to use the enzyme at its optimum pH. At this pH (pH6), we are below the pI (7.3) of the enzyme and the enzyme would carry a net positive charge. The resin, being above its pKa value, would be in its unprotonated form (ie R - COO$^-$) and would carry a negative charge. Thus the enzyme would bind.

At pH6, the R - $^+NH_3$ resin would be protonated because it is below its pKa value. Thus the resin would be positively charged and would repel the positively charged enzyme at this pH. Thus the enzyme would not be retained.

2) It would not be suitable.

Responses to SAQs

Let us assume that we can load the enzyme onto the resin at a pH below 7 but above 4 (for example pH 5.5). At this pH, the ion exchange resin will be mainly in the negatively charged form R - COO⁻ as it is above its pKa value. The enzyme, being below its pI value, will be in its positively charged form and will, therefore, bind to the resin. But, if we would like to use the enzyme at its optimum pH (pH8). This pH would be above the pI value of the enzyme and thus the enzyme would carry a net negative charge. It would, therefore, elute from the resin.

4.4 You should have had little difficulty with this question. Trypsin is itself a protein. If the pores of the gel are to be made sufficiently small to entrap trypsin within the gel, they will also prevent the substrate protein molecules from entering the gel. Therefore, trypsin is not a suitable candidate for entrapment.

4.5 From what you have learnt so far, you should have chosen only 4).

In 1) the negatively charged support would repel the substrate and thus [S] in the micro-environment of the support would be lower than in the bulk liquid. The product would tend to be retained by the support and thus [P] would be higher than in the bulk liquid. Thus with a lower $\frac{[S]}{[P]}$ ratio in the micro-environment of the support, we would expect the net reaction in the direction S → P to be slower.

In 2), again the substrate would tend to be repelled by the support and the product to be retained. Thus the ratio of $\frac{[S]}{[P]}$ would be lower in the support than in the bulk liquid and the reaction would be slower.

In 3) both S and P would accumulate in higher concentrations than in the bulk liquid. If they were retained by the same relative extent, we would expect the $\frac{[S]}{[P]}$ ratio to be similar to that in the bulk liquid and thus the net reaction rate to be largely unaltered.

In 4), S would be attracted and P repelled. Thus we might expect a higher $\frac{[S]}{[P]}$ ratio than in the bulk liquid and thus a faster reaction rate.

In reality, however, the situation is not as straightforward as this. As we will learn in the next section, since diffusion within the support may be limited, the product concentrations will tend to build up in the support at the beginning of the incubation when the rate of formation of P is greatest and will only decline when the rate of formation of P declines to a value below the diffusion rate out of the support.

4.6 There are many factors which will increase the probability that diffusion will limit the overall rate of a reaction catalysed by an immobilised enzyme. The main ones are:

- low $[S_b]$ value;

- low substrate diffusion rates (K_L will be low);

- high enzyme loading;

- high specificity constant $\left(\frac{V_{max}}{K_M}\right)$;

- low K_M (K_M');

- low rates of mixing, this will increase δ and thus will decrease $K_L = \frac{D_s}{\delta}$.

4.7 1), 2) and 3).

In the text we showed that at steady state:

$$K_L ([S_b] - [S_s]) = \frac{V_{max} [S_s]}{K_M' + [S_s]}.$$

1) - increased agitation would reduce the depth (δ) of the stagnant layer. This would effectively increase K_L and thus the value of $V_{max}[S_s]$ (ie the reaction rate).

2) - replacement of the bulk liquid would mean that product P would not build up in the bulk liquid (that is $[P_b] = 0$). Thus a step (maximal) gradient of product concentration would be maintained between the surface of the support and the bulk liquid. Thus P would diffuse at maximum rates away from the surface thereby enhancing the rate of the forward reaction (law of mass action) and reducing product inhibition.

3) - increasing $[S_b]$ would increase the rate of flux of substrate to the surface. This would increase the rate of reaction provided that $[S_s]$ was not too high and caused substrate inhibition.

4) - increasing the viscosity of the solution would deicrease the diffusion coefficient (D_s) and thus reduce the mass transport coefficient (K_L) of the substrate to the surface. Thus at steady state there would be a reduction of the rate of the forward reaction.

4.8

1) Since $D_s = 10 \times 10^{-10}$ m^2 s^{-1} and $\delta = 5 \times 10^{-6}$ m:

$$K_L = \frac{10 \times 10^{-10}}{5 \times 10^{-6}} = 2 \times 10^{-4} \text{ ms}^{-1}.$$

This is much lower than the V_{max}/K_M ratio (20 ms^{-1}).

Thus we must assume that since K_L is much lower than V_{max}/K_M and the system will be diffusion limited.

Under these conditions:

$v_A = K_L [S_b]$

$S_b = 1$ mol l^{-1} = 1000 mol m^{-3}. Therefore;

$v_A = 2 \times 10^{-4}$ m s$^{-1} \times 1000$ mol m^{-3} s^{-1} = 0.2 mol m^{-2} s^{-1}.

2) In this case K_L still = 2×10^{-4} ms^{-1} but V_{max}/K_M ratio is reduced to 20×10^{-6} m s^{-1}.

Thus V_{max}/K_M is now much smaller than K_L. Thus the amount of enzyme becomes limiting and the relationship we use is:

$v_A = [S_b] (V_{max}/K_M') = 1000 \times 20 \times 10^{-6}$ mol m^{-2} s^{-1}

$= 2 \times 10^{-4}$ mol m^{-2} s^{-1}.

3) In this case K_L and V_{max}/K_M have relatively similar values. Thus in this case we use the relationship:

$$v_A = \frac{[S_b]}{1/K_L + 1/(V_{max}/K_M')}$$

$$v_A = \frac{1000}{5 \times 10^3 + 10^4} \frac{\text{mol m}^{-3}}{\text{m}^{-1} \text{ s}} = \frac{1000}{1.5 \times 10^4} = 0.067 \text{ mol m}^{-2} \text{s}^{-1}.$$

4.9

1) The substrate does not penetrate the whole biocatalyst. We know this from the following calculation.

The total volume of the biocatalyst is area x thickness = 100×10^{-3} m^3 = 0.1 m^3

But the zeroth order reaction constant = 1 mol m^{-3}h^{-1}. Thus 0.1m^3 of membrane should consume 0.1×1 mol h^{-1}. However, it only consumes 0.08 mol h^{-1}. Thus it is working below capacity and this must reflect substrate depletion.

2) 0.8 mm.

Since the substrate is consumed at 0.08 mol h^{-1} and the zeroth order reaction constant = 1 mol m^{-3}h^{-1}, this must mean that 0.08 m^3 of the biofilm is consuming substrate.

Responses to SAQs

The area of the biofilm is 100 m^2, so the effective working depth = $\frac{0.08}{100}$ m = 0.8 mm.

4.10 1) Volume = 1m^3, area = 500 m^2.

The total amount of biocatalyst used was 100 kg at a concentration of 100 kg m^{-3}. Thus the volume of the biocatalyst must be 1m^3. But since the thickness of the biocatalyst was 2mm (= 2 x 10^{-3}m), the area must be $\frac{1}{2 \times 10^{-3}}$ m^2 = 500 m^2.

2) Substrate only penetrates to 1.3 mm.

Since the reaction rate constant = 1 x 10^{-3} mol s^{-1} kg^{-1}, 100 kg of the biocatalyst should consume 100 x 1 x 10^{-3} mol s^{-1} = 0.1 mol s^{-1}.

However it only consumes 0.06 mol s^{-1}. Thus only $\frac{6}{10}$ of the biocatalyst is consuming substrate. Thus the substrate has only penetrated $\frac{6}{10}$ of 2 mm = 1.2 mm.

3) Effectiveness factor = 0.6.

Since $\eta = \frac{r}{R}$ and R = 2 mm and r = 1.2.

4) 0.327 mol kg^{-1}.

Since $\eta = \sqrt{(2[S_b] D/R^2 K_0)}$ (see Table 4.1).

Then $0.6 = \sqrt{(2[S_b] \times 2.2 \times 10^{-9}/(2 \times 10^{-3})^2 \times 1 \times 10^{-3})}$.

Note we converted R to meters to have consistent units.

Thus $0.36 = \frac{2[S_b] \times 2.2 \times 10^{-9}}{4 \times 10^{-6} \times 1 \times 10^{-3}}$

$0.36 = 1.1 [S_b]$

Thus $S_b = 0.327$ mol kg^{-1}.

Note the rather strange units of concentration. This is because the reaction rate constant was expressed in mol s^{-1} kg^{-1}.

4.11 These are all true.

1) Product will accumulate in the immobilised matrix and thus inhibit the reaction. Thus if v is lower, it means that the value of η will be lower.

2) The positively charged inhibitor will partition into the negatively charged support. Thus its concentration in the micro-environment of the enzyme will be higher than it would be in the vicinity of the enzyme in free solution. The enzyme would, therefore, be more inhibited.

3) If the system is not diffusion limited, the substrate concentration at the liquid:immobilised enzyme interphase would be high and the enzyme molecules at the surface would be inactivated. As these surface molecules are inactivated, substrate would continue to diffuse into the enzyme matrix. Thus, progressively the enzyme molecules deeper in the matrix would be inactivated. Ultimately all enzyme molecules would be inactivated. If, however, the system was diffusion limited, the enzyme molecules within the matrix would tend to keep the substrate concentration low and this would reduce or prevent inactivation.

4) Consider the following graph displaying reaction velocity against enzyme loading (this is similar to Figure 4.15 in the text) for an immobilised enzyme system.

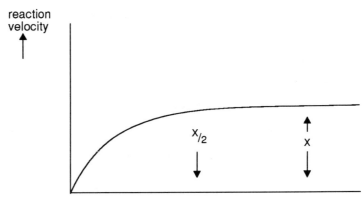

At high enzyme loadings (eg at X), then the reaction velocity is independent of enzyme loading (is diffusion limited).

Even if we halved the amount of enzyme (eg at $\frac{X}{2}$), there would be very little effect on the reaction velocity.

Now, returning to the question, if we had a high enzyme loading (as at X) but choose a pH in which each enzyme molecule could only work at 50% efficiency, then this would only have a marginal effect on the observed reaction velocity. In other words, in diffusion limited systems, enzymes *appear* to be less sensitive to pH. Similarly in diffusion limited systems enzymes appear to be less sensitive to temperature, ionic strength of the solution, inhibition or activation.

5) See 4). We could significantly inhibit enzyme supplied at a concentration of X without any noticeable effect on reaction velocity.

Responses to Chapter 5 SAQs

5.1

1) 60.1 min. We have used the relationship:

$$V_{max}t = K_M \ln \frac{1}{1-F} + F[S_b]$$

From the question V_{max} = 10 mol min^{-1} m^{-3}, K_M = 1 mol m^3, $[S_b]$ = 1000 mol m^{-3}, F = 0.6.

Thus:

$$10 \times t = 1 \ln \frac{1}{1-0.6} + 0.6 \times 1000$$

$$10t = 1 \times 2.303 \log 2.5 + 600 = .916 + 600$$

$$t = \frac{601}{10} = 60.1 \text{ min.}$$

The assumption we have made is that throughout the process diffusion of substrate into the biocatalyst has not become rate limiting.

2) 69.2 min.

By analogy with 1):

$$10t = 100 \ln \frac{1}{1-0.6} + 0.6 \times 1000$$

Responses to SAQs

$$t = \frac{91.6 + 600}{10} = 69.2 \text{ min.}$$

5.2 1) 11.38 m.

2) 17.8 m.

1) This is quite a difficult calculation, so we will break it down into stages.

First we can calculate the residence time we need to achieve the desired conversion.

Using $V_{max}\theta = K_M' \ln \frac{[S_0]}{[S_\theta]} + ([S_0] - [S_\theta])$

where θ = maximum residence time; $[S_\theta]$ = substrate at θ; $[S_0]$ = substrate supplied to the column.

In this case $[S_\theta] = 500$ mmol l^{-1}; $[S_0] = 1000$ mmol l^{-1}; $K_M' = 100$ mmol l^{-1} and $V_{max} = 10$ mmol l^{-1} min^{-1}.

Thus $10 \theta = 100 \times 2.303 \log 2 + 500 = 569$ m.

So the residence time required to achieve the desired conversion = 56.9 min.

We know that the cross sectional area of the column is 0.1 m^2 and that half the volume of the reactor is occupied by the enzyme-support matrix. The flow rate (ϕ) = 1 litre min^{-1}.

The residence time θ = :

$$\frac{(\text{cross-sectional area} \times L) - (\text{volume occupied by support})}{\phi}$$

We need to bring the values to the same units, $\phi = 1$ litre min^{-1} = 0.001 m^3 min^{-1}.

Substituting in the values for θ and ϕ:

$$56.9 = \frac{0.1 \times L - 0.5(0.1 \times L)}{0.001}$$

$$56.9 = \frac{0.05L}{0.001}$$

$$L = 1.138 \text{ m.}$$

2) We use a similar calculation except to get 75% conversion $[S_\theta] = 250$ mmol l^{-1}.

Using this value, the residence time θ required can be calculated from:

$$V_{max}\theta = K_M' \ln \frac{[S_0]}{[S_\theta]} + ([S_0] - [S_\theta]).$$

Thus $\theta = 88.9$ min.

The flow rate, ϕ, is the same as before ($\phi = 0.001$ m^3 min^{-1}).

$$88.9 = \frac{0.1 \times L - 0.5(0.1 \times L)}{0.001}$$

Thus $L = 1.78$ m.

5.3 80%. The batch process is operated for $\frac{20}{25}$ of the time available = 0.8 of the total time.

Since the PFR operates all of the time, the ratio of the yield will be 1:0.8 (PFR:batch) if the same amount of enzyme is used in each case. In other words, the yield from the batch system would be 80% of that of the PFR.

5.4

1) We need to use sufficient enzyme to give a V_{max} of 25 mmol l⁻¹ h⁻¹.

   We can use the relationship:

   $$D\,([S_0] - [\tilde{S}_r]) = \frac{V_{max}[\tilde{S}_r]}{K_M' + [\tilde{S}_r]}.$$

   In this case $D = \frac{5}{100}$; $[S_0] = 1000$ mmol l⁻¹; $[\tilde{S}_r] = 500$ mmol l⁻¹.

   $K_M' = 1$ mmol l⁻¹.

   Thus:

   $$\frac{5}{100}(1000 - 500) = \frac{V_{max}[500]}{1 + 500}.$$

   $V_{max} = 25$ mmol l⁻¹ h⁻¹.

2) We need to use sufficient enzyme to give a V_{max} of 50 mmol l⁻¹ h⁻¹.

   Since:

   $$D\,([S_0] - [\tilde{S}_r]) = \frac{V_{max}[\tilde{S}_r]}{K_M' + [\tilde{S}_r]}$$

   $$\frac{10}{100}(1000-500) = \frac{V_{max}500}{1 + 500}$$

   $V_{max} = 50$ mmol l⁻¹ h⁻¹.

3) We need to use sufficient enzyme to give a V_{max} of 60 mmol l⁻¹ h⁻¹.

   $$D\,([\tilde{S}_0] - [\tilde{S}_r]) = \frac{V_{max}[\tilde{S}_r]}{K_M' + [\tilde{S}_r]}$$

   $$\frac{10}{100}(1000-500) = \frac{V_{max}500}{100 + 500}$$

   $V_{max} = 60$ mmol l⁻¹ h⁻¹.

   If we compare 2) and 3) we can see that with lower affinity enzymes (high K_M') we need to use more enzyme to achieve the same conversion rate.

5.5

1) a) We need to add sufficient enzyme to give a V_{max} 0.12 mol h⁻¹ per litre of reaction.

   Since:

   $$D\,([S_0] - [\tilde{S}_r]) = \eta \left(\frac{V_{max}[\tilde{S}_r]}{K_M' + [\tilde{S}_r]} \right)$$

   $$0.07 = 0.6 \left(\frac{V_{max} \times 0.3}{0.01 + 0.3} \right)$$

   $V_{max} = 0.12$ mol l⁻¹ h⁻¹.

   b) 1.4 mol h⁻¹.

   The output per unit volume of the reactor = $D\,([S_0] - [\tilde{S}_r]) = 0.07$ mol l⁻¹ h⁻¹.

   But the total volume of the reacter is 20 litres.

Thus the total output $= 0.07 \times 20 = 1.4$ mol h^{-1}.

2) We need to add sufficient enzyme to give a $V_{max} = 0.18$ mol h^{-1} per litre of vessel.

In this case, we still need to convert 0.07 mol h^{-1} per litre of the reactor.

$$\eta\left(\frac{V_{max}[\tilde{S_r}]}{K_M + [\tilde{S_r}]}\right) = 0.07$$

$$0.4\left(\frac{V_{max}\, 0.3}{0.01 + 0.3}\right) = 0.07$$

$V_{max} = 0.18$ mol h^{-1} per litre of the reactor.

5.6

1) We could apply Michaelis-Menten kinetics as these would approximate the reaction kinetics.

Since we intend to stop the reaction when 80% of the substrate is converted, the ratio of [P]/[S], at this stage will be 4. This is considerably lower than the K_{eq} value ($= 10^4$). When [P]/[S] = 4, the reverse reaction (P + E → ES) would be insignificant compared to the forward reaction. Thus Michaelis-Menten kinetics would apply.

2) We could not apply Michaelis-Menten kinetics in this case. At the completion of the incubation $\frac{[P]}{[S]} = \frac{60}{40} = 1.5$. Since $K_{eq} = 3$, we would be approaching the equilibrium position. Thus the reverse reaction would be quite significant compared to the forward reaction.

3) Again we could not apply Michaelis-Menten kinetics for the whole process. For quite a considerable portion of the incubation time, the ratio [P]/[S] would be well below 10^2 and thus the kinetics of the reaction would approximate to the Michaelis-Menten relationship. However, towards the end of the incubation, with the build up of [P] and the decrease of [S], the ratio [P][S] would approach 10^2 (its final value would be 0.66×10^2). Thus the kinetics of the reverse reaction would become significant and the Michaelis-Menten relationship would no longer hold.

What this SAQ has emphasised is that providing a reaction does not approach the equilibrium position, then the reaction kinetics are approximated by the Michaelis-Menten relationship. If, however, we approach the equilibrium, and the rate of the reverse reaction becomes significant, then the **net** forward reaction is slowed and the Michaelis-Menten relationship no longer holds. In designing an industrial process this must be borne in mind. Obviously, we would like to achieve a high substrate conversion but if this means that we are approaching K_{eq}, then this will significantly increase the reaction time required. The maximum substrate conversion is achieved when $\frac{[P]}{[S]} = K_{eq}$, but, in theory, this is only achieved after incubation for infinite time.

In practice, therefore, a compromise is reached between substrate conversion and incubation time.

We can represent this situation graphically:

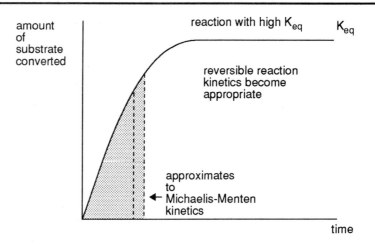

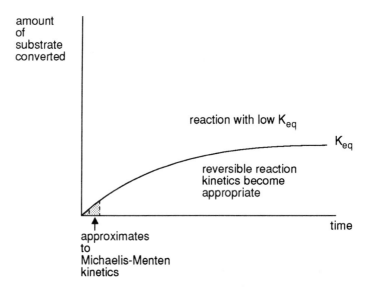

5.7 1) transaminase, 2) glucose isomerase.

These two enzymes catalyse the reactions:

$$R^I-\underset{O}{\overset{\|}{C}}-COOH + R^{II}-\underset{H}{\overset{NH_2}{\underset{|}{C}}}-COOH \rightleftharpoons R^I-\underset{H}{\overset{NH_2}{\underset{|}{C}}}-COOH + R^{II}-\underset{O}{\overset{\|}{C}}-COOH$$

and:

$$glucose \rightleftharpoons fructose$$

Neither reactions involve large energy changes (ie ΔG is small) and the equilibrium constants are close to unity. In other words the reactions are freely reversible.

With both ATPase and acetyl CoA thiolase catalysed reactions there is a large energy change:

Responses to SAQs 359

$$\text{ATP} + \text{H}_2\text{O} \xrightleftharpoons{\text{ATPase}} \text{ADP} + \text{Pi}$$

$$\text{acetyl-SCoA} + \text{H}_2\text{O} \xrightleftharpoons[\text{thiolase}]{} \text{acetate} + \text{HSCoA}$$

Thus the equilibrium lies far to the right (ie K_{eq} is large). Thus these reactions behave as though they are virtually irreversible.

5.8   642 mmol l^{-1}.

We use the relationship:

$$\ln \frac{\left\{([S_0] - [\tilde{S}_r^0]) + K_M \left(\frac{[S_0] - [\tilde{S}_r^0]}{[\tilde{S}_r^0]}\right)\right\}}{\left\{([S_0] - [\tilde{S}_r^t]) + K_M \left(\frac{[S_0] - [\tilde{S}_r^t]}{[\tilde{S}_r^t]}\right)\right\}} = k_d t.$$

Substituting in the values given in the question (converting to mmol l^{-1}) and using $x = [\tilde{S}_r^t]$.

$$\ln \left\{(1000-100) + 10\left(\frac{1000-100}{100}\right)\right\} - \ln \left\{(1000-x) + 10\left(\frac{1000-x}{x}\right)\right\} = 0.1 \times 10.$$

$$\ln(990) - \ln\left\{(1000-x) + \left(\frac{10000}{x}\right) - 10\right\} = 1$$

$$6.898 - 1 = \ln\left(1000 - x + \frac{10000}{x} - 10\right)$$

$$5.898 = \ln\left(1000 - x + \frac{10000}{x} - 10\right)$$

$$364.1 = 1000 - x + \frac{10000}{x} - 10$$

therefore $-625.9 = -x + \dfrac{10000}{x}$

and $x^2 - 625.9x - 10000 = 0$.

Solving this quadratic equation gives $x = 642$ mmol $l^{-1} = [\tilde{S}_r^t]$.

Clearly, the decay of the enzyme over the 10 day incubation means that less substrate is converted in the reactor. Thus the output substrate concentration will be higher.

5.9
1) This should not present a problem since the cofactor is attached to the enzyme and it is first modified (forms a Schiffs base) by interaction with the first substrate (the keto acid) and is converted back to its original form (ie regenerated) by interaction with the second substrate.

2) Choice b) looks most appropriate, a) is too expensive and c) is inappropriate because we need intact mitochondria to generate ATP from the oxidation of substrates and electron transport coupled to oxidative phosphorylation.

b) is a good choice since acetyl phosphate is relatively cheap and we can use the enzyme to catalyse the reaction:

$$\text{acetyl} \ⓟ + \text{ADP} \rightleftharpoons \text{acetate} + \text{ATP}$$

Notice we use the term 'relatively' cheap. It is still quite expensive and can only be used on a small scale with high-valued products.

5.10 This is a fairly open ended question. In order for the technologist to determine the effectiveness factor for various configurations of the biocatalyst, he specifically needs to have kinetic data relating to the enzyme.

He can determine the rate of diffusion of substrate into the biocatalytic matrix and the rate of diffusion of product out of the biocatalytic matrix from purely physical data. But, to determine the actual substrate/product profiles within the matrix, he will need to know the rates of the catalysed reaction over a range of substrate concentrations and the effect the presence of product has on these rates. Thus he will need to know the relationship between v and [S]. For example, does the system follow Michaelis-Menten kinetics? If so, he will need to know the value of the K_M' of the enzyme in the matrix. Similarly if the product is an inhibitor then he will need to know the extent to which the reaction is inhibited. We will not go into specific details here, but mathematical relationships between inhibitor concentrations [I] and reaction rate are well established for various types (competitive, non-competitive, uncompetitive) inhibitors. (These are dealt with in the BIOTOL text, 'Principles of Enzymology for Technological Applications'). He will also need to be alerted to the appropriate relationship(s) in order to properly predict substrate consumption rates and effectiveness factors. He will also need to be aware of the stability of the enzyme within the matrix as this will influence calculations concerning the need to re-charge the reactor with enzyme (ie the life expectancy of the reactor).

Responses to Chapter 6 SAQs

6.1 The correct answer is 3). Amylose is a straight chain of α1-4 glucose residues; α-amylase (option 1) splits amylase into maltose not into glucose; pullulanase (option 2) cleaves α-1-6 links and will not hydrolyse amylose. Option 4 is a waste of enzyme, pullulanase will not have any effect on α-amylose and β-amylase will act slowly on amylose as it will only hydrolyse the amylose from the non-reducing end. It is, however the only enzyme of those listed which releases glucose. Although not offered as a choice, a mixture of α-amylase and maltase would be optimal.

6.2 The factor favouring the production of HFCS rather than invert sugar is 1). Factor 2) also applies to invert sugar. Factor 4) does not favour HFCS since invert sugar production requires only one enzyme. Factor 3 favours invert sugar, since a product with higher fructose content is produced.

6.3 The appropriate enzymes are:

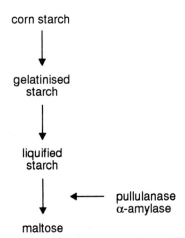

Note that: glucose isomerase would convert glucose to fructose; invertase hydrolyses sucrose; β-amylase releases glucose not maltose as its main product.

Both pullulanase and α-amylase are needed to hydrolyse α1-6 and α1-4 links respectively.

Responses to SAQs

6.4 You should have labelled the diagram as shown below.

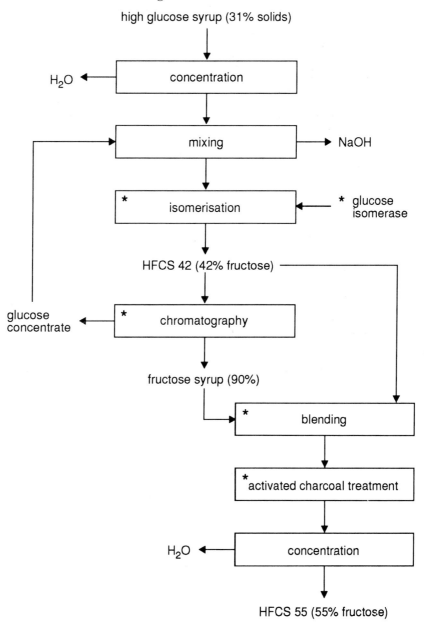

6.5
1) False. α-amylase increases loaf volume by decreasing amylose concentrations leading to the production of low molecular mass carbohydrates which can be metabolised by yeast to produce CO_2.

2) True.

3) True.

6.6 The factors favouring the use of lipases are 1, 3 and 4. Lipases which are non-specific (option 2) give no advantage over the chemical processes.

6.7 You should have recalled, from the text, that immobilised aspartase (from *E. coli*) can be used to convert fumarate to L-aspartate:

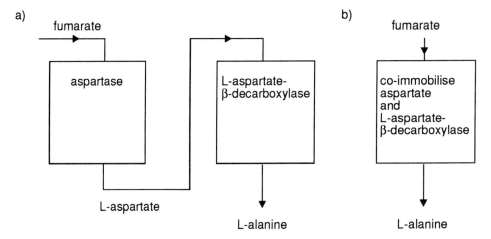

We could, therefore visualise either a two-stage system or a single-stage system in which aspartase and L-aspartate-β-decarboxylase are used in immobilised forms. This utilises the two enzymes. Thus:

a) fumarate → aspartase → L-aspartate → L-aspartate-β-decarboxylase → L-alanine

b) fumarate → co-immobilise aspartate and L-aspartate-β-decarboxylase → L-alanine

In practise a two-stage system would probably be used because:

- it gives the flexibility of producing either L-aspartic and or L-alanine;

- there will probably be differences in the stabilities of the two enzymes (for example *E. coli* aspartase appears to be stable for 3 months. The *Xanthomonas oryzae* L-aspartate-β-decarboxylase is less stable). Using two separate stages enables us to replenish the enzymes separately. Note that, commercially, aspartase is used within *E. coli* cells, as the extracted enzyme is unstable. The stability of extracted L-aspartate β-decarboxylase from *Xanthomonas* would have to be evaluated.

6.8 1) The enzyme you require is penicillin acylase. This could be in an immobilised form in a PBR or in solution in a STR.

Incubation of the enzyme with penicillin G at pH8 causes hydrolysis of the amide link releasing 6-aminopenicillanic acid and phenylacetate. The 6-aminopenicillanic acid is then collected, the pH lowered to pH5, phenyl glycine is added and the mixture incubated with penicillin acylase. It is perhaps easiest to visualise this in the form of a two-stage PBR:

Responses to SAQs

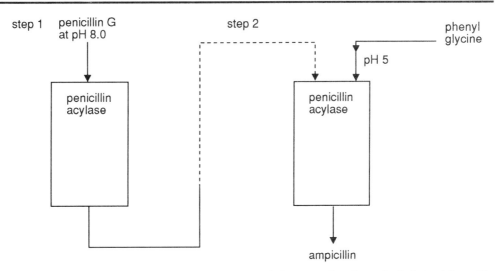

The critical point to remember is the change in pH. Alkaline pH values favour hydrolysis of the amide links. Acid pH values favour formation of the amide link.

2) a) For the chemical synthesis of methicillin, a reactive species is required. We would suggest using the acid chloride:

This would react quite vigorously with the amino group of 6 aminopenicillanic acid to form methicillin.

b) In this case, we would add the acid and penicillin acylase at pH5.

This acid has the structure:

Responses to Chapter 7 SAQs

7.1

1) You should have chosen glucose oxidase. From the data given, this enzyme appears to be much more specific then hexokinase. Note, however, it is not completely specific - enzymes rarely are! This must be borne in mind whenever an enzyme is used for analysis. A key question to ask is whether the sample being assayed is likely to contain other compounds which can be used by the enzyme and which will interfere with the estimation.

2) Not very conveniently. The assay of glucose-6-phosphate dehydrogenase depends on monitoring (at 340 nm) the reduction of NAD^+. A crude extract (cell homogenate) almost certainly contains the enzyme NADH oxidase. Thus, if oxygen was present, the NAD^+ reduced by the activity of glucose-6-phosphate dehydrogenase would be re-oxidised.

The points we are making in this question are that we need to consider the specificity of the enzyme and the possible effects of other compounds which may have an effect on the enzyme (eg alternative substrates, inhibitors etc). We also need to consider the consequences of other enzymes that may be present in the enzyme preparation or sample.

7.2

1) a) A is predominantly first order.

    b) C is predominantly zeroth order.

    c) B is a mixture of zeroth and first order.

2) Region A, since rate proportional to [S].

3) Region C, since here the rate is independent of [S].

4) The K_M will influence the range in which first order and mixed order kinetics are followed. The lower the K_M, the lower the [S] range which will give first order kinetics.

5) We must use the same amount of enzyme and incubation conditions (temperature, pH, ionic strength) as in the test samples. We must also be alert to interference to the activity of the enzyme from interfering compounds (see SAQ 7.1).

6) If this was not an enzyme-catalysed reaction, we would expect the reaction to be first order with respect to [S] at all [S] values. In other words, a plot of rate of reaction against substrate concentration would be linear.

7.3

0.68 mmol l^{-1}.

This can be determined graphically. A plot of initial velocity against [pyruvate] gives us a calibration curve. An initial velocity of 18.4 µmol min^{-1} is given by using a pyruvate concentration of 0.68 mmol l^{-1}.

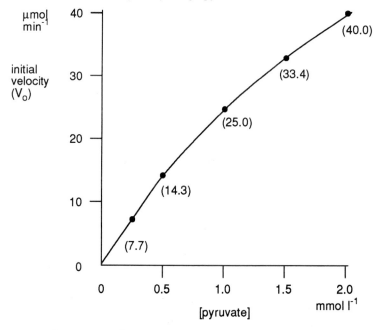

Alternatively a plot of $\frac{1}{v_o}$ against $\frac{1}{[pyruvate]}$ could have been produced (ie a Lineweaver-Burk plot).

This has the advantage of producing a straight line.

7.4

1)

The substrate concentration (0.1×10^{-4} mol l^{-1}) is one quarter of the stated K_M value for the enzyme. From the Michaelis-Menten equations we can calculate that the initial velocity will be 20% of V_{max} (see Section 3.3.1 if you are in any doubt over how to do this). If you think of the hyperbolic relationship between v_0 and [S] (shown below), this is within the region which approximates to first-order kinetics. The observed rates in this region can readily be used to discriminate between different [S]. Thus a kinetic approach would be suitable and accuracy should be good.

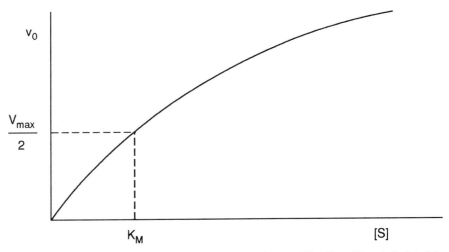

2) [S] is now well above (20 x) the K_M value, and the initial velocity will be 95% of V_{max} (calculated from the Michaelis-Menten equations). This is now in the region of the v_0 against [S] plot where zero-order kinetics occur, ie alterations in [S] has negligible effect on v_0. Hence estimation of [S] from the observed v_0 will be highly inaccurate and a kinetic method will be most unsuitable, unless the sample is diluted before assay. An end-point approach could still be used.

7.5

1) Yes.

We can use the relationship.

$$t = \frac{2.3 K_M}{V_{max}} \log_0 \frac{[S_o]}{[S_t]} + \frac{[S_o] - [S_t]}{V_{max}}$$ (see Equation 7.1).

In this case $\log_{10} \frac{[S_o]}{[S_t]} = 2$ and $V_{max} = 6$ IU ml^{-1} = 6×10^{-6} mol ml^{-1} min^{-1}.

If we substitute in (converting all units to mol ml^{-1}).

$$t = \frac{2.3 \times 4 \times 10^{-6} \times 2}{6 \times 10^{-6}} + \frac{10^{-6}}{6 \times 10^{-6}}$$

= 3.1 + 0.167 min = 3.3 min (approx).

The reaction would reach 99% completion within 3.3 min.

2) 3.88 IU ml^{-1}.

We use the same relationship as in 1) but this time to calculate V_{max} we get:
$$\frac{2.3 K_M}{V_{max}} \log \frac{[S_o]}{[S_t]} + \frac{[S_o] - [S_t]}{V_{max}} = 5 \text{ min.}$$

Thus $5 = \dfrac{2.3 \times 4 \times 10^{-6} \times 2}{V_{max}} + \dfrac{10^{-6}}{V_{max}}$ (ignoring [S$_t$] = 10^{-8})

$$V_{max} = \frac{18.4 \times 10^{-6} + 10^{-6}}{5} = 3.88 \text{ µmol min}^{-1} \text{ ml}^{-1} = 3.88 \text{ IU ml}^{-1}.$$

7.6 3.3 μmol ml^{-1} (approx).

First we convert activities into % of the uninhibited enzyme and also % inhibition.

Thus:

Inhibitor concentration (μmol ml^{-1})	% activity	% inhibition
0	100	0
1	85.9	14.1
2	71.8	28.2
4	57.6	42.4
8	43.5	56.5
16	29.4	70.6
x	62.0	38.0

Now we need to produce the calibration curve. If we plot % inhibition against inhibitor concentration on linear scales, the result is a hyperbolic curve, which is not particularly satisfactory as a calibration curve. If however we plot % inhibition against [inhibitor] plotted on a log scale we obtain a linear plot.

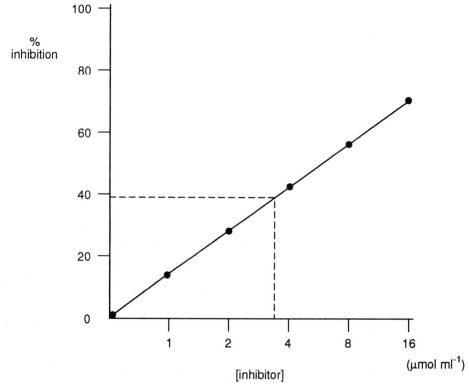

Using this calibration curve, the concentration of inhibitor in X = 3.3 μmol ml^{-1} (approx) (note that the inhibitor axis is logarithmic).

You should note that the data we have provided in this SAQ are somewhat stylised. The data generated in practice rarely give such good calibration curves as that produced here.

7.7 There are several features. The main ones are:

Responses to SAQs 367

- antibodies usually bind their target antigens avidly and will, therefore, bind them at low concentration;
- the enzymes used act as an amplification device.

Thus for each enzyme: antibody complex bound (ELISA) or enzyme: antigen complex left unbound (EMIT), the enzyme molecule converts many substrate molecules into products. This acts as an amplification factor. The higher the turnover number of the enzyme, the greater the amplification. Thus for every molecule of antigen we are attempting to detect, we may produce many thousands of product molecules from the activity of the enzyme used in the system. The sensitivity of ELISA will be further enhanced if the product of the enzymatic reaction has a high absorption coefficient.

7.8

1) Approximately 7.5 nmol ml^{-1}.

First we use the data to produce a calibration curve.

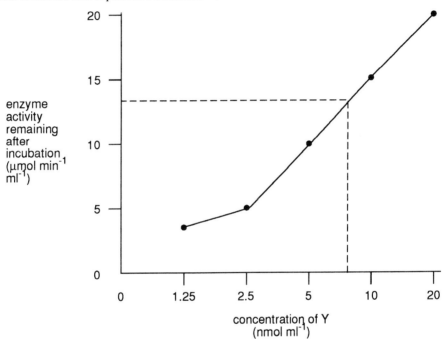

From the calibration curve, we can see that the concentration of Y in the unknown was about 7.5 nmol ml^{-1} (Again note that the concentration of Y is plotted on a logarithmic scale).

2) a) If we used more antigen:enzyme complex in each reaction mixture, we would expect that more enzyme would be left in the active form even when Y was not added to the system. This would, in effect, shift the calibration curve to give higher values of enzyme activity for each calibration point. The slope of the line would also be shallower.

b) If more antibody was used in each reaction mixture, the calibration curve would be moved to the right.

In the extreme case, where we had a vast excess of antibodies, inclusion of Y at low concentrations in the reaction mixture would result in hardly any increase in activity.

The important point to remember is that to carry out this type of assay successfully the antigen: enzyme/antibody ratio must be in balance.

Thus we can visualise a range of curves:

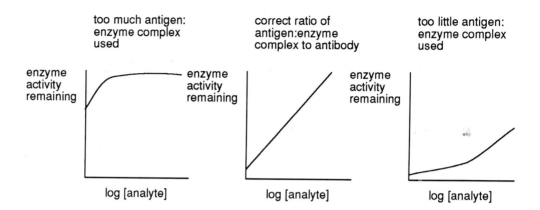

Responses to Chapter 8 SAQs

8.1 1) 0.193 coulombs.

Since n = 2, F = 96.487 coulombs mol^{-1} and N = 1 µmol.

Then $10^{-6} = \dfrac{Q}{2 \times 96\,487}$. Therefore, Q = 0.193 coulombs.

2) 400 µmol.

Since $N = \dfrac{77.2}{2 \times 96\,487}$ mol (from Faraday's Law). Hence $N = \dfrac{77.2 \times 10^{6}}{2 \times 96\,487}$ µmol

= 400 µmol.

3) 0.2 mol l^{-1}.

Since 400 µmol are present in 2 ml. Therefore in 1 litre there are $\dfrac{400 \times 1000}{2}$ µmol

= 0.2 mol.

8.2 The potential advantages of using ferricyanide is that it is a relatively small molecule and will be quite motile allowing it to diffuse into the active site of the enzyme to receive electrons and to diffuse across to the electrode where it will discharge its electrons. This same high motility does mean, however, that it will be difficult to maintain high ferricyanide concentrations at the enzyme - electrode interface. In other words, ferricyanide would have to be added to the galactose solutions being tested.

Free ferrocene has many similar properties to ferricyanide but it can be more readily polymerised thereby reducing its motility and leakage from the enzyme - electrode interface. The real advantage of using ferrocene is, however, that a lower operating potential can be used (181 mV compared with a standard calomel electrode) than with ferricyanide (450 mV compared with a standard calomel electrode). Thus using ferrocene, we can use lower electrode potentials. This means that a ferrocene mediated biosensor will probably exhibit less interference from other analytes that may be oxidised or reduced at the electrode surface.

8.3 Trivalent.

We determined this in the following way.

Slope of the graph = 0.02 V.

From the Nernst equation 0.02 V = 2.303 RT/zF.

Thus $z = \dfrac{2.303\ R\ T}{0.02 \times 96\ 487} = \dfrac{2.303 \times 8.314 \times 300}{0.02 \times 96\ 487} = 2.976 = 3$.

8.4 For this type of device, we would suggest the co-immobilisation of the pyruvate kinase and firefly luciferase. By including luciferin, Mg^{2+}, K^+ ADP, phosphate and O_2 in the system we would anticipate that the following sequence of reactions would take place.

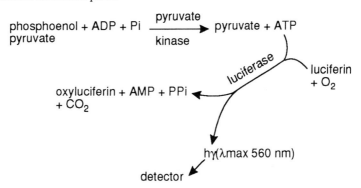

The two enzymes could be co-immobilised onto membranes which would be placed on the end of a glass fibre-optic. The light generated in response to phosphoenol pyruvate (via ATP generation) could be passed along the fibre-optic to a luminometer.

8.5 This is a fairly open ended question. The key point is to think how you would detect the reaction catalysed by the enzyme.

Hydrolysis of the ester is unlikely to lead to significant changes in light absorption. There might be a change in conductivity as the:

$$R'-O-\overset{\overset{\displaystyle O}{\|}}{C}-R''$$

is replaced by:

$$R'-OH + HO-\overset{\overset{\displaystyle O}{\|}}{C}-R''$$

but this would be small. There will however be the heat of hydrolysis so you might consider using a thermal detection system.

This might be made even more efficient by using an organic phase enzyme electrode especially as the compounds of interest are not very water soluble. A key question that would need resolution would be whether or not the esterase would be active in a non-aqueous environment. Remember that some water would be needed for the reaction:

$$R'-O-\overset{\overset{\displaystyle O}{\|}}{C}-R'' + H_2O \longrightarrow R'-OH + HO\overset{\overset{\displaystyle O}{\|}}{C}-R''$$

so the system would have to include some water.

The sort of device we anticipate you would arrive at would be similar to that shown in Figure 8.4.

There are however a number of other possibilities. For example you might have considered using additional enzymes, such as alcohol dehydrogenase to oxidise the alcohol released in a system like this:

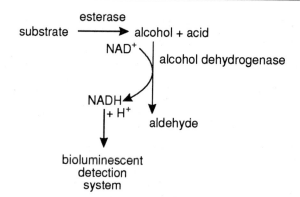

As a general rule, the smaller the number of the enzymes used, the more reliable the system.

A key point to remember is that the specificity of the biosensor is dependent upon the specificity of the enzymes used. In our case here, esterases are often not very specific, so we may well get interference from other esters in test samples.

8.6

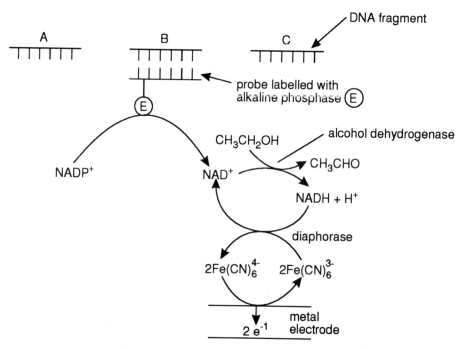

8.7 There are several possible ways to tackle this. Here we will deal with a genetic solution. The basic problem can be represented diagrammatically as:

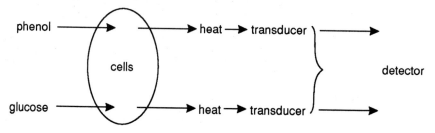

Responses to SAQs

8.8 1) By using diaphorase instead of glucose oxidase, the reaction sequence would be:

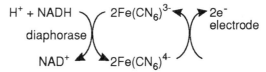

Thus NADH would be determined coulometrically.

2) This would be similar to 1) except that lactate dehydrogenase would also be included.

Thus the reaction sequences in the biosensor would be:

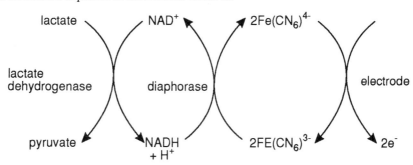

Responses to Chapter 9 SAQs

9.1 1) Phage not previously grown up in E. coli K will only infect this strain with low efficiency as the DNA will not be protected from restriction by methylation of the recognition sequences. In a few instances, bacterial methylases will modify the DNA before it is exposed to the restriction enzymes, allowing phage replication and the production of a small number of plaques.

2) Replating on the same strain will give a very high efficiency infection as all particles from the plaques in 1) will be methylated and hence protected from restriction enzymes found in this host.

3) If these phage are used to infect a different strain such as E. coli C, only very low efficiency will be obtained. This is because this strain will contain different restriction modification systems recognising different DNA sequences. The methylation obtained in E. coli K will not protect the DNA from digestion by the enzymes in E. coli C.

4) Any phage particles produced following infection in 3) will now be methylated at positions dictated by the restriction-modification system of E. coli C, and will have no protection at sites recognised by the system of E. coli K. Hence, the result obtained will be as in 1) and 3) - a low efficiency infection.

9.2

Property	Class I	Class II	Class III
Size	≥ 400 000 D	20 - 100 000 D (endonuclease)	250 000 D (endonuclease)
Complexity	Endonuclease and methylase activities in same molecule	Separate endonucleases and methylases	As Class II except that the enzymes share a common subunit
Recognition sequence	Long, containing two discrete sequences separated by a long variable region	Short, palindromic	Short, not palindromic
Cleavage site	1000 - 7000 bp from recognition sequence. Direction depends on enzyme	Within or very close to recognition sequence	24 - 26 bp downstream of recognition sequence
Methylation site	A or C residues within the recognition sequence		
Cofactor requirement for endonuclease	Mg^{2+}, SAM, ATP	Mg^{2+}	Mg^{2+}, SAM, ATP
Usefulness in recombinant DNA work	None	Yes	None

9.3

Cloning in pUC19 and similar vectors is carried out by inserting the required DNA into one of the unique restriction enzyme sites in the multiple cloning site. This insertion causes the *lac* Z gene to be inactivated. If bacteria transformed with a ligation mixture containing pUC19 and cloned DNA are plated on to agar containing X-gal, IPTG and ampicillin two types of colony will be observed.

One type will be blue. This will consist of bacteria which have taken up vector molecules that have rejoined without an inserted piece of DNA. Hence the *lac* Z gene will be intact, become induced by IPTG and the β-galactosidase will metabolise the X-gal into a blue product, colouring the colony.

The other type will be white. These will contain recombinant DNA molecules, vectors into which the DNA to be cloned has been inserted. This insertion inactivates the *lac* Z gene such that no functional β-galactosidase is produced and hence no blue product formed from X-gal.

Bacteria which have not taken up vector sequences will not grow on plates which contain ampicillin.

The basis of recombinant clone selection when using pUC19 is hence a colour reaction.

9.4

To achieve this, digest a sample of DNA with the enzyme preparation and determine the size of the fragments produced by comparison to known standards on agarose gel electrophoresis. This will narrow down the recognition sequence to one of two areas on the DNA:

The arrows indicate the two possible cut sites for the enzyme to produce 2 and 5 kb fragments from a 7 kb piece of DNA.

Study of the sequences around these positions may indicate the presence of a palindromic sequence - a likely candidate as the recognition site. To confirm this identification, site-directed mutagenesis may be used to alter the sequence, one base at a time. The effects of these alterations on the ability of the enzyme to cleave the DNA are easily monitored by gel electrophoresis and should allow accurate definition of the recognition site. Mutations outside the recognition site are unlikely to have major effects on enzyme activity.

Responses to Chapter 10 SAQs

10.1 1) Starting from cDNA:

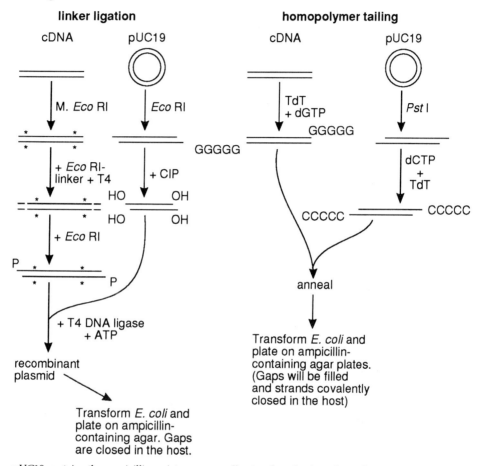

pUC19 contains the ampicillin resistance gene, allowing for selection of transformed bacteria.

2) If the vector is cut with *Pst* I, and poly dT tails are added using TdT, the following situation will be observed:

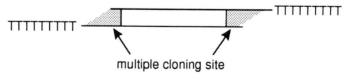

One of the tails may be removed by digestion with another restriction enzyme cutting within the multiple cloning site, to give:

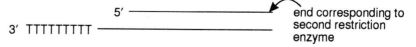

This molecule may then be annealed to mRNA to act as the primer for reverse transcriptase (RT) as follows:

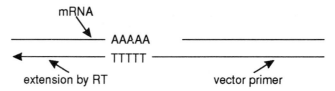

Following second strand DNA synthesis, addition of T4 DNA ligase and ATP allows cyclisation of the molecule to produce recombinant plasmid for introduction into the host.

This method circumvents the need for modifying the cDNA ends for cloning.

10.2 1) If the region of DNA sequence encompassing a disease-linked RFLP is known it should be simple to identify two unique oligonucleotide sequences suitable for use as primers. These should be about 20 nucleotides long, flanking the polymorphic site and complementary to opposite strands of the DNA eg:

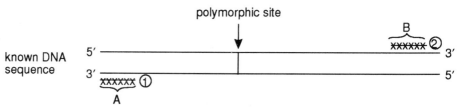

Primer 1) is complementary to sequence A, and primer 2) to sequence B. These primers are used in a PCR to amplify the genomic DNA. The amplified material is then subjected to digestion with the restriction enzyme for which the site is polymorphic, and the fragments are run on an agarose gel and stained with ethidium bromide. Comparison of the resulting pattern of bands from the foetal tissue with those from other affected and unaffected members of the family should allow its status as regarding the disease to be ascertained. This method is just a quicker, less hazardous, alternative to that described in Section 9.9 and illustrated in Figures 9.8 and 9.9.

2) To sex a foetus by PCR, use must be made of sequences on the Y chromosome which have no counterpart on the X chromosome. Amplification of such sequences will reliably distinguish male from female by the presence or absence, respectively, of amplified material. On electrophoretic gels it is vital that positive and negative controls are carried out in such analyses, as a false negative result could be disastrous in that a potentially affected male foetus could be missed.

10.3 *Hin*d III recognises the sequence 5'AAGCTT3', cutting DNA to produce fragments with 5' overhangs as follows:

5' ...A
3' ...TTCGA5'

The Klenow fragment will 'fill in' the top strand of this end if supplied with all of the nucleoside triphosphates and Mg^{2+} ions. If one of the nucleotides used is radioactively labelled (eg (α-^{32}P-dCTP) this will be incorporated into the fragments, which will then show up on autoradiography of gels etc. This method labels the fragments without significantly altering their sizes, and hence their mobility on electrophoretic gels. So, requirements for this labelling reaction are the enzyme, all four nucleotides (one being radioactively labelled on the α-phosphate group), Mg^{2+} ions and a buffer of suitable pH.

10.4 1) For a dideoxynucleotide chain termination method of RNA sequencing similar to that used for DNA sequencing, reverse transcriptase (RT) would be the obvious enzyme of choice as it copies RNA, and, being a DNA polymerase, will use dideoxynucleotides as substrates in the same way that the Klenow fragment may.

2) As for DNA sequencing, a short complementary oligonucleotide specific to the 3' end of the RNA to be sequenced must be used as a primer. It is not appropriate to use an oligo (dT) primer as this will bind to any mRNA, and will not be specific to the mRNA under study unless a completely pure preparation is available.

3) To overcome such problems, the only good solution is to label only the last nucleotide in each strand, ie the dideoxynucleotide. This may be achieved by using fluorescent-labelled dideoxynucleotides.

Responses to Chapter 11 SAQs

11.1 The reason this approach cannot be generally applied to enzymes is that most enzymes are large molecules and contain methionine residues. Thus when we come to treat the chimeric protein with cyanogen bromide, we not only break the link between the two parts of the chimeric protein, but we also disrupt the structure of the enzyme we hope to isolate and use. In general terms, the larger the target enzyme is, the lower the probability of success with this strategy.

11.2
1) Type I (site-directed mutagenesis). A lot is known about the structure-activity relationship in lysozyme, so we could use this technique.

2a) The generalised approach must be to use random mutagenesis (that is Type IV protein engineering). We have no knowledge of which part of the of enzyme is functionally involved in the binding of substrate, catalytic function or thermal stability. Thus we cannot successfully apply site-directed mutagenesis (Type I protein engineering) or localised random mutagenesis (Type II). Nor are we aware of a homologous enzyme from other sources so we could not use Type III protein engineering.

2b) Obviously after carrying out random mutagenesis we need to screen for mutant colonies of *E. coli* carrying genes coding for modified enzyme molecules which are stable at 37°C and to compare their activities with those of colonies carrying the wild type gene.

Note it would be best if we could find an assay that could be used *in situ*. For example, the substrate, being an aldehyde, might be toxic to *E. coli* and, therefore, if the enzyme is present and active, the host cells might be protected from aldehyde toxicity.

11.3
1) Although the enzyme catalyses the same type of reaction (the hydrolysis of peptide bonds) irrespective of the protein substrate used, the bond to be hydrolysed has to be brought to the active site. The side chains and conformation of the protein will influence this. Thus the rate of hydrolysis need not be the same with different substrate proteins.

2) No. The data presented in Table 11.3 show that replacement of the methionine at position 216 by many different amino acids has only limited effect on the affinity (as measured by K_M) of the enzyme for its substrate. This suggests that the methionine at position 216 is not primarily involved with substrate binding. We have, however to be a little cautious, it could be that methionine is not involved in the binding of SAAPFpNA (the substrate used to measure K_M) but could be involved, in a minor way, with the binding of large molecules. The evidence, however, from Table 11.3 is that methionine 216 is more closely associated with the catalytic activity rather than the binding activity of the enzyme, since substitution of methionine at position 216 by other residues has much more marked effects on k_{cat}.

11.4 The more thermostable the enzyme, the more rigid the enzyme becomes. This loss in flexibility usually reduces the catalytic efficiency of an enzyme, especially in those circumstances in which movement of the enzyme is required in order to bring the amino acid residues involved in catalysis into the correct conformation (Koshland's induced fit model of enzyme action).

11.5
1) You should have selected b).

The β-*gal* promoter normally controls the expression of the β-galactosidase gene and other genes coding for enzymes involved in lactose metabolism. Glucose represses the expression of genes controlled by this promoter (catabolite repression). Thus if we used medium a), the gene coding for enzyme A would not have been expressed. Similar, nutrient broth contains readily metabolisable substrates and again, by catabolite repression, would switch off expression of the desired gene. Glycerol is a poorly metabolised substrate and the cells would grow only slowly although some enzyme A might be anticipated to be produced. Lactose, in the absence of other easy to catabolise substrates, such as glucose, is readily metabolised and supports good growth rates and will switch on genes controlled by β-*gal* promoters. Thus option b) is the option of choice.

2) There are several possibilities here. First the β-*gal* promoter and the gene(s) it control must be 'in step' with each other to produce the correct amino acid sequence(s). Consider for example the nucleotide sequence.

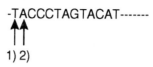

1) 2)

We have indicated two different start points for transcription. If 1) was used the RNA transcript would be:

AUGGGAUCAUGUA------
aa_1 aa_2 aa_3 aa_4

If 2) was used, the RNA transcript would be:

UGGGAUCAUGGUA------
aa_1 aa_2

Since AUG is the initiation codon, we should anticipate that the second transcript would produce a peptide missing some amino acids and be made up of a completely different sequence of amino acids. This would, of course, not lead to the successful production of enzyme A.

An alternative, explanation is that the original organism (*Salmonella typhi*) carried out some post-translational modification of the protein product derived from gene A and that this is essential to convert this protein into an active form of enzyme A.

Responses to Chapter 12 SAQs

12.1

1) Slightly more activity should be retained in hexane. Hexane is slightly less polar than cyclohexane (log P = 3.5 for hexane compared with log P = 3.1 for cyclohexane, see Table 12.1). The higher the value of log P, the greater the retention of activity (see Figure 12.1).

2) This proposal is somewhat illogical. Presumably the idea is that by using an organic solvent the concentration of cortisol that could be used would be much higher than that which could be used using water as the solvent. You should note however that cortisol is quite soluble in water. (Note that it contains a number of polar groups which makes it quite soluble in water).

The key point is that the dehydrogenase must have an electron/hydrogen accepting cofactor. With dehydrogenases this is usually $NAD(P)^+$ or FAD. The key question is how these can be supplied to the enzyme in a non-aqueous system. Of course if we did add NAD^+ (or FAD) this would be attracted to the aqueous interphase and then tend not to diffuse away. We could therefore only use a rather limited amount of these cofactors. Unless we could find a way of re-oxidising these cofactors with some acceptor compatible with the organic phase, the reaction would soon come to a stop. You might have thought that a solution to this might be to use whole bacterial cells rather than the extracted enzyme. The cofactors required for the reaction might be retained inside of the cells and be capable of re-oxidation. It is however questionable whether such cells would remain intact as the organic solvent would tend to dissolve the plasma membrane.

12.2

The key to successfully converting palm oil and olive oil into a cocoa butter substitute is to replace some of their unsaturated fatty acid moieties by saturated fatty acids.

One way of achieving this substitution is to incubate the oils with lipase in an organic solvent (eg hexane) containing just sufficient water to activate the lipase. Included in the incubation mixture are saturated fatty acids such as stearic acid ($C_{17}H_{35}COOH$) or palmitic acid ($C_{15}H_{31}COOH$).

Under these conditions some of the fatty acid moieties of the oils are replaced by these saturated fatty acids producing triglycerides with higher melting points. The reaction can be allowed to proceed until triglycerides with the required melting temperature are obtained. The products may be recovered from the organic solvent.

Suggestions for further reading

Atkinson, B. and Mavituna, F., Biochemical Engineering and Biotechnology Handbook, 2nd Edition, Macmillan, 1991.

Bailey, J.E. and Ollis, D.F., Biochemical Engineering Fundamentals, McGraw-Hill, 1986.

Bioprocess Technology: Modelling and Transport Phenomena, BIOTOL series, Butterworth-Heinemann, 1992.

Biorector Design and Product Yield, BIOTOL series, Butterworth-Heinemann, 1992.

Chaplin, M.F. and Bucke, C., Enzyme Technology, Cambridge University Press, 1990.

Dordick, J.S. (Ed), Biocatalysts for Industry, Plenum Press, 1991.

Fogarty, W.M. and Kelly C.T. (Eds), Microbial Enzymes and Biotechnology, 2nd Edition, Elsevier, 1990.

Hall, E.A.H., Biosensors, Open University Press, Biotechnology series, 1990.

Operational Modes of Bioreactors, BIOTOL series, Butterworth-Heinemann, 1992.

Oxender, D.L. and Fox, C.F. (Eds), Protein Engineering, Alan R. Liss, Inc., 1987.

Palmer, T., Understanding Enzymes, 3rd Edition, Ellis Horwood, 1991.

Principles of Enzymology for Technological Applications, BIOTOL series, Butterworth-Heinemann, 1993.

Product Recovery in Bioprocess Technology, BIOTOL series, Butterworth-Heinemann, 1992.

Appendix 1

Abbreviations used for the common amino acids

Amino acid	Three-letter abbreviation	One-letter symbol
Alanine	Ala	A
Arginine	Arg	R
Asparagine	Asn	N
Aspartic acid	Asp	D
Asparagine or aspartic acid	Asx	B
Cysteine	Cys	C
Glutamine	Gln	Q
Glutamic acid	Glu	E
Glutamine or glutamic acid	Glx	Z
Glycine	Gly	G
Histidine	His	H
Isoleucine	Ile	I
Leucine	Leu	L
Lsyine	Lys	K
Methionine	Met	M
Phenylalanine	Phe	F
Proline	Pro	P
Serine	Ser	S
Threonine	Thr	T
Tryptophan	Trp	W
Tyrosine	Tyr	Y
Valine	Val	V

Index

1,4β-cellobiosidase, 160
1,4β-glucosidase, 160
3' overhang, 282
3' recessed DNA molecules, 282
3'-OH groups, 280
3'5,5'-tetramethyl benzidine (TMB), 199
3'→5' exonuclease activity, 290
5'-phosphate groups, 280
5'→3' elongation, 290
5'→3' exonuclease activity, 290
5,6-dihydroxycyclohexa-1,3-dione, 175
5-methyl cytosine, 270
6-aminopenicillanic acid, 168, 169
6-phosphoglucone-d-lactone, 195
7-amino-cephalosporanic acid (7-ACA), 171

A

α-amylase, 61, 80, 89, 92, 94, 155
α-ketoglutarate, 183
α-L-aspartyl-L-phenylalanine methyl ester, 334
A. oryzae, 161
α1-4 links, 151
α1-6 links, 151
absolute temperature, 15
abzymes, 49
acetyl cholinesterase, 197
acetylcholine receptor, 240
acid-base catalysis, 7
acid-base catalysts, 48
Actinoplanes missouriensis, 157
activated charcoal, 92
activation energy barrier, 6
active site, 98, 308
activity:pH profile, 63
adenosine, 242
adenosine nucleotides, 250
adipic acid production, 176
advantages and disadvantages of
    site-directed mutagenesis, 311
affinity, 133
affinity chromatography, 42
affinity exchange chromatography, 44
agar, 80
agarose, 80
agarose gel electrophoresis, 266, 273
air traps, 46
alanine, 315, 335
alcohol, 151, 241
alcohol dehydrogenase, 101, 142, 143, 144, 204, 236
alginate, 94
alkaline isomerisation, 156
alkaline phosphatase, 204, 236
alkaline protease
    storage stability, 317
alleles, 267
allergenic properties, 169
allosteric effector, 252
allosteric enzymes, 12
alumina, 37, 80, 243
amino (NH$_2$) groups, 90
amino acid substitution/modification, 303

amino acids, 164
aminoacylase, 94, 164
ammonia, 241
ammonia gas selective electrodes, 222
ammonium, 223
ammonium fumarate, 165
ammonium persulphate, 95
ammonium sulphate precipitation, 42, 321
amoxycillin, 172
amperometric analyses
    instrumentation of, 213
amperometric enzyme biosensors, 212
amperometric enzyme electrodes, 214
amperometric transduction, 212
amperometric transduction systems, 231
amperometry, 186
ampholyte buffers, 45
ampicillin, 169, 170, 172, 259, 262
amplification of forensic material, 289
amylases, 153, 333
amylopectin, 151
amylose, 151
anaemias, 233
ancient DNA samples, 289
animal cell cultures
    as sources of enzymes, 23
animal viruses, 204
animals, 20
anion exchange chromatography, 47
anion exchange resins, 92
anisole, 336
anti-antibodies, 203
anti-freezing agents, 69
anti-microbial agent, 64
antibiotics, 64, 165, 241
antibodies, 200, 230
antibody-based biosensors, 230
antioxidants, 62
apo-enzyme, 6, 322
apparent K_M, 99
application of biphasic systems, 332
application of ELISA, 204
aqueous biphasic partition, 29
arabans, 161
arabinases, 161
arabinose, 161
Archimedes screw, 31
arginine, 321
Arrhenius constant, 15
Arrhenius relationship, 15
artificial enzymes
    production of, 48
ascorbate oxidase, 48
asparaginase, 94, 174
aspartame, 165, 314, 334
aspartame synthesis, 334
aspartate, 334
aspartate ammonium lyase, 165
aspartic acid, 80
aspartic acid production, 165
aspartyl proteases, 314
Aspergillus niger, 153, 161, 162
Aspergillus oryzae, 153, 164
Aspergillus proteases, 24
Aspergillus spp, 161
ATP, 142, 195, 252, 256, 257, 259
ATP-specific system, 227
ATPase activity, 256

Index

attached enzyme, 111
attaching enzymes to the membranes, 141
attenuated total reflection (ATR), 225
automatic pH titration system, 63
Ava II, 251
avian myeloblastis virus (AMV), 278

B

β-amylase, 89
β-D-galactosidase, 89
β-gal promoter, 304
β-glucanases, 153
B. licheniformis, 153
B. subtilis, 153
β1-4 links, 160
Bacillal proteases, 314
Bacillus amyloliquefaciens, 153, 162, 313
Bacillus coagulans, 157
Bacillus globigii, 251
bacitracin, 172
back pressure, 130
bacterial acylases, 169
bacterial luciferase, 227
bacterial peroxidase, 223
bacterial proteases, 182
bactericide, 161
bacteriophage, 259
bacteriophage attack, 248
bacteriophage promoters, 323
bacteriophage T4, 248, 257
baking, 151, 266
baking industry, 153, 161
Bam HI, 251, 261
batch mode, 140
batch systems, 129
batch-fed systems, 122
bead mills, 36
beans, 61
bed height, 130
beef products, 204
Beer's Law, 223
benezene cis-glycol production, 175
benzene cis-glycol (BCG), 175
benzoquinone, 216
benzyl chloroformate, 334
benzyloxycarbonyl (BOC) derivative, 334
Bgl I, 251
Bgl II, 251
bicarbonate ions, 223
bifunctional reagents, 89, 90
binding capacity, 45
binding target cells, 303
biochemical oxygen demand, 241
biological amplification, 258
bioluminescent ATP monitoring, 227
bioluminescent NAD(P)H monitoring, 227
bioluminescent systems, 227
bioreactor design, 24
bioreceptor-based biosensors, 240
biosensors, 198, 210, 242
biospecific interaction analysis (BIA), 234
biotin, 142, 265
biotinylated, 291
biphasic partition, 34
biscuit manufacture, 162
bisdiazobenzidine-2, 2-disulphonic acid, 91

bisimidate, 91
bisoxirane, 91
bleach-containing detergent, 317
bleaching agents, 314, 315
blood, 198
blood clots, 174
blood glucose, 215
blood glucose-lowering therapeutics, 215
blood groups, 204
blunt ended DNA, 261
blunt-ended cDNA, 257
BOC-L-aspartame, 334
bone marrow transplants, 270
bread, 162
brewing, 151, 153
bridge formation, 32
brittle bone disease, 264
bromide, 221
buffer, 63
ulk liquid, 103
bulk wave devices, 239
butanol, 329, 330

C

C-terminal carboxyl group, 80
Ca-alginate, 95
Ca^{2+}, 64, 142
calcium, 221
calcium alginate, 230
calcium phosphate gel, 92
calf chymosin, 22
calf intestinal alkaline phosphatase (CIP), 280
calibration curve, 198
calorimetry, 186
calve stomachs, 20
Candida albicans, 239
canned meat products, 162
capital cost of equipment, 122
carbenicillin, 170
carbohydrates, 43
    relative sweetness, 156
carbonic anhydrase, 197
carboxyl group (-COO), 80
Carica papaya, 162
carriers, 80
catabolic enzymes, 24
catabolite repression, 24
catalase, 94, 161
catalytic activity, 80
catalytic efficiency, 8
catalytic efficiency of enzymes, 301
cation exchange resins, 92
cDNA, 21, 295, 304, 323
    random priming, 280
    specific priming, 280
cDNA copies, 278
celite, 32
cell debris, 43
cellobiose, 160
cells, 114
    disruption of, 35
cellular location of enzymes, 27
cellular RNase H, 279
cellulase, 160, 333
cellulose, 80, 151, 160
centrifugal vacuum filters, 32

Index

centrifugation, 29
centrifuges, 30
cephalexin, 172
cephalosporin, 171, 172
cephalosporin C, 171, 172
cephalothin, 172
cetyltrimethylammonium bromide = CTAB, 328
changes in the primary sequence, 302
changing the pI of an enzyme, 321
charge, 321
cheese, 161
chem FETs, 222
chemical approaches to protein engineering, 303
chemical conversion to glucose to fructose, 156
chemical methods, 146
chemical mutagen, 22
chemically crosslinking enzymes, 79
chemically modified form of T7 polymerase, 294
chimeric gene constructs, 304
chitin, 80
chitosan, 95
chloridised silver electrode, 212
chloroform, 64, 220
choice of bacterial strain as host, 263
choice of conditions for PCR, 286
cholesterol, 199, 217
cholesterol esterase, 217
cholesterol oxidase, 218
cholesterol sensor, 217, 220
choline oxidase, 89
chromatofocusing, 45
chromatography, 45
chromatography columns, 46
chromogenic, 199
chromogenic substrates, 204
chymosin, 314
chymotrypsin, 20, 89, 94, 314
chymotrypsin-like synzymes, 48
CIP, 281
Clark, 214
Class I restriction-modification systems, 252
Class II enzymes, 255
Class II restriction-modification systems, 253
Class III restriction-modification systems, 255
cleavage site, 256
clinical analysis, 198
clinical biochemistry, 204
clogging, 127, 140
clogging of membranes, 97
cloned cDNA, 288
cloned gene, 311
cloning, 258
cloning site, 295
cloxacillin, 170
co-enzymes, 303
co-surfactants, 329
CO_2, 151, 153
coagulating particles, 31
coagulation, 31
coagulation of milk, 161
coated wire electrodes, 222
coatings, 175
cobalt, 282
cobalt ions, 282
cocoa butter, 163
coding regions, 277
coenzyme A, 142

coenzymes, 6
cofactor requirements, 64, 142
cofactors, 6, 62, 142
Col E1, 259
collagen, 80
collagenase, 23, 174
colour blindness, 264
colourimetry, 186
column bed height, 46
column volume, 46
comparison of CSTRs and PBRs, 134
competitive ELISA, 230
competitive ELISA technique, 201
competitive inhibition, 10
complementary DNA (cDNA), 277
complex deactivation kinetics, 69
compressibility, 32
compulsory order reactions, 9
computer aided graphic systems, 307
concentration gradients, 103
concentration profiles, 105
conductimetric enzyme biosensors, 223
conductivity, 46
conductivity measurements, 43
conjugation, 248
consequences of immobilisation, 97
continuous centrifugation, 30
continuous culturing, 25
continuous flow stirred tank reactors (CSTRs), 130
continuous mode, 140
continuous process, 130
control module, 47
conversion of low levels of pollutants, 126
convert porcine insulin into human insulin, 303
core regions, 268
cosmids, 259
cost, 45
cost of operation, 122
cosubstrate requirements, 142
cosubstrates, 142
coulometric transduction, 213
counter (or auxiliary) electrode, 212
coupled enzymes, 143, 197
coupled substrates, 144
covalent binding, 80
covalent binding of an enzyme to a support, 79
covalent bond formation, 7
creatine amidinohydrolase, 227
creatinine amidohydrolase, 227
creatinine in blood, 227
cross-linked dextrans, 204
crosslinking, 89
crossover points, 320
CSTRs
    performance of, 131
    productivity of, 138
culture medium, 26
cultured animal cells, 96
cyanide poisoning, 174
cyanogen bromide (CNBr), 305
cyclodextrin-based synzymes, 48
cyclohexane, 176
cyclohexane hydroxylase, 176
cyclohexanol, 176
cysteine, 64, 80, 242
cysteine proteases, 314
cystic fibrosis gene, 272
cytidine nucleotides, 270

Index

D

D-amino acid oxidase, 218
D-amino acids, 165
dA tails, 283
dC tails, 283
de-differentiate, 23
dehydrogenase, 204
delivery module, 46
denaturation, 69, 70
depreciation, 122
desalting, 43, 47
detection of transcriptionally active genes, 270
detergent, 52
detergent proteases, 313
dG tails, 283
diabetes, 199
diagnostic dipstick tests, 198
diaphorase, 236
diatomaceous earth, 163
dideoxy chain termination method, 292, 294
dideoxynucleotide ddATP, 292
dideoxynucleotide ddCTP, 292
dideoxynucleotide ddGTP, 292
dideoxynucleotide ddTTP, 292
dielectric constant, 97
dielectrics, 243
different pulse voltammetry, 236
differentially methylated regions, 271
diffusion, 55, 102, 103, 105, 111
diffusion barrier, 123
diffusion limitation, 133, 134
diffusion limited, 125
diffusion of substrates, 122
difluoro-2, 4-dinitrobenzene, 91
diglycerides, 162
digoxin, 234
dilution, 43
dilution rate (D), 131
dimethyl ferrocene, 216
diperoxydodecanoic-di-acid, 315
diploid organisms, 265
dipstick tests, 198
disadvantages of PBRs, 130
disc bowl centrifuge, 31
disc-stack centrifuge, 31
disease phenotype, 267
diversification of penicillins, 167
divinylsulphone, 91
dizygotic, 270
DNA fingerprint analysis, 289
DNA fingerprinting, 267
DNA hybridisation experiments, 291
DNA ligase, 248, 257, 279, 295
DNA ligase in cloning, 258
DNA ligation, 281
DNA polymerase, 311
DNA polymerase I, 290
DNA polymerase II, 290
DNA polymerase III, 290
DNA polymerase of *Thermus aquaticus*, 276
DNA probes, 265, 268
DNA replication *in vivo*, 290
DNA sequencing, 291, 292, 294
DNA synthesis, 257
DNA synthetically, 310
DNA-dependent DNA polymerase activity, 277

dNTPs, 286, 291, 292
donation/acceptance of protons, 7
Donnan equation, 100
dopamine, 242
double antibody ELISA technique, 202
double reciprocal plots, 8
double-priming, 311
double-stranded heteroduplex molecule, 310
down-time, 130
downstream processing, 44, 122, 123, 157
drug breakdown products in serum samples, 205
dT tails, 283
dual-sensor configuration, 223
Dyno-Mill, 36

E

E. coli, 22, 25, 169, 248, 257, 290
E. coli DNA pol I, 290
E. coli DNA polymerase, 276, 279
E. coli host, 311
E. coli K12, 263
E. coli ligase, 258, 259
E. coli R245, 251
E. coli thioredoxin, 294
Eco B, 252, 263
Eco K, 252, 263
Eco RI, 251, 258
Eco RI linker, 261
Eco RII, 251
effectiveness factor, 104, 112, 124, 133, 134
efficient (turbulent) mixing, 126
egg white, 161
elastase, 314
elective abortion, 264
electrochemical detection of nucleic acids, 236
electrochemical immunosensors, 231
electrochemical methods, 144, 146
electroconductive pastes, 243
electrodes, 198
electromotive force (emf), 220
electron transfer mediator molecules, 215
electronic signal
    generation of, 211
electrophoretic pH-gradient investigations, 47
ELISA, 232
    applications of, 204
    competitive, 230
    sandwich, 230
ELISA method, 230
ELISA technique, 201
elution buffers, 46
embryonic differentiation, 271
emulsions, 61
end-point (equilibrium) analysis, 197
end-point determination method, 192
end-point method, 190
endo-1,4β-glucanase, 160
endonuclease, 253, 255, 277
Endothia parasitica, 161
entrapment, 93, 95
entrapped enzyme, 111
environmental monitoring, 240
environmentally damaging, 177
enzymatic methods of cofactor generation, 143
enzyme, 122
enzyme activators

assay of, 197
enzyme activities, 241
enzyme amplification, 236
enzyme immobilisation, 76, 77
enzyme inhibitors
    assay of, 197
enzyme kinetics, 7
    physical environment, 15
enzyme labelling of antibodies, 231
enzyme multiplied immunoassay technique EMIT, 204
enzyme multiplied immunoassays (EMIT), 198
enzyme optrodes, 226
enzyme protectants, 64
enzyme re-use, 122
enzyme reaction kinetics, 213
enzyme reactions in non-aqueous system, 328
enzyme stabilisation in non-aqueous solvents, 330
enzyme stability, 80, 138
enzyme structure
    changes to, 98
enzyme use, 70
enzyme-inactivators, 61
enzyme-ligand complexes, 69
enzyme-linked immunosorbent assay (ELISA), 198, 200, 230
enzymes, 4, 6
    artificial, 48
    contamination of, 43
    evolution of, 300
    inhibition of, 10
enzymes extracellular, 25
enzymes immobilisation
    key issues, 79
enzymes in affinity biosensors, 230
enzymes in analysis, 182
enzymes in large-scale industrial application, 150
enzymes in the food industry, 151
enzymes in the leather industry, 177
enzymes in the petrochemical industries, 174
enzymes in the pharmaceutical
    and healthcare industries, 166
enzymes processed, 322
enzymes used as therapeutic agents, 173
enzymes used in ELISA, 204
enzymes used to modify carbohydrates, 151
equilibrium method, 190
Escherichia coli, 161, 165
    See also *E. coli*
Escherichia coli RY13, 251
ethanol, 61, 144, 333
ethyl acetate, 333
ethyl cellulose, 97, 199
eukaryotic genes, 322
eukaryotic genomic DNA, 263
eukaryotic proteins, 27
evanescent wave, 225
evanescent wave sensors, 234
ExacTech, 217
exo-1,4β-glucosidase, 160
exo-cellobiohydrolase, 160
exon deletions in the dystrophin gene, 290
exons, 277
exoribonuclease, 277
expression vector, 295, 311
extracellular microbial lipases, 163

F

FAD (FMN), 142
Faraday's Law, 213
fats, 162
fatty acid, 333
fatty acids soluble, 97
feedback repression, 24
feedstock concentration, 46
ferricyanide, 236
ferritin, 233
ferrocene, 216, 218
ferrocene polymers, 218
Fersht, 313
FETs, 231
fibre entrapment, 94
fibre-optic oxygen systems, 226
fibres, 175
ficin, 20
Fick's law, 106, 111
film, 111, 175
filter (guard) layer, 130
filter aid, 32
filter blockage, 32
filter paper, 204
filter press, 33
filters, 46
filtration, 29, 32
firefly luciferase, 227
firefly *Photinus pyralis*, 227
first order kinetics, 111
flat-bed membranes, 96
flocculating particles, 31
flocculation, 31
flour starch, 153
flow, 127, 139
flow rate, 45, 46, 130
flow resistance, 45
fluid flow
    laminar, 123
    turbulent, 123
fluidised bed reactors (FBRs), 139
fluidises the bed, 140
fluorescein dyes, 232
fluorescence, 232
fluorescence capillary fill device (FCFD), 234
fluorescent dyes, 265
fluorescently-labelled nucleotides, 291
fluoride, 221
fluoride ion-selective electrode, 222
fluorimetry, 186
fluorophenol, 223
fluorosceinated alpha bungarotoxin, 240
flush or blunt ends, 254
food, 165, 204
food and health care industries, 93
food industry, 60, 333
food supplements, 164
force fields, 308
forensic analysis, 270
fractionation module, 46
free fatty acids, 162
free radicals, 294
Frohman, 289
fructose, 151, 155, 156
fruit extracts, 161
fruit juice, 161

Index

fruits, 156
functional groups within the enzyme, 7
fungal α-amylases, 153
fungal acylases, 169
fungal enzymes, 24

G

galactanases, 161
galactans, 161
galactose, 156, 160, 161
galactosidase, 94
gapped duplex approach, 311
gapped duplex technique, 311
gas constant, 15
gel entrapment, 94
gel filtration, 43, 48, 321
gelatinisation, 60, 152, 161
genetic engineering, 301
    objectives of, 300
genetic engineering techniques, 161
genetic fingerprint, 268
genetic fingerprinting, 267
geraniol, 144
germinate, 153
Gibbs Energy Function change, 15
glass, 37, 80
glass pH electrode, 221
glucans, 153
glucoamylase, 94
gluconic acid, 161, 214
glucose, 151, 153, 156, 160, 161, 183, 194, 199, 214, 228
glucose hexokinase, 194
glucose isomerase, 63, 93, 155, 156
glucose oxidase, 89, 161, 183, 199, 218, 223, 226, 230
glucose oxidase (GOD), 214
glucose syrups, 153
glucose-6-phosphate, 195
glucose-6-phosphate dehydrogenase, 194, 204
Glucosensor, 215
glucosidase, 194
glutamate, 242
glutamic acid, 80
glutaminase, 174
glutamine mutant, 315
glutardialdehyde, 89, 91
gluten, 162
glycerol, 69, 97
glycosidic links, 334
glycosylation, 322
gold, 243
Gram negative bacteria, 25
graphite, 243
grinding, 35
growth rates of the host cells, 323
guard layer, 130

H

Hae III, 251
Haemophilus influenzae Rd, 251
half the maximum velocity, 56
half-life ($t_{\frac{1}{2}}$) of an enzyme, 16
hazes, 161
hazes in sucrose solutions, 155
heart, 162

heart attack, 174
heavy duty powder detergent, 314
hemi-cellulose, 160
hemimethylated DNA, 271
hepatitis, 239
herpes simplex, 239
heterogeneity of immobilised systems, 99
heterogeneity of the system, 122
heterogeneous soluble enzyme reactors, 60
hexadecane, 330
hexamethylene bis-isocyanate, 91
hexamethylene bis-maleimide, 91
hexanol, 329
hexanucleotides, 291
hexokinase, 194, 195
high alkaline protease from
    Bacillus alcaligenes, 315
high copy number vectors, 323
high transmission fibres, 224
Hin dIII, 251
histidine, 80
histidine ammonia lyase, 165
HIV, 239
hollow fibres, 95
holo-enzymes, 6
homogenisation, 61
homogenisers, 38
homologous chromosomes, 267
homologous enzymes, 307
hormones, 204
horse meat, 204
host cell RNA polymerase, 295
Hpa II, 272
HPLC systems, 45
human chorionic gonadotrophin (HCG) in urine, 234
human immunoglobulins, 239
human serum, 233
human superoxide dismutase, 47
hyaluronidase, 174
hybrid (chimeric) gene, 320
hybrid enzymes, 320
hybrid formation, 305
hybrid genes (chimeras), 302
hydrocarbons, 328
hydrogen bonding, 92
hydrogen bonds, 319
hydrogen peroxide, 161, 214, 242
hydrogenation, 334
hydrolases, 25
hydrolytic enzymes, 24
hydrophillic enzymes, 330
hydrophobic binding-sites, 48
hydrophobic enzymes, 330
hydrophobic interactions, 92
hydrophobic residues, 319
hydrophobic substrate, 101
hydrophobic supports, 101
hydroxyl ion, 223
hydroxylapatite, 92
hypervariable regions (HVRs), 267

I

identification and purification
    of restriction enzymes, 273
identify target regions for mutagenesis, 320
imidazole group, 80

immobilisation of enzymes, 302
immobilisation process, 123
immobilisation technique, 169
immobilised enzyme, 52, 122
immobilised enzyme particles, 140
immobilised properties, 169
immobilised systems, 171
immune precipitation, 200
immunosensors, 230
impervious immobilised enzyme systems, 105
improving yield by genetic engineering, 322
inclusion bodies, 27
increase the yield, 301
indirect ELISA technique, 202
induction, 22, 24
industrial applications of
       protein engineering, 313
industrial reactor broths, 182
inertia of liquids, 123
influence of ionic strength, 63
inheritance of
       disease phenotypes, 264
       genes, 264
inhibition, 109
inhibitor effects
       of high substrate and product concentrations, 134
inhibitors, 61
initial substrate concentrations, 99
ink-jet printing, 245
inorganic carriers, 80
inorganic cofactors, 142
insecticides, 165, 197
insulin, 204, 304
insulin mRNA, 304
insulin-dependent diabetic patients, 215
inter-esterification of lipids, 333
interconversion of glucose and fructose, 63
interdigitated micro-electrodes, 223
interferometric sensors, 224
intracellular enzymes, 25
introns, 21, 259, 276, 277, 322
inverted micelles, 328
iodide, 221
ion exchange chromatography, 44, 321
ion-exchange, 42
ion selective electrodes, 221
ion-selective field effect transistor (ISFET), 222
ionic interactions, 101
ionic strength, 15, 62, 92, 97, 99, 100, 302
irreversible inhibition, 10
isoamyl acetate, 333
isoamyl alcohol, 333
isocitrate dehydrogenase, 197
isocyanate derivatives, 90
isoelectric focusing, 321
isolation and purification of enzymes, 301

J

Jeffries, 267

K

K^+, 142
κ-casein, 161
Kaiser, 303

kaolinite, 92
k_{cat}, 55
kidneys, 162
kinetic method, 187
kinetics of coupled reactions, 197
kinetics of immobilised enzymes, 103
kinetics of the electrode, 213
k_L = mass transfer coefficient, 133
Klenow fragment, 279, 284, 290, 291, 292, 311
Klenow fragment of *E. coli* DNA polymerase, 294
Kluyveromyces fragilis, 161
Kluyveromyces lactis, 161
Kluyveromyces spp., 22
K_M, 10, 56, 99, 302
K_M', 102
Krebs cycle, 183

L

L-alanine, 165
L-amino acids, 164
L-aspartate, 165
L-aspartate b-decarboxylase, 165
L-aspartic acid, 334
L-histidine, 165
L-lactate in biological fluids, 228
L-phenylalanine methyl ester, 334
labels, 198
lactase, 160
lactate, 189
lactate dehydrogenase, 189, 228
lactose, 151, 156, 160
lactose intolerance, 160
laminar flow, 134
LAP sensor, 236
LAPS, 236
large-scale extraction, 29
large-scale purification, 42
laundry detergents, 313
Lawrence, 303
leakage of enzyme from the support, 93
leather, 177
leukaemia, 174
lidocaine, 234
light emitting diodes (LEDs), 224, 236
lignin, 160
linear flow rate, 46
Lineweaver-Burk plots, 8
linkage analysis, 267
linkers, 257, 261
lipase, 97, 162, 230, 330, 332, 333
lipids, 43, 60, 97, 328
liposomal vesicles, 97
liquefaction, 152
liquid/solid separations, 29
liver, 162
localised (directed)-random mutagenesis, 318
localised heat output, 130
localised or directed random mutagenesis, 307
localised overheating, 72
localised random mutagenesis, 305, 306, 309, 311, 320
luciferase, 227
luciferin, 227
luminescent systems, 226
lyophilisation, 35
lysine, 80
lysozyme, 35

Index

M

M. *Ava* II, 251
M. *Bam* HI, 251
M. *Eco* RI, 251
M. *Hae* III, 251
M. *pusillus*, 161
macroporous matrices, 92
magnesium, 197
magnesium ions, 252, 253
magnetite, 80
maintenance methylation, 271
maize, 156
maltase, 194
malting, 153
maltose, 151, 156, 194
maltose groups, 153
manufacturing technologies, 242
marine bacteria *Vibrio* species, 227
marker gene, 259
mass load, 45
mass transfer coefficient, 107
mass transfer to the electrode surface, 212
matrix, 23
mature mRNA, 278
Maxazyme GI Immob, 93
maximum velocity V_{max}, 187
Mbo I, 251
measuring enzyme reactions, 186
meat extracts, 54
mechanisms of enzyme catalysis, 6
media formulation for soluble enzyme systems, 62
mediated amperometric enzyme electrodes, 215
medicine, 303
melt in the mouth feel, 163, 333
melting temperature of the triglycerides, 333
membrane, 140
membrane reactors, 95, 140
mercapto group (-SH), 80
mercaptoethanol, 64
mesophilic enzymes, 318
metabolic end-products, 24
metabolic inhibitors, 241
metabolites, 184
metalloproteases, 314
methicillin, 170
methionine, 304, 315
methyl donor, 252, 256
methyl ester of L-phenylalanine, 334
methylase, 251, 252, 255
methylase enzyme (M. *Eco* R1), 262
methylase enzyme activity, 248
methylases, 250, 253
methylation in eukaryotic DNA, 271
Mg^{2+}, 64, 142, 227, 256, 282, 286, 291
Mg^{2+} titration curve, 286
Michaelis-Menten equation, 55
    integrated form, 190
Michaelis-Menten kinetics, 8, 54, 56, 61
Michaelis-Menten relationship, 111, 187
Mickle shaker, 36
micro-encapsulated system, 143
micro-encapsulation, 79, 93, 94, 97
micro-organisms, 20, 175
microbial biosensors, 240
microbial contamination, 328
microbial culture, 24
microbial cultures, 21
microprocessors, 210
milled grain, 153
mini-wash tests, 318
minisatellite DNA, 267
minisatellites, 268
mis-priming, 284
mixing, 60
mixing in stirred tank reactors, 123
mixing of the bulk liquid, 212
mixing procedure, 24
Mn^{2+}, 64
modal dispersion, 224
modified nucleotides
    in the sequencing reactions, 294
modified T7 DNA polymerase, 294
modifying enzymes to simplify purification, 320
molecular oxygen, 161
monitoring module, 46
monoclonal antibodies, 96, 200
monoglycerides, 162
monozygotic, 270
Moraxella bovis, 251
mRNA, 21, 288
Msp I, 272
Mucor miehei, 161, 162, 163
Mucor spp, 333
multi-subunit enzymes, 323
multifunctional reagents, 89, 90
multimode fibre, 224
multiple copies, 323
multiple enzymes
    assays using, 194
multiplex PCR, 290
multistage reactions, 175
murine leukaemia virus (MuLV), 278
Muscular Dystrophy, 290
mutagenesis, 22
mutagenic primer, 311
mutagenicity, 241
mutant strain, 22
mutants, 24, 241
mutated bacterial strains, 263
mutating gene, 302
mutation 'hot-spots', 290
mutation frequency, 311
mutations, 277, 308
myoglobin, 48

N

N-terminal amino group, 80
$NAD(P)^+$, 204
$NAD(P)H$-specific system, 227
NAD^+, 64, 259
NAD^+ (NADH), 142
NAD^+ (nicotinamide adenine dinucleotide), 257
NADH, 187, 189
NADH oxidase system from *Escherichia coli*, 144
$NADP^+$, 195
$NADP^+$ (NADPH), 142
NADPH, 187, 195
natural carriers, 80
negative control samples, 289
Nernst equation, 220
nickel oxide, 80

See also NAD and NADP
nicotinic acetylcholine, 240
nikkomycin, 172
nitrate, 221, 241
nitrobenzene, 204
nitrocellulose, 97
nitrocellulose filters, 204
non-aqueous solvent, 328
non-aqueous systems, 328
non-coding intervening sequences, 277
non-coding regions, 259
non-competitive inhibition, 10
non-ionogenic detergent, 204
non-proliferating whole cells, 52, 175
non-specific proteases, 162
non-specific recombination, 256
Northern blotting, 235
nucleases, 43
nucleic acid biosensors, 235
nucleic acids, 43
nucleotide analogues, 294
nutrient concentrations, 24
nylon, 80, 97, 176

O

o-phenylene-diamine, 204
O_2, 140, 227
octanol, 329
oestrogen, 204
oils, 162
oligo(dT) primer, 288
oligonucleotide primer, 286, 292
oligonucleotide-directed mutagenesis, 296
optical enzyme biosensors, 223
optical fibre, 224
optical immunosensors, 232
optical transduction, 223
optimisation of operating potential, 213
organ transplantation, 270
organic acids, 241
organic cofactors, 64, 142
organic phase enzyme electrodes, 230
organic phase enzyme electrodes (OPEEs), 219
organic polymers, 80
organofluorine compounds (X-F), 223
orientation effects, 7
origin of replication, 259
osmotic shock, 25, 28
overall yield, 25
oxaloacetate, 183
oxidase, 226
oxidation, 314
oxidation resistancy, 315
oxidation resistant proteases, 314
oxidoreductase activity, 303
oxygen, 24
oxygen optrodes, 226
oxyluciferin, 227

P

p-nitrophenol, 204
packed bed reactors (PBRs), 107, 127
palindromes, 254
palm oil, 163

pancreatic lipase, 330
pancreatic proteases trypsin, 314
papain, 20, 89, 162, 303
paper pulp, 160
parathion, 198
partial synthesis (semi-synthesis), 303
particulate substrate, 122
partitioning, 110
partitioning effects, 102
patent applications, 313
pathogenic bacteria in clinical samples, 239
patulin, 172
PBR reactors, 157
pBr322, 21, 259, 262
PBRs, 164
  productivity of, 138
PCR, 296
  and mutations, 312
PCR (polymerase chain reaction), 236
peas, 61
pectin, 161
pectin esterase, 161
pectin lyase, 161
pectinase, 61, 70, 161
penicillin acylases, 168
penicillin amidase, 94
penicillin G, 167, 172
penicillin-G-amidase, 22
penicillin-G-amidase from *E. coli*, 21
Penicillium spp, 161
pepsin, 63
peptide bonds, 90, 334
peptides, 241
perfluoroalkyl groups, 222
performance of PFRs (PBRs), 127
periplasmic proteins, 28
periplasmic space, 25
peroxidase, 199, 204, 222, 223
peroxidase-positive bacteria, 223
petroleum ether, 163
pH, 4, 24, 43, 46, 62, 97, 99, 122, 130, 134, 302, 330, 334
pH optimum, 302
pH probe, 63
phage M13 vector, 309
pharmaceuticals, 165
Pharmacia BIAcore instrument, 234
phenol group, 80
phenol sensors, 220
phenols, 220
phenyl-napthol phosphates, 236
phenylacetic acid, 169
phenylalanine, 165
phenylhydrazine, 183
phenylhydrazone, 183
phosphate, 241
phosphatidylcholine, 328
phospholipids, 97
photodiode, 224
phycobiliproteins, 232
physical adsorption onto a support, 92
physiological fluids, 182
pI isoelectric point, 321
piezoelectric detection of nucleic acids, 238
piezoelectric-based detection
  of nucleic acids, 239
pigments, 43
pilot scale, 44
ping pong reactions, 9

Index

pKa's, 101
plant cell cultures
    as sources of enzymes, 23
plant cell walls, 160
plant viruses, 204
plants, 20
plasmid
    linearisation, 259
plasmids, 24, 259
plug flow reactors (PFRs), 92, 126
plug flow systems, 129, 134
pO_2, 24
poly A tail, 278
polyacrylamide, 95
polyacrylamide beads, 204
polyethyleneimine, 48
polyethyleneimine agarose, 45
polygalacturonase, 161
polymerase chain reaction (PCR), 284
polymeric carbohydrates, 160
polymeric matrices, 95
polyoxirane, 80
polypyrimidine sequences, 289
polysaccharide hydrolases, 333
polystyrene, 80, 97
polystyrene microtitre plates, 204
polyurethanes, 80
polyvinyl alcohol, 95
porcine insulin, 335
pore sizes in the entrapping gel, 95
porous glass, 92
porous immobilised enzyme systems, 111
porous membranes, 95
positive control samples, 289
post-translational modification, 22, 297
potassium, 221
potassium ferricyanide, 216
potatoes, 156
potentiometric enzyme biosensors, 220
potentiometric enzyme electrodes, 222
potentiometric enzyme sensors, 222
potentiometric transduction, 220
potentiometric transduction systems, 231
potentiometry, 186
power of impellers, 124
practical aspects of the ELISA technique, 204
precautions in using PCR, 289
precipitation of nucleic acids, 43
prenatal diagnosis, 264
preparation of double-stranded cDNA, 279
pressure, 32, 42, 46
pressure drop across PBRs, 130
primary structure, 302
primer, 277
primer-dimers, 286
probes for hybridisation, 291
producing more thermostable enzymes, 318
product, 101, 122
product accumulates, 58
product inhibition, 60
product yield, 54
production of labelled DNA probes, 291
    using the Klenow fragment, 291
profile of reaction rates, 127
promoter, 311
proof-read DNA, 290
proof-reading, 277

prostatic acid phosphatase, 231
protamine sulphate, 43
proteases, 177, 333
proteases in the food industry, 161
proteases with a higher wash performance, 317
protein engineering, 22, 295, 301
    objectives of, 300
protein engineering by the mutagenesis cycle, 307
protein hydrolysates, 161
protein inhibitor, 317
protein kinase, 89
proteins, 60
proteolytic enzymes, 27, 54
Providencia stuartii, 251
provirus, 277
proximity effects, 7
pseudo first order, 9
pseudo-single substrate reactions, 55
Pseudomonas putida, 175
Pst I, 251, 282
pUC19, 259
pullulanase, 153
pure (cloned) strains, 21
purification
    large-scale, 42
purification by chromatography, 43
purine nucleotides, 282
pyridine nucleotide, 197
pyridoxal coenzymes, 49
pyridoxal phosphate, 142
pyrimidine, 282
pyruvate, 183, 189, 242
pyruvate dehydrogenase, 143

Q

quality assurance, 270
quality control, 204
quartz crystal microbalance (QCM), 239

R

racemic mixture, 335
radioactive immune assays, 200
radioactive phosphorus, 265
radioactively labelled antibodies, 204
radioactively labelled antigens, 204
radioactively labelled denatured DNA probe, 266
radioactively labelled markers, 292
radioimmune assays (RIA), 204, 232
radiolabelled, 291
radiometry, 186
raemic mixture, 164
random cloning of genomic DNA, 259
random libraries, 263
random localised mutagenesis, 317
random mutagenic procedures, 302
random order reactions, 9
randomly mutated genes, 307
rate of diffusion, 111, 213
reaction mechanism of
    enzyme catalysed reactions, 313
reaction times, 123
reactions involving gases, 140
reactor, 157
reactor kinetics and reversible reactions, 136
reactor maintenance and control, 93

receptors embedded in bilipid membranes, 240
recognition site, 248, 256
recombinant DNA techniques, 27, 295
recombinant DNA technology, 22
recombinant yeast, 47
recombination at meiosis, 267
recombination at mitosis, 267
reducing viscosity, 42
reduction in number of non-recombinant clones, 280
reference electrode, 212, 220
regenerating the required cofactor
    cosubstrate within the reactor, 143
regeneration buffers, 46
regeneration of
    oxidised pyridine nucleotides, 143
regions conserved variable, 307
regulation of enzyme synthesis, 22
removal of hair fromskin, 178
rennet, 161
repair of single strand breaks, 257
repeated sequences, 267
replica plating, 263
repression, 22
reproducibility, 126
research, 303
residence time, 122, 130, 133, 157
resistance to antibiotics, 167
resistant against an antibiotic, 259
resistant to penicillins, 167
resolution, 45
resonating silicon bridges, 229
restriction
    modification defence system, 248
restriction endonucleases, 248
restriction enzyme sites, 311
restriction enzymes, 248, 258, 265
    in diagnosis of genetic disorders, 264
    See also specific restriction enzymes
restriction fragment length polymorphism, 265
restriction-modification system nomenclature, 251
retroviruses, 277
reverse transcriptase (RT), 21, 276, 277
Reynolds number (Re), 123
RFLPs, 267
Rhizopus spp, 333
rhodanase, 174
ribonuclease (RNase H)-type activity, 277
RNA hybridisation experiments, 291
RNA-dependent DNA polymerase activity, 277
RNase, 174
rotary vacuum filters, 32
RSF 2124, 259
RT, 279
Rubella antibody in serum, 234
running costs, 122

S

S-adenosyl methionine (SAM), 252
S1 nuclease, 279
saccharification, 151
Saccharomyces cerevisiae, 155
salt concentration, 4
SAM, 256
sample, 46
sample concentration, 46
sample load, 46

Sanger, 292
sarcosine oxidase, 227
Satellite G system, 217
Sau 3A, 261
scale-up, 126
Schiff base, 90
screen printing, 243
scroll (decanter) centrifuge, 31
SDS-PAGE, 47
secondary structure, 286
secondary structure of the enzyme, 319
selectable markers of plasmids, 262
selection of primers, 286
selection test, 307
self-buffering substrates, 63
self-priming for second strand synthesis, 279
semi-permeable microcapsules, 97
semi-solid culture, 24
semi-specific PCR, 288
semi-synthetic antibiotic products, 171
semi-synthetic antibiotics, 166
semi-synthetic penicillins, 167
semiconservative replication, 296
sensing element, 220
sensitivity, 182
separation module, 46
Sephacryl S-200, 48
Sequenase, 294
sequence data from regions of DNA, 294
sequence diversity, 265
serial deactivation scheme, 69
serine, 80, 315
serine mutant, 315
serine proteases, 313
serine proteases from *Bacillus* species, 313
SH groups, 62
SH protectants, 64
Sharples centrifuges, 31
shelf-life detergents, 317
shuttle vectors, 22
side-chain modification of proteins, 303
signal peptide, 322
silica gel, 80, 92
silver, 221
silver pastes, 243
single mode fibre, 224
single priming, 311
single stranded DNA, 236
single stranded DNA to be sequenced, 292
single stranded DNA vector, 309
single stranded form
    of the recombinant vector, 295
single stranded template, 311
single substrate reactions, 55
single-batch, 25
site-directed mutagenesis, 295, 306, 307, 309, 310, 311, 318
site-directed mutations, 320
site-specific mutagenesis, 302, 305
site-specific mutation of enzymes, 276
site-specific recombination, 256
site-specifically mutated cDNA, 296
skin ulcers, 174
slippage during replication, 267
small-scale
    high-priced operations, 140
sodium ion-selective electrodes, 221
sodium perborate, 315
solid phase, 204

Index

solid support, 111, 204
solubility, 60, 321
soluble enzymes, 52, 54
soluble enzymes in industrial processes, 52
solvent extraction, 35
soups, 162
sources of enzymes for industry, 20
Southern blots, 268
Southern blotting, 235, 266
soya protein, 204
specific washing performance, 318
specificity, 7
specificity constant, 8
specificity constant of enzymes, 55
specificity of antibodies, 200
specificity of sensors, 210
spectrophotometric assay, 200
spectrophotometry, 186
SPR-based immunosensors, 234
stability, 302
stability of enzymes in vitro, 301
stability of the enzyme, 68, 122
stability:pH profile, 63
stagnant layer, 107, 111, 212
stagnant liquid layer, 103
starch, 151, 156, 160
starch in barley, 153
starch suspensions, 60
stearic acid, 163
stereochemistry of steroids, 166
steric hindrance, 98
steroid hydroxylations, 166
steroids, 166, 328
stirred tank batch reactors, 122
    productivity of, 124
stirred tanks, 134
stirring, 107
stomach of calves, 161
storage, 69
strain, 7
strains selection, 22
strength of the promoter, 323
streptokinase, 174
Streptomyces spp., 157
streptomycin sulphate, 43
stringently controlled vectors, 323
strong promoter, 295
structure:activity relationships of enzymes, 301
structure:function analysis, 320
structure:activity relationships, 307
structure:function relationships, 313
submerged culture, 24
substrate, 122, 199
substrate analogue, 317
substrate concentration, 97
substrate limitation, 133
substrate specificity, 182
substrate specificity of enzymes, 301
substrate-binding, 80
substrates, 101, 122
subtilisin, 303
subtilisin derived from
    Bacillus licheniformis, 315
subtilisin from *Bacillus amyloliquefaciens*, 315
subtilisins, 313
sucrose, 151, 155, 156
sugar beet, 155, 156

sugar cane, 155, 156
sugars, 241
sulphide, 221
sun-screening agent, 165
surface acoustic wave (SAW), 239
surface plasmon, 226
surface plasmon resonance, 234
surface plasmon resonance (SPR), 226, 232, 238
surfactants, 25, 315, 328
sweetness ratings, 155
synthesis of polymers, 334
synthesis of single-stranded cDNA, 278
synthetic human insulin, 335
synzymes, 48
    with transaminase activity, 49

T

t-butanol, 336
T4 DNA ligase, 259, 261
T4 ligase, 258
T7 DNA polymerase, 276
tank module, 46
Taq polymerase, 284, 286
techniques available
    for modifying protein structure, 301
techniques for immobilising enzymes, 78
temperature, 4, 24, 27, 42, 122, 130, 134, 302
temperature quotient Q_{10}, 68
tenderisation of meat, 161
terminal deoxynucleotidyl transferase, 282
terminal transferase (TdT), 282
tertiary structure of the enzyme, 319
tetracyanoquinodimethane, 216
tetracyclin, 259, 262
tetrathiafulvalene, 216
therapeutic agents, 166
thermal denaturation, 68
thermal enzyme biosensors, 228
thermal protein denaturation, 69
thermal stability of an enzyme, 16
thermal-optical biosensor, 229
thermistors, 228, 230
thermolysin, 314, 334
thermometers, 228
thermophilic organisms, 318
thermopiles, 228
thermostability, 330
thermostable enzymes, 155
Thermus aquaticus, 284
thiamine pyrophosphate, 142
thin film techniques, 242
thiol anti-oxidants, 69
thiolprotease activity, 303
thiosulphate sulphur transferase, 174
three dimensional structure of the enzyme, 302
threonine, 80, 335
threshold system, 237
tissue section biosensors, 241
tissue transplantation, 270
tissue-specific inactivation of genes, 271
toluene, 64
total internal reflection fluorescence (TIRF), 225
total protein content, 42
total reaction time, 68
total synthesis of enzymes, 303
toxins, 204

transacylases, 333
transcriptase, 304
transcription termination signal, 295
transesterification of triglycerides, 333
transformation, 248
tributyrin, 330
trichloro-s-triazine, 91
Trichoderma reesei, 160
trifluoroacetic acid, 336
triglycerides, 162, 333
tris (1,10-phenanthroline) ruthenium (II), 226
tryp E promoter, 304
trypsin, 20, 89, 335
tubular bowl centrifuge, 31
turbidity, 186
turbulence, 124
turbulent flow, 134
turnover number, 8, 55
Tween 80, 25
twins, 270
two-phase systems, 34
tylosin, 172
tyrosine, 80, 242
tyrosyl tRNA-synthetase, 313

U

ultrafiltration, 42
ultrasonication, 35
unsaturated fatty acid residues, 333
urate oxidase, 89
urea, 182, 223, 242
urease, 94, 182, 223, 236
urine, 198
urocanic acid, 165
urokinase, 174
use of homopolymer tails in gene cloning, 282
use of pilot scale trials, 44
use of reverse transcriptase to produce cDNA, 278
use of the Klenow fragment in DNA sequencing, 292
usefulness of sticky (cohesive) ends, 255
uses of genetic fingerprinting, 270
UV absorption, 46
UV irradiation, 266

V

van der Waal's forces, 92
vector DNA, 259
viral infections, 174
viscosity, 32, 42, 60, 130, 213
viscous forces, 123
vitamins, 241
V_{max}, 10, 99
volume load, 45
volumetric flow rate, 46

W

washing powder, 54
water film, 140
water insoluble solutes, 141
water-substrate interfacial area, 60
whole cell sensors, 240
whole cells, 143

wine manufacture, 161
working (or sensing) electrode, 212

X

X-press, 35
X-ray film, 266
X-rays, 22
xanthine oxidase, 89
Xanthobacter, 176
xylose isomerase, 157

Y

yeast, 22, 153
yields from PCR, 286

Z

zeroth order kinetics, 111
zinc metalloenzyme, 277
zirconium oxide, 37
zirconium silicate, 37
Zn^{2+}, 314